Information and Living Systems

Information and Living Systems
Philosophical and Scientific Perspectives

Edited by George Terzis and Robert Arp

A Bradford Book
The MIT Press
Cambridge, Massachusetts
London, England

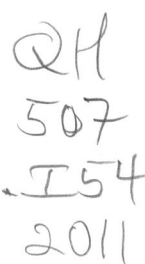

© 2011 Massachusetts Institute of Technology

All rights reserved. No part of this book may be reproduced in any form by any electronic or mechanical means (including photocopying, recording, or information storage and retrieval) without permission in writing from the publisher.

For information about special quantity discounts, please email special_sales@mitpress.mit.edu

This book was set in Sabon by Toppan Best-set Premedia Limited. Printed and bound in the United States of America.

Library of Congress Cataloging-in-Publication Data

Information and living systems : philosophical and scientific perspectives / edited by George Terzis and Robert Arp.
   p. cm.
Includes bibliographical references (p.) and index.
ISBN 978-0-262-20174-2 (hardcover : alk. paper)
1. Information theory in biology. I. Terzis, George, 1951– II. Arp, Robert.
QH507.I54   2011
570—dc22

                                                                          2010026309

10 9 8 7 6 5 4 3 2 1

For George, Katherine, Chris, and Alexis

Contents

Preface ix
Introduction xi

I The Definition of Life 1

1 The Need for a Universal Definition of Life in Twenty-first-century Biology 3
Kepa Ruiz-Mirazo and Alvaro Moreno

2 Energy Coupling 25
Yaşar Demirel

II Information and Biological Organization 53

3 Bioinformation as a Triadic Relation 55
Alfredo Marcos

4 The Biosemiotic Approach in Biology: Theoretical Bases and Applied Models 91
João Queiroz, Claus Emmeche, Kalevi Kull, and Charbel El-Hani

5 Problem Solving in the Life Cycles of Multicellular Organisms: Immunology and Cancer 131
Niall Shanks and Rebecca A. Pyles

6 The Informational Nature of Biological Causality 157
Alvaro Moreno and Kepa Ruiz-Mirazo

7 The Self-construction of a Living Organism 177
Natalia López-Moratalla and María Cerezo

8 Plasticity and Complexity in Biology: Topological Organization, Regulatory Protein Networks, and Mechanisms of Genetic Expression 205
Luciano Boi

III Information and the Biology of Cognition, Value, and Language 251

9 Decision Making in the Economy of Nature: Value as Information 253
Benoit Hardy-Vallée

10 Information Theory and Perception: The Role of Constraints, and What Do We Maximize Information About? 289
Roland Baddeley, Benjamin Vincent, and David Attewell

11 Attention, Information, and Epistemic Perception 309
Nicolas J. Bullot

12 Biolinguistics and Information 353
Cedric Boeckx and Juan Uriagereka

13 The Biology of Personality 371
Aurelio José Figueredo, W. Jake Jacobs, Sarah B. Burger, Paul R. Gladden, and Sally G. Olderbak

Contributors 407
Index 409

Preface

This volume is the product of talks we had about the important ways in which information shapes our understanding of biological organization, and more generally, of the difference between living and inanimate matter. It later occurred to us to broaden our conversation to include other researchers, both philosophers and scientists, who could write on this theme at a level of biological organization that reflects the interests of their respective disciplines. With this in mind, we solicited contributions, through both invitation and a global call for papers. As a result, we were able to assemble a group of papers that addresses our informational theme at levels of organization that range from the genetic and epigenetic, at one end of the biological continuum, to the cognitive and linguistic, at the other. These papers also exemplify the deeply interdisciplinary nature of our theme.

In preparing this volume, we were helped and encouraged by several of our colleagues and friends to whom we wish to express our deep appreciation: to former MIT Press senior editor, Tom Stone, for his friendliness and belief in the value of our project; to current senior editor Philip Laughlin and his assistant, Marc Lowenthal, for guidance during the project's final stages; to MIT Press editor Kathleen Caruso who, together with Nancy Kotary and Michael Sims, oversaw the final edit of our project; to Theodore Vitali, CP, Philosophy Department Chair, Saint Louis University, for lightening our load so that our project could be completed and for moral support; to the university's College of Arts and Sciences for a Mellon Faculty Development Grant that helped us kick-start our project at the beginning; and to Ronald Belgau and Matthew Piper, for helping us prepare the introduction to our volume.

Finally, we are also grateful to Kathy Stone, Melissa Kinsey, and Judy MacManus for their helpful editorial advice, and especially to Aileen Keenan, who took time from her own responsibilities as *African American Review*'s managing editor to do an excellent job in preparing the final version of the manuscript.

Introduction

The idea of information has become increasingly important in our efforts to understand the nature of biological organization. It is now generally recognized that information-related ideas play a key role in characterizing life at every organizational level. Familiar examples of these ideas include: how DNA codes for specific amino-acid sequences, how gene expression shapes biological development, how chemical signaling enables the immune system to combat pathogens, and at the cognitive end of the biological continuum, how populations of neurons represent features of our environment. But although pervasive, the contribution that information-related ideas make to our understanding of these processes is not yet well understood. Indeed, only fairly recently have researchers begun to explore their deeper theoretical significance.

A similar point can be made about the philosophical and scientific controversies that inevitably arise from our understanding of these information-dependent biological processes. For example, because organisms use information to construct, maintain, repair, and replicate themselves, it would seem only natural to include information-related ideas in our attempt to understand the general nature of living systems; the causality by means of which they operate; whether they could exist and evolve in nonbiological matter; and how, in certain species, they give rise to cognition, value, and language. But again, philosophers and scientists have only gradually incorporated information-related ideas in the methodology they employ to discuss these long-standing scientific and philosophical controversies.

Our volume thus seeks to respond to what we believe is a neglect of the many contributions that information-related ideas play in shaping our understanding of the nature of biological organization, as well as the controversies to which this understanding inevitably gives rise. Admittedly, this response will in some cases just add fuel to the flames.

For example, as there is an abundance not just of information-related ideas but also of broader information-theoretic perspectives, ranging from the purely quantitative to the functional and semantic, an important difficulty concerns how closely or loosely these ideas and perspectives are related to one another. In other words, to what extent, if any, can they be meaningfully unified? Also, although the description of certain biological processes may require an information-based vocabulary—one that includes terms such as code, signal, messenger, transmission, transcription, translation, correction, and the like—the deeper scientific and philosophical meaning of this vocabulary is still unclear. Can there, then, be a literal philosophical or scientific understanding of the bioinformational processes these terms describe, or is their meaning merely heuristic or metaphorical? Although they are deeply puzzling, we believe that any aversion toward addressing these questions is overshadowed by the fact that information is indeed integral to our understanding of the organization of life. For this reason, these more recent questions are just as much in need of scientific and philosophical examination as are the more traditional ones.

The chapters in this volume are the work of an international community of researchers who seek to shed light on these issues, and more generally, on the informational nature of biological organization. Their respective disciplines, which reveal the deeply interdisciplinary nature of these issues, include: anthropology, biology, biosemiotics, chemistry, cognitive science, computer science, information theory, linguistics, mathematics, medicine, paleontology, philosophy, physics, psychology, and systems theory. The contributions of our authors are intended to be useful not only to fellow researchers, but also to advanced undergraduate and graduate students in both science and philosophy and perhaps to any thoughtful person who has been deeply struck by the fact that information is in some sense crucial to our understanding of the difference between living and inanimate matter.

The organization of our authors' contributions follows a simple, logical progression. Part I introduces the idea of an organism or living system, and characterizes some of its basic physical, chemical, biological, and informational properties. Part II introduces the idea of an informational perspective on biological organization, which it gradually and intuitively expands to include philosophical, evolutionary, and developmental or epigenetic considerations. Finally, Part III extends our bioinformational theme to questions concerning the biological basis of cognition, value, language, and personality.

I Defining Life

In chapter 1, "The Need for a Definition of Life in Twenty-first-century Biology," Kepa Ruiz-Mirazo and Alvaro Moreno remark that despite its many successes, progress in twentieth-century biology was also limited by its attempt to give a reductive account of the nature of life based on the gene concept (Westerhoff and Palsson 2004). In response to this impediment, researchers in biology have recently begun to move away from a reductive, molecular approach toward a more synthetic, functional one (Benner and Sismour 2005). This shift in their approach to biology has led researchers to reconsider the nature of life itself, including its definition. Now, the definition of life can be attempted in a number of ways, one of which is to give a list of necessary and sufficient conditions for saying that something is alive. Possible entries on such a list might include development, metabolism, homeostasis, genetic program, adaptation, evolution, and the like. But apart from the obvious difficulty of trying to produce and defend a single, correct list of necessary and sufficient properties, there is the deeper problem of trying to understand how the many properties are related to one another. This is especially important in regard to life, where properties need to be understood in terms of how they fit together in a hierarchy in which some are more basic than others. Thus, to avoid this difficulty, the authors seek a definition of life that articulates its fundamental nature or principles of organization—in traditional philosophical terms, an "essentialist definition." These, of course, include the principles of physics and chemistry as well as our best account of how life might have emerged from these more basic levels of organization.

Unfortunately, current essentialist definitions, Ruiz-Mirazo and Moreno point out, have been only partly successful in achieving these objectives. A recurring weakness found in the different definitions is that they fail to integrate two distinct types of features found in living phenomena: the individual and metabolic, on the one hand, and the collective and evolutionary, on the other. Instead, current essentialist definitions tend to focus on one type of feature to the relative exclusion of the other. An example of this definitional one-sidedness is found in the account of the nature of life defended in autopoietic theories (Maturana and Varela 1973; Varela 1994). Though autopoietic systems admit of reproduction and structural modification, they do not also admit of Darwinian evolution. Conversely, definitions according to which life is a chemical system that evolves through natural selection overlook the organizational structure that such a system requires in order to be capable of such evolution (Sagan 1970; Joyce 1994).

The goal, then, of Ruiz-Mirazo and Moreno's discussion is to discover a definition of life that effectively integrates both types of features. In constructing this definition, the authors begin from the individual-metabolic, rather than the collective-evolutionary, side of life. In this respect, their work is influenced by Tibor Gánti's (1975) well-known notion of a *chemoton*: an organizational entity whose components include a boundary, metabolism, and information network. But unlike Gánti's more formal notion, Ruiz-Mirazo and Moreno want a definition with enough biological particularity to shed light on the origins of life. Of course, this is not an easy condition to fulfill, because we do not know, in regard to the conditions in which terrestrial life emerged, which organizational features are essential (or necessary) and which are accidental (see Keller 2007). For this reason, Ruiz-Mirazo and Moreno admit that their definition must be set forth in a provisional manner. Still, they are hopeful that advances in astrobiology, computational models, and experiments—both in vitro and in silico—will help sort out the difference between the essential and the accidental over time. In other words, a satisfactory "twenty-first-century" definition of life must be responsive to these theoretical and experimental considerations.

There must, then, be stable organizational structure of the kind suggested by Gánti. This must also be true in a thermodynamic sense: that is, the structure's components must be organized so that it harnesses energy to continuously reproduce these components (Kauffman 2000, 2003). But although important, the authors claim that this autonomous or recursive self-maintenance will not be sufficiently robust, unless a further condition is met. Such self-maintenance, they conjecture, depended on life having evolved in such a way that the informational domain gradually became relatively *decoupled* from the metabolic domain. Eventually, this decoupling allowed the informational domain to generate, store, and even rearrange its informational vocabulary, namely, its one-dimension sequence of nucleotides, relatively independently of its protein (enzymes) controlled metabolism (Pattee 1982; see also Neumann 1949). Of course, such decoupling is entirely compatible with the different domains remaining deeply interdependent. After all, the DNA sequence of nucleotides codes for proteins some of which play a central role in transcription (DNA to mRNA) and translation (mRNA to amino-acid sequence) processes. Later in this discussion (see Expanding the Functionalist Account), we will explore in greater depth the authors' decoupling thesis, including the informational causality that, they maintain, connects the two domains. Our present objective, however, is to note that this thesis produces a definition of life that appropriately combines the collective-evolutionary and the individual-metabolic aspects. According to Ruiz-Mirazo and Moreno's definition, "Life is a complex network of self-reproducing autonomous agents whose basic organization is instructed by material records generated through the open-ended process in which that collective network evolves" (16).

As we have observed, although Ruiz-Mirazo and Moreno object to a reductionistic account of biological phenomena, they still insist that the fundamental principles of physics and chemistry play a key role in their efforts to define life. They also play this role in chapter 2, Yaşar Demirel's "Energy Coupling," which further explores the thermodynamic properties of living systems. Central to this perspective is the idea that organisms are open, dynamic systems that create and maintain themselves by continuously exchanging matter, energy, and information with their environment. Because organisms maintain and create order, their behavior may at first appear to be inconsistent with the Second Law of Thermodynamics, according to which any energy exchange leads to an increase in the entropy of the universe. As we know, this inconsistency is apparent, not actual: any gain in an organism's order and complexity will be offset by a greater amount of disorder—for example, in the form of heat loss—that the organism will dissipate into its surroundings (Schrödinger 1945). However, this response leaves unanswered the important question of *how*, in a manner consistent with the Second Law, organisms are able to create and maintain order. The answer to this question, explained in depth by Demirel, is that organisms meet their energy needs by *coupling* spontaneous processes with nonspontaneous ones. Familiar examples of such coupled processes include photosynthesis, oxidative phosphorylation, and the hydrolysis of adenosine triphosphate (ATP). Intuitively, the distinction between spontaneous and nonspontaneous processes is evident. A spontaneous process is one that proceeds on its own, whereas the initiation of a nonspontaneous process depends on an external energy source. Thus water running down hill is spontaneous, but the turning of a turbine to create electricity is not, although the latter can be coupled to the former to perform work. But a more precise way of making the distinction is to say that a spontaneous process is one in which there is a loss of *free energy*, and in a nonspontaneous process the amount of free energy increases. A system's free energy, G, is related to its total energy, H, by means of the equation

$G = H - TS$, where T refers to absolute temperature and S refers to entropy. This equation says that the system's free energy is the amount of its energy that is available to perform work once we've subtracted the system's entropy, which is amplified by its absolute temperature. Finally, because the amount of free energy is not fixed but changes once a process is underway, we can measure the change in G, denoted by ΔG, by subtracting the amount of free energy the system possesses in its final state from the amount it possesses in its initial state. In other words, $\Delta G = G_{\text{final state}} - G_{\text{initial state}}$. Thus, a spontaneous change is one where $\Delta G < 0$, whereas in a nonspontaneous change, $\Delta G > 0$. Finally, combining this result with the previous one gives us the Gibbs free energy equation: $\Delta G = \Delta H - T\Delta S$. In other words, the change in the system's free energy equals the change in its total energy minus the change in its entropy, again amplified by its absolute temperature.

The idea of free energy thus enables us to understand how, consistent with the laws of thermodynamics, organisms effectively couple spontaneous with nonspontaneous processes. Consider, for example, the hydrolysis of adenosine triphosphate (ATP) to adenosine diphosphate (ADP). ATP is an energy-rich molecule with an unstable triphosphate tail, which—when hydrolyzed—transfers a phosphate group to another molecule, enabling it to perform work. In this case, adding a molecule of water to a molecule of ATP yields a molecule of ADP plus an inorganic phosphate, thus producing a decrease in free energy where $\Delta G = -7.3$ kcal/mol. Conversely, the regeneration—namely, the phosphorylation—of ADP to ATP, which is nonspontaneous, depends on an external energy-releasing process, such as cellular respiration, or—in the case of plants—the absorption of sunlight. In this case, adding an inorganic phosphate molecule to a molecule of ADP yields a molecule of ATP and $\Delta G = +7.3$ kcal/mol. Therefore, these results are consistent with the Gibbs free energy equation.

In addition to deepening our understanding of energy coupling, Demirel suggests how an understanding of the thermodynamic properties of living systems, including free energy, begins to shed light on the idea of information. Drawing on Wicken (1987), Demirel observes that "without information the inflow of energy would not lead to self-organization" (47). Information in this sense, Demirel argues, is more than information in the Shannon and Weaver (1949) sense; it is functional and can be thought of as information in both an "instructional" (Brooks and Wiley 1988) and "control" sense (Corning and Kline 1988; McIntosh 2006), as it requires information that creates complex structures—for example, enzymatic proteins—and metabolic pathways that productively channel the flow of energy both within an organism and between the latter and its environment.

II Information and Biological Function

These last remarks hint at the central question of the present volume: what *is* information as it pertains to biological organization? Or, more succinctly, what is bioinformation? At the beginning of chapter 3, "Bioinformation as a Triadic Relation," Alfredo Marcos identifies a conceptual difficulty we need to confront before attempting to answer this question. This difficulty is that the contemporary concept of information is really a "family of concepts whose members lack any clear interconnection" (55). In the first part of his discussion, Marcos briefly outlines the history that produced this plurality of meanings. The main points of his outline are (1) that, in ancient and medieval philosophy, a central meaning of the Latin verb *informo* was "to inform or shape" and that this meaning influenced that of the Latin noun *informatio* (the word from which the English word "information" is derived) so that it could mean not just "idea" or "representation" but the "action whereby a thing is shaped" (this last meaning was then applied to biological phenomena, including, embryological development); (2) that early modernism, in rejecting scholasticism, also rejected this meaning, substituting in its place a noncausal meaning of information as idea or representation; (3) that the development of communications technology in the nineteenth and twentieth centuries eventually led to the idea of information as a measurable commodity—a development that found expression in the communication theory of Shannon and Weaver (1949); (4) that this theory further evolved in a way that enabled it to be formulated independently of the particulars of the Shannon-Weaver telegraphy-based paradigm; and (5) that the idea of information also gradually acquired a semantic and functional/pragmatic meaning, and that these meanings were later applied to both biological phenomena and biologically based accounts of cognition.

In spite of the fact that the contemporary idea of information has become a plurality of ideas, Marcos believes that a significant degree of unity can be found within this plurality and that a distinction originally made by Weaver (Shannon and Weaver 1949) hints at how we might find it. This distinction concerns the basic types of information-related problems that communication theory confronts. As already suggested, these are (1) syntactic problems that concern the amount of information a message can contain (see Shannon and Weaver 1949; Kolmogorov 1965; Solomonoff 2003; Li and Vitányi 1997; Grünwald and Vitányi 2003); (2) semantic problems that concern the meaning of information, that is, the system that a message conveys information about (see Barwise and Seligman 1997); and (3) pragmatic problems that concern whether a message successfully influences the receiver's behavior in a desired way. A central thesis of Marcos's discussion is that the third of these three types of problems is more fundamental than the other types and that they can be derived from it through "abstraction, ellipsis, or addition." As Marcos observes, "the mere material transfer of what has been reproduced would not be information, while reproduction itself is worthless unless it refers to what has been reproduced. The receiver of information can be so-called only if it can relate what is received to what was emitted" (106). For these reasons, Marcos holds that the fundamental concept of information is relational in nature; that is, a message (m) informs a receiver (R) about some system of reference (S). He also holds that this concept can be modified to meet the requirements of bioinformation, where a message's effect on a receiver can be construed to include its capacity to cause a functional or adaptive response in an organism.

Marcos derives from this relational definition of information two important results. First, a recurring controversy concerning the nature of information is its location. Just where does it reside? Is it a property of things, like order, complexity, and diversity are? Or is it itself a thing or, more precisely, a basic reality on a par with matter and energy? As should now be apparent, it belongs to neither of these categories but is instead relational in nature (see also Barwise 1986; Dennett 1987; MacKay 1969; Küppers 1990; Queiroz, Emmeche, and El-Hani 2005). Consequently, the traditional problem of the specific location of information need not arise.

A second consequence of Marcos's relational definition is that it can be used to derive a concept of the measurement of information that can be applied to biological functionality. This is noteworthy, because traditional concepts of information measurement capture genetic content without also capturing biological functionality, yet concepts that purport to capture such functionality produce counterintuitive results, such as the identification of maximum information with maximum redundancy. As we have seen, Marcos maintains that the core meaning of information is pragmatic or functional in nature. Thus the aim of a concept of the measurement of information is to determine the magnitude of this functional effect. According to Marcos, this magnitude reflects the difference in the recipient's knowledge of the system of reference before and after receipt of a message (see Marcos's discussion for the formal presentation of this concept; see also Peirce 1932–1935; Popper 1990; MacKay 1969; Dretske 1981; Wicken 1987, 1988). The greater this difference, the greater the amount of information contained in such a message.

As we have seen, Marcos employs a distinction made by Weaver in order to defend a triadic-relational account of information, including bioinformation. In chapter 4, "The Biosemiotic Approach in Biology: Theoretical Bases and Applied Models," João Queiroz and his colleagues use the resources of biosemiotics—that is, the study of sign processes in biological systems—to defend a similar account. An important part of the methodology of biosemiotics is that it seeks to integrate the findings of a variety of biological sciences as well as those of chemistry and physics. But although an understanding of sign-related processes in biological systems depends on an understanding of the physics and chemistry of those processes, it is not reducible to the latter understanding. Thus, to try to capture the special nature of sign-related processes, biosemiotics employs the further methodological tenet of interpreting these perspectives from a Peirce-inspired theory of signs—a theory that although inherently formal, can be fruitfully applied to biological processes. Central to this theory is Peirce's idea of the triadic relationship of sign, object, and interpretant (CP 2:171, CP 2:274). According to this theory, a sign is anything that stands for something other than itself. But in order for a sign to convey information *about* an object, it must also communicate that information *to* something or someone, which is why this relationship must be triadic. In this way, the sign creates an effect upon this thing or person—this effect being the interpretant—thereby causing it to refer to the original object in the same way in which the sign itself does (CP 2:274). Moreover, this effect can itself become a sign, which means that it can represent the same object to another interpretant, and so on. Thus, the sign-object-interpretant relation can generate an indefinitely large chain of triads (CP 2:303).

Because Peirce adhered to a pragmatic theory of meaning, the effect that the sign has on an interpreter—that is, the interpretant—must be understood in pragmatic terms. This is especially important where the interpretive system is not a conscious mind, but, say, a cell or a part of a cell—an organelle—such as a ribosome. To express this fact, Peirce states that the object, as mediated by the sign, conveys a *form* to the interpretant (MS 793:1–3). This form, according to Peirce, is not a thing or simple feature of a thing, such as its shape, but a regularity or rule-governed tendency (CP 5:397). The sign, then, in conveying this regularity to the object, changes the state of the interpretive system, thereby constraining its own rule-governed tendencies. Understood in this way, the idea of form allows us to grasp the connection between Peirce's theory of signs and a Peirce-inspired biosemiotic account of information. According to this account, information is the conveyance of a form from the object to an interpretant through the mediation of a sign (Queiroz, Emmeche, and El-Hani 2005; El-Hani, Queiroz, and Emmeche 2006; Queiroz and El-Hani 2006a, 2006b). What is conveyed, as we have now seen, is a rule-like regularity that produces a further regularity, namely, the system's own interpretive response.

In the remainder of their discussion, Queiroz and his colleagues try to show that their Peircean theory of signs can be used to model two well-known biological processes: the first, the process by which DNA codes for an amino acid sequence (see also Queiroz, Emmeche, and El-Hani 2005; El-Hani, Queiroz, and Emmeche 2006); the second, signal transduction in B cell activation. In regard to the first process, it is sometimes thought that genetic information resides in the stretch of DNA that (indirectly) codes for a specific sequence of amino acids. But according to the authors, this is only a part of the triadic process, so it is at most information in a potential sense. (Note that this conclusion overlaps with Marcos's triadic-relational account of information.) For example, in transcription, a segment of DNA is converted to mRNA, via RNA polymerases, which function as one interpretive system, and ribosomes function as another in the translation of mRNA to amino-acid sequence. In addition, specific forms of aminoacyl-tRNA synthetase recognize the link between an amino acid and its corresponding tRNAs. Ultimately, therefore, genetic information in the actual sense is the entire triadic process—a process that ensures that the form of a protein (the object represented by a segment of DNA) is preserved from one cell generation to the next.

In the second model, the authors' biosemiotic perspective is used to shed light on signal transduction in B cell activation (see also Bruni 2003; Queiroz and El-Hani 2006a, 2006b; El-Hani, Arnellos, and Queiroz 2007). In signal transduction, an extracellular signal initiates a signaling process that is continued within the cell along various signaling pathways (see Reth and Wienands 1997). In the case of B cell activation, the extracellular signal is an antigen, which signals the presence of a pathogen within the organism. The response to the antigen is initiated when a ligand binds to a transmembrane protein, which in turn activates its associated transducers, resulting in a complex signaling process that can ultimately influence gene expression. Thus, the goal of the authors' model is to show how the different material bases of this complex signaling process are, nevertheless, responses to the same element—namely, the original pathogen. For example, the antigen, which is a sign of the pathogen, produces an interpretant: the phosphorylated state of the B cell receptor, which itself can be a sign that represents the pathogen to yet another interpretive system, and so on. In this way, the different material bases—including, ultimately, second messengers that influence B cell production—can be understood as parts of one and the same signaling process.

Expanding the Functionalist Account

Evolutionary Considerations One way in which to deepen our understanding of chemical signaling in the immune system is to make the evolutionary dimension of such processing somewhat more prominent than it is in biosemiotics. In chapter 5, "Problem Solving in the Life Cycles of Multicellular Organisms: Immunology and Cancer," Niall Shanks and Rebecca A. Pyles develop this response by showing that Darwinian evolutionary principles can be used to explain how problem-solving adaptations can occur at the cellular level of the immune system. They call this form of Darwinian evolutionary explanation "ontogenetic Darwinism" (Shanks 2004) to distinguish it from "phylogenetic Darwinism," which is the form of evolutionary explanation that Darwin originally defended. In phylogenetic Darwinism, problem-solving adaptations occur over the span of many generations, whereas in ontogenetic Darwinism, these adaptations can occur within days, weeks, and months. Again, in phylogenetic Darwinism, the authors explain, adaptations are possible because (1) a population of an organism exhibits variation in regard to an inherited characteristic possessed by its members; (2) the variations result in differential rates of reproductive success in their respective members; and (3) heritable variations that contribute to their members' differential success rate will be passed to their progeny. Of course, a variant characteristic counts as a genuine adaptation, not merely because it is inherited by successive generations, but because it can be shown to confer a "fitness advantage" on its members (Sober 2000). But as Shanks and Pyles argue, these conditions are also met at the cellular level—that is, where variations in a molecular characteristic are present in members of a population of cells. Here, too, different, heritable variations will lead to differential rates of reproductive success in members of the population of cells that exhibit such variations, with the result that variants that contribute to their members' reproductive success will usually be passed on to successive generations. Finally, as in the case of phylogenetic Darwinism, a variant characteristic counts as a problem-solving adaptation, because it confers a "fitness advantage" on its possessors.

To appreciate this more unusual version of Darwinian explanation, consider, again, the role played by B cells in adaptive immune response. As we know, the B cell is a Y-shaped molecule, each of whose two tips function as a separate binding site. The B cell's molecular shape is formed from four polypeptide chains: two identical light and two identical heavy chains. The stem of the Y, whose components are relatively constant, allow the molecule to be classified as one of a small number of immunoglobulins. On the other hand, the tips of the Y admit of enormous variation, which is why they here function as the relevant variant characteristic. Because of this variation, it is likely that one or a few members of the population of B cells will successfully bind to some degree to their respective antigen. This binding will result in a signal being sent to a specific group of T cells—namely, helper T cells—which in turn signal the B cell to reproduce itself. Initially, the degree of binding of the B cell to its respective antigen may still be relatively weak. However, daughters of the B cell are likely to undergo somatic hypermutation (Parham 2000) so that, over time, the affinity between the B cell and its respective antigen will significantly improve. As Shanks and Pyles observe, this improved fit will similarly receive "preferential signaling" from T cells, causing its subpopulation of B cells to enjoy greater reproductive success and to compete more effectively for its antigen than other B cell subpopulations.

Of course, the other side of the coin is that the principles of ontogenetic Darwinism apply to pathogens that threaten an organism no less than they do to the organism itself. For example, another specialized group of T cells, regulator T cells, help ensure immunological homeostasis by appropriately reducing the population of B and T cells, as, for example, when a pathogen has been destroyed. Cancer, however, uses chemical misinformation to cause regulator T cells to reduce the population of other T and B cells when it is not beneficial to the organism (Knutson et al. 2007). For this reason, the idea of ontogenetic Darwinism can contribute to a medical as well as a more abstract theoretical understanding of molecular informational mechanisms.

Earlier we saw how, in biosemiotics, a Peirce-inspired theory of signs enables us to appreciate extended sign-related processes, such as signal transduction. The principal reason for this is that the effect of a sign on an interpretive system—namely, the interpretant—can itself be a sign. For this reason, a multitude of signs—including first and second messengers—can stand in relation to one another as parts of a single chemically based information process. But by viewing this process within the perspective of ontogenetic Darwinism, we enjoy the further advantage of understanding it as an *evolving*, or, alternatively, an *adaptive* process. As we have just seen, the reason why the process can be understand in this second way is that it can be shown to meet the same evolutionary explanatory criteria met by phylogenetic Darwinism, though of course on a much different spatial and temporal scale.

An evolutionary perspective also informs Moreno and Ruiz-Mirazo's "The Informational Nature of Biological Causality" (chapter 6). The starting point of the authors' discussion is the open-ended nature of Darwinian evolution. This means that the trajectory of a population of an organism transcends the ontogenetic trajectory of its individual members in a way that is not predetermined. A main objective of the authors' discussion is to try to explain how organisms originally acquired this evolutionary capacity. According to Moreno and Ruiz-Mirazo, a partial answer to this question depends on an understanding of the organizational structure of the precursors of organisms. As we saw in our discussion of their account of definition, the authors maintain that although the evolution of a form of recursive self-maintenance is necessary to account for such structure (Bickhard 2004), it is not sufficient. More specifically, protocellular organisms began to develop an "infrastructure" whose processes operated on a much slower time scale than its remaining processes, so that the former processes were "partially decoupled" from the latter ones. This decoupling was strengthened by the emergence of a simple template—for example, one-polymer RNA—that could store and reproduce the chemical structure of the organism's processes. Although some argue that the template depended on an early form of natural selection rather than on the organism's organizational structure, the authors maintain that without a sufficiently robust form of self-maintenance, natural selection could not result in increasing organizational complexity (see Wicken 1987). Both conditions, then, are needed to account for the emergence of the sort of template that can maintain and propagate an organism's organizational structure.

Although the combined effect of these conditions could explain how organisms could begin to evolve, it is insufficient, the authors argue, to explain evolution in the previously noted open-ended sense of the term. This is because the separate aims of storing and copying biological organization and of translating this organization into functional structure can interfere with each other when a single type of molecular component is required to perform both tasks (Moreno and Fernández 1990; Benner 1999). For this reason, an even stronger form of decoupling is needed: on the one hand, enzymes that carefully regulate metabolic function; on the other hand, templates that specialize in storing and transmitting information. Because an organism's template, now encoded in DNA's one-dimensional sequence of nucleotides, can function independently of the time-scale of enzyme-controlled chemical reactions (Pattee 1977), the template can not only produce the enzymes that regulate lower-level chemical activities, but can also produce variant sequences, thereby making open-ended evolution possible.

This stronger sense of decoupling also sets the stage for the central thesis of the authors' discussion, namely, the special ways in which the two decoupled domains interact with each other. The authors' answer to this question—an answer in which each domain influences and is influenced by the other—is articulated in broadly Aristotelian (1995) terms. According to this answer, sequences of nucleotides code for specific sequences of amino acids that define the enzymes that regulate chemical activity in the individual organism domain. At the same time, the coding process is carried out by means of some of these structures. Thus, in a recursively circular way, the translation from DNA to amino acid sequence is achieved by means of structures that are also the product of this translation. In broadly Aristotelian terms, the amino acids that are the building blocks of protein structures can be said to constitute the *material cause* of the process of protein synthesis, while the enzymes that polymerize them are its *efficient cause*. However, the process also reveals a *formal cause*, because DNA is a template for producing a particular protein in virtue of its *sequence* of nucleotides—in other words, because of the specific way in which the nucleotides are arranged. Finally, this cause is also significant from an informational perspective. As we have seen, the DNA template occupies a stable domain that has been effectively decoupled from the organism's enzyme-regulated metabolic activity. This stability ensures not only the creation and preservation of

DNA's nucleotide sequence, but also its rearrangement in potentially functionally significant ways—a rearrangement that makes possible a truly open-ended evolution (see Pattee 1977, 1982).

Which, Where, and When: The Role of Epigenetic Information As stated earlier, both Shanks and Pyles and Moreno and Ruiz-Mirazo rely on various evolutionary considerations to defend their respective accounts of chemical signaling. But such an account can also derive support from recent work in developmental biology. This work deserves our attention, because it relies on a combination of genetic and epigenetic information. In chapter 7, "The Self-construction of a Living Organism," Natalia López-Moratalla and María Cerezo examine the role that these two types of information play in regulating an organism's capacity for self-development. As the authors observe, once an organism is constituted, its genetic information is sufficient to explain its biological identity. But this information is not sufficient to explain its individuality. According to the authors, organisms are spatiotemporal entities that undergo continuous self-development, so their individuality must be understood in terms of the overall trajectory of their development. Furthermore, the countless changes that shape this trajectory depend on specific genes expressed at specific space and time coordinates. This, then, is the role of epigenetic information. Of course, it is not the only source of biological information, as portions of the DNA sequence also regulate development. The main point, though, is that epigenetic information plays a crucial role in regulating the continuous changes that enable an organism to be a self-developing member of its species.

Now, epigenetic information depends on epigenetic mechanisms that generate such information. As noted previously, an organism's self-development depends on gene expression at specific space and time coordinates. Gene expression, in turn, depends on the processes of transcription (DNA to mRNA) and translation (mRNA to amino-acid sequence). Epigenetic mechanisms, therefore, generate information by selectively influencing both transcription and translation processes. Two familiar mechanisms are methylation and demethylation, which influence gene repression and expression, respectively, during the transcription process. For example, in demethylation, methyl groups can be attached or removed from DNA segments, thereby increasing epigenetic information (Jenuwein and Allis 2001; Jones and Takai 2001; Strahl and Allis 2000; Torres-Padilla et al. 2007; Turner 2000). In the case of translation, alternative splicing—a mechanism highly responsive to environmental factors—recombines mRNA exons, enabling a single gene to code for different proteins.

Because an organism is continuously developing, the epigenetic information that shapes this development is similarly changing. Thus, the phenotypic expression of the organism's development at any given moment will be the outcome of its current epigenetic information, current environmental factors, including its intra- and intercellular and its external environment, and its genetic information. But because this moment is itself the outcome of a prior phenotypic expression, the state of these causal factors will be constrained by their state at an earlier time, and so on. For this reason, the organism's epigenetic information is far more dynamic than its genetic information, which is not influenced by the previous environmental factors and which, once fully constituted, remains essentially fixed.

An important consequence of López-Moratalla and Cerezo's epigenetic account of an organism's self-development is that it enables them to present an interesting middle-ground position in the contemporary version of the classic debate between preformationism and vitalism: namely, the debate between genetic determinism and extreme contingentism or environmental plasticity (Maynard Smith 2000; Oyama 1985; Oyama, Griffiths, and Gray 2003). This compromise position, which (like Moreno and Ruiz-Mirazo) they use Aristotelian terminology to describe, is that the form of an organism is inherent in the organism in one sense, though not in another. It is inherent in the organism because genetic information is an inherent feature of the organism. It is not inherent, however, in the sense that form is also generated through the developmental process—that is, through mechanism-generated epigenetic information—and because this information is co-constituted by various types of environmental factors.

In chapter 8, "Plasticity and Complexity in Biology: Topological Organization, Regulatory Protein Networks, and Mechanisms of Genetic Expression," Luciano Boi further contributes to our understanding of epigenetic information. As we have seen, such information depends on mechanisms that selectively influence transcription and translation processes. However, these and other epigenetic mechanisms are in turn influenced by the topological and dynamic properties of the organizational structures in which they occur. Of special importance in this regard are the properties of chromosome and chromatin. As we know, the enormous amount of DNA contained within a eukaryotic cell must be stably packaged within its chromosomes—a goal achieved by means of specialized proteins that fold and compact the DNA. However, this complex of DNA and proteins, referred to as *chromatin*, must also be flexible enough to accommodate the needs of DNA replication and repair as well as gene expression (Vogelauer et al. 2000; Lutter, Judis, and Paretti 1992; Fraga et al. 2005; Almouzni et al. 1994).

For example, in chromatin remodeling, certain proteins use the energy of ATP hydrolysis to modify chromatin structure so that other proteins can gain access to relevant DNA segments. Further evidence of chromatin's dynamic nature is found at the level of the nucleosome, the level in which DNA is wound around a core of histone molecules. For example, patterns of histone tail modification can attract other proteins and can signal the initiation or completion of information-related processes, including gene expression. For these reasons, an understanding of the topology and dynamics of chromosome and chromatin reveals a level of information over and above that of the linear DNA sequence (Abbott 1999; Coen 1999; Cornish-Bowden and Cárdenas 2001; Jaenisch and Bird 2003; Misteli 2001).

Because chromatin and chromosome dynamics shape gene expression, they not only deepen our understanding of the developmental processes examined by López-Moratalla and Cerezo, but also those explored in Shanks and Pyles's discussion of the relevance of ontogenetic Darwinism to the study of cancer. For example, chromosome aberrations can influence transcriptional processes so as to silence tumor suppression genes. One mechanism that plays a role in these processes is CpG island hypermethylation. CpG islands often appear at promoter regions of a gene, that is, untranscribed regions that turn transcription processes "on" or "off." As Boi explains, abnormal chromatin structure hypermethylates normally unmethylated CpG islands, thereby silencing tumor suppression genes.

Because the study of chromosome and chromatin dynamics promises to shed light on an array of topics, including gene expression, embryology, cell regulation, and tumorigenesis, Boi's discussion points to a unified vision of the biological sciences. Essentially, biology must move away from its previous genetic and molecular approach toward one that is epigenetic and highly integrative. Boi also predicts that future insights into the nature of epigenetic information will most likely occur at areas where mathematics (especially topology), physics, and biology overlap.

III Information and the Biology of Cognition, Value, and Language

Until now, our discussion has focused on information-related ideas and processes at a purely biological level. As we know, however, these ideas also play a role in our efforts to understand the biological basis of cognition, value, language, and even personality. Thus, a more complete appreciation of information-related perspectives must take into account relevant findings in neuroscience, psychology, linguistics, philosophy, and other related disciplines. This is what we do in the remainder of our volume, where our procedure is to begin with cognitive and evaluative capacities that apply to a broad range of organisms, then gradually proceed to more complex versions of these capacities—ultimately, to those that define our human nature.

In chapter 9, "Decision Making in the Economy of Nature: Value as Information," Benoit Hardy-Vallée outlines a theory of biological decision making that applies to all "brainy animals." In doing so, he departs from long-standing traditions in both economics and philosophy, in that the target of his explanation is not the theoretical, linguistic, rationality of *Homo sapiens*, but the much wider class of non–linguistically mediated decisions that we and other animals share in common. To this end, Hardy-Vallée locates the idea of biological decision making within the context of Darwin's theory of evolution, both at a general level and in regard to an important refinement of that theory—namely, Darwin's notion of the economy of nature. First, natural selection suggests that brainy animals are inherently goal-oriented, value-based information processors. Without the tacit goals of survival and reproduction, it is unlikely that an animal's genes will be represented in future generations, as mating can be a difficult and costly biological activity (Darwin 1859/2003). But second, Darwin's concept of the economy of nature provides a framework for assessing natural rationality in a more agent-relative way than the statistical features of natural selection permit. That is, natural selection describes population dynamics over long periods of time. In contrast, a theory of natural rationality—of biological decision making—requires a more refined assessment of the merit or demerit of individual animals' actions. The economy-of-nature concept facilitates such a refinement, in that it is a methodological principle for modeling animals as scarce-resource maximizers and their brains as decision-making organs (Cartwright 1989).

For example, optimal foraging theory (OFT) models animals as net caloric intake maximizers (MacArthur and Pianka 1966; White et al.

2007). Animals are thus viewed as seeking to find ecologically situated nutrients as efficiently as possible. To make such models ecologically realistic, a host of parameters must be included, such as where it forages, what prey it seeks, and how it allots its time. The quantitative expression of these and other parameters can be used to derive calorie-maximizing algorithms that can predict an animal's behavior.

Because the valuation mechanisms are instantiated in animals' neural systems, the theoretical framework appropriate for modeling such decisions is the nascent discipline of neuroeconomics, which studies the neural basis of cost/benefit computations (Glimcher 2003). Here, Hardy-Vallée argues that "much of goal-oriented neural computation is realized by midbrain dopaminergic systems activity" (Montague et al. 2004). The importance of work on dopaminergic systems for a theory of naturalistic rationality stems from the fact of their ubiquity in nature, operating even in fruit flies (Zhang et al. 2007). Hardy-Vallée further suggests that temporal-difference (TD) learning algorithms (Niv et al. 2005) can successfully model dopamine-mediated decision making (Montague 2006).

As experiments by Antonio Damasio (1994) and his colleagues reveal, effective decision making also requires well-functioning affective states. Hardy-Vallée articulates three different affective mechanisms roughly differentiable by neurobiological and phylogenetic complexity. First, there are core affect mechanisms, subserved by phylogenetically older brain structures that encode the magnitude and valence of stimuli. Such structures include the amygdala for fear responses, the anterior insula for disgust responses, and the nucleus accumbens for pleasure responses (Griffiths 1997; Lieberman 2007). Second, there are emotion-monitoring mechanisms that integrate naturally selected affective dispositions with agent-relative (memory-dependent) learning. Areas such as the orbitofrontal cortex and anterior cingulate cortex are particularly important for such integration. Third, there are control mechanisms that allow affect mechanisms to be overridden by high-level representation of the sort that occurs in the dorsolateral prefrontal cortex. Given the complexity of the foregoing picture of value regulation, Hardy-Vallée advocates an "evolutionary mechanistic functionalism" according to which older neural structures can be redeployed in novel contexts through the activity of phylogenetically newer neural structures.

The philosophical implications of the foregoing view of biological decision as regards the idea of information are striking. If perception and decision making are guided by evolutionary prescribed and inherently value-based goals, then animals filter the world relative to the utility of the world in order to affect the attainment of their agent-relative goals. If this account is correct, then the traditional view of semantic information as preceding evaluative information is mistaken; rather, we should explain the first kind of information within the broader and more basic context of the second.

In chapter 10, "Information Theory and Perception: The Role of Constraints, and What Do We Maximize Information About?," Roland Baddeley, Benjamin Vincent, and David Attewell explore recent efforts, including their own, to use Shannon's (1948) information-theoretic perspective to understand early visual processing in human and nonhuman primates. In early visual processing, varying patterns of light are first detected by the retina's rods and cones, then transmitted via retinal ganglion cells to the lateral geniculate nucleus, then transmitted again to neurons in V1. Since the work of Hubel and Wiesel (1962), we have learned a great deal about the function of these neurons, which is essentially that of coding for the orientation of edges or lines segments. How V1 neurons extract this information is a challenging problem, as it is not explicit in the varying patterns of light that are normalized and coded for prior to neurons receiving information in V1. There is a further problem, in that information encoded in huge numbers of rods and cones must be compressed when it reaches the retinal ganglion cells, which can devote only about 1 million cells per eye—in other words, an information bottleneck problem. The researchers' objective, then, is to use information theory to shed light on these issues.

As the authors observe, researchers who address these issues agree that there is a need to understand how early vision maximizes information about the world subject to certain constraints. They also agree that input signals must reflect realistic images—those that animals are likely to encounter—rather than random patterns of light, and that these images should be modeled by means of linear filters. But researchers disagree about additional constraints that are needed to understand such visual processing. For example, Baddeley and his colleagues argue that attempts to understand these constraints simply in terms of the number of filters needed to represent images fail to capture accurately features of receptive fields when perception occurs for any significant length of time (see Foster and Ward 1991). To overcome this difficulty, they propose a model that takes into account not only the number of linear filters, but also energy-related considerations. As the authors remind us, the brain is a fairly prodigious consumer of energy (Rolfe and Brown 1997), so it is likely that it evolved to handle perception efficiently. But the task of properly formulating energy-related constraints is difficult. For example, in a model that addresses the previously described information bottleneck, reducing the average firing rate of V1 neurons produces negligible energy savings. Instead, Baddeley and his colleagues propose a model according to which there is a reduction in synaptic activity, a reduction that yields significant energy savings and that accords with our understanding of the properties of retinal ganglion cells (Vincent and Baddeley 2003; Vincent et al. 2005).

Despite these promising results, the researchers are candid about the limitations of their energy-constrained information-maximizing model. Simply put, this is because perception is not about light patterns, but about the world, especially its behaviorally significant features. Their most recent work, therefore, tries to extend their previous research so that it can determine when light variation is indicative of such features. But to make this determination requires statistics about behaviorally relevant features of the world, not just about patterns of light in natural scenes, which is why their information-theoretic approach to this problem must be enhanced—for example, by incorporating Bayesian techniques.

What is interesting about this last response is that it hints at a basic difference between information in a causal sense and information in an intentional sense, between information as a pattern of light and information as representing behaviorally significant features of the world. In chapter 11, "Attention, Information, and Epistemic Perception," Nicolas J. Bullot further develops this distinction, using a methodology that, like that of Baddeley and his colleagues, includes information theory and neurobiology, but that also combines these disciplines with an epistemological analysis of perception. As Bullot reminds us, modern information theory began with Claude Shannon's mathematical theory of information (Shannon 1948). However, as this theory has been assimilated into biological psychology, philosophy, and cognitive science, it has sparked a wide diversity of approaches (Cherry 1953; Miller 1956; Broadbent 1958; Moray 1969; Dretske 1981, 1994; Adams 2003). In order to bring clarity to this discussion, Bullot distinguishes two concepts of information beyond the formal, mathematical sense developed by Shannon: *causal information* and *semantic information*. Causal information is an objective property of things in the world. If As always lead to Bs, or have a propensity to lead to Bs, then As carry causal information about Bs. For example, because an increase in the number and intensity of earthquakes in the vicinity of a volcano tends to be linked with impending volcanic activity, seismic monitoring of the volcano can give scientists information about the likelihood of an impending eruption. On the other hand, semantic information differs from causal information in that it is intentional. The earthquakes are *caused* by movements of magma deep beneath the volcano, but the volcano does not *intend* to communicate anything by them. However, if a scientist detects the earthquakes and sends a message to civil authorities that the volcano may soon erupt, his or her message is an example of semantic information, because the scientist intends to communicate. Because we express our knowledge to ourselves in the same way that we would report it to others (the scientist would think to himself or herself, "The volcano may erupt soon"), Bullot thinks we must explain how the mind moves from the causal information that bombards our sense organs to the semantic information that expresses knowledge.

To explain this transition, Bullot focuses on the nature of perceptual *attention*, which, he believes, is the fundamental cognitive means by which humans attain knowledge of objects and agents in their environment. According to Bullot, there are two kinds of attention, both of which are essential to human cognitive processes. First, there is *overt* attention, for example, directing the eyes toward an object in the world. Without overt attention, human beings cannot attend to the relevant causal information about a particular target. However, pointing to research showing that overt attention is not sufficient to explain cognitive access (Posner 1978, 1980; Posner, Nissen, and Ogden 1978; Posner, Snyder, and Davidson 1980), Bullot argues that we must also consider *covert* information processing. That is, in addition to directing gaze, one must actually process the causal information obtained (instead of, e.g, staring blankly). Bullot argues that overt and covert attention are distinct, but jointly necessary, conditions for cognitive access.

The required information processing, Bullot argues, is performed by a hierarchical system of sensory-motor *routines* and *procedures* that are necessary to complete a given epistemic task. The procedure for completing each kind of task will make use of a series of attentional routines (e.g., focusing on a particular object) performed in order, and each routine may be used by a variety of different epistemic tasks. It is these procedures that allow the human mind to "navigate" the causal structure of the world, identifying particular objects within it, and making judgments about the truth or falsity of causal propositions about those objects. From this, Bullot concludes that knowledge is constituted by the attentional faculty's ability to focus on particular objects and extract semantic information from their causal information.

The capacity to make perceptual judgments about objects and agents, that is, to ascribe general properties to such individuals enables us to understand the need for a language that enables us to encode and articulate such judgments. In chapter 12, "Biolinguistics and Information," Cedric Boeckx and Juan Uriagereka use information-theoretic concepts to examine the biological basis of this cognitive-based language. As in the case of Bullot, information theory is the authors' point of departure, but, unlike Bullot, they not only distinguish the Shannon and Weaver (1949) account of information from their own semantic account, but they actually draw inspiration from the former in their effort to understand the latter. Here is how. According to Shannon, information reduces uncertainty regarding the receiver's knowledge of the world. This means that the amount of information reflects the difference between what the receiver already knows and what he or she learns from the signal. But such a view suggests that in order to learn something new, the receiver must already know something. Stated another way, without any expectations about how the world is, a signal would have to convey an infinite amount of information in a finite time. As this cannot occur, the receiver must possess an inherent structure—for Boeckx and Uriagereka, an inherent linguistic structure—that limits the range of possible learning.

In defending the idea that language has a partly innate structure, Boeckx and Uriagereka acknowledge their indebtedness to Chomsky in the following ways. First, as already suggested, they endorse Chomsky's basic insight (1956, 1957, 1968) that there are innate features of the brain that constrain the kinds of language that are possible ("universal grammar"). To illustrate this point, they use an example that Chomsky (1968) drew from Peirce (1931–1966). As Peirce notes, the little chicken does not, once hatched, work through all the possibilities for action, then somehow settle on pecking for food. Rather, it has an innate instinct to peck, an instinct that rules out any other possibility. Instinct thus eliminates uncertainty about how the chicken will act. Peirce claims that humans have similar structures, and Chomsky argues that some of these structures restrict the types of grammar that are possible for human languages.

Second, Boeckx and Uriagereka point out that as Chomsky's theories developed, he recognized that it is not just genetic factors (innate structures in the brain) or environmental factors (the linguistic environment in which the child develops) that shape the development of human languages. Chomsky (1995, 2005) argued that the amount of genetic information needed to specify the brain structures that make language possible may be minimal. Instead, a third set of factors, epigenetic principles of growth and form, could play a critical role in shaping the development of linguistic capacities (see Gould 2002).

Finally, Boeckx and Uriagereka focus on the unique contribution that language makes to human cognitive capacities. They take up the suggestion of Fodor (1983) and Pylyshyn (1984, 1999) that the mind is composed of modules that encapsulate the cognitive processes inside the module, thereby preventing them from having access to outside cognitive influences. Contemporary research (e.g., Spelke 2003; Carruthers 2002, 2006; Pietroski 2005; Mithen 1996; Arp 2008; Boeckx, forthcoming) suggests that language is able to combine information from various modules that would otherwise remain isolated. Thus, for example, nonhuman animals seem only able to use either relative or absolute positional information, though humans are able to use both kinds of information (Gallistel and Cramer 1996). Language is, therefore, a crucial ingredient that contributes to the flexibility and creativity of human cognitive processes.

Finally, in chapter 13, "The Biology of Personality," Aurelio José Figueredo, W. Jake Jacobs, Sarah B. Burger, Paul R. Gladden, and Sally G. Olderbak offer us a way of integrating the previously discussed biological, cognitive, emotional, and linguistic levels. The authors liken the biology of personality to the hierarchical organization of life (Mayr 1982), where there are multiple levels of organization—for example, molecules, macromolecules, organs, organ systems, and the like—and where there is causal interaction both within and among the many levels. Because the authors deny that this causal interaction is deterministic, they require a somewhat weaker notion of constraint. To this end, they appeal to the Shannon-Weaver model of "information or reduction of uncertainty, as the degree of *constraint* that each level of hierarchical organization exerts upon another" (Shannon and Weaver 1949, 698–699; see also Dretske 1981). According to the authors, the greater the number of identified constraints a given level exerts on another, the more we are able to make predictions about the latter's behavior. The authors are also attracted to the Shannon-Weaver model, because we can apply it across different hierarchical levels without having to worry about questionable semantic or functional interpretations of biochemical processes.

In the body of their discussion, the authors point to several illustrations of the sort of mutual causal constraint that makes the Shannon-

Weaver model appealing to them. For example, in their account of the genetics of personality, they remind us that because genes code for amino-acid sequences, they constrain but do not determine the development of behavior (West-Eberhard 2003). Because genes code for enzymes that regulate chemical reaction, they play a causal role in determining reaction rates that exert a continuous bottom-up effect on the organism. But gene function is, in turn, constrained by a variety of cellular environmental factors, including hormones, that at an epigenetic level influence gene expression (see also the discussions of López-Moratalla and Cerezo and Boi).

This complex picture of mutual constraint may also be found in the relationship between neurotransmitters and the stable pattern of personality traits they support. Here the starting point of the authors' discussion is the model of Zuckerman (1991, 1995), according to which the monoamine neurotransmitters dopamine, seratonin, and norepinephrine respectively influence approach, inhibition, and arousal, which in turn shape the personality traits, extraversion, impulsivity, and neuroticism. But, as the authors explain, the model quickly becomes complicated, as these neurotransmitters are influenced by one another, by other neurochemicals, and even by the way in which the behaviors that they respectively constitute interact. Furthermore, neurotransmitters and other neurochemicals must not only be able to support these stable traits, but they must also allow them to be flexibly applied to environmental particulars; otherwise, they would have little survival value for an organism. For example, both the LC-NE (locus coeruleus–norephinephrine) and dopaminergic systems' neurons admit of both phasic and tonic activity (see also the discussion of Boeckx and Uriagereka). In the case of LC neurons, the different levels of activity are thought to provide feedback that can appropriately support the organism's performance in the following way. When activity appears to the organism to be successful, neural bursts can create a memory of the activity, enabling the organism to call upon it in similar future situations, whereas when such activity appears unsuccessful for any relatively prolonged period, tonic neural activity allows the organism to disengage, freeing it to consider alternative strategies (Aston-Jones and Cohen 2005). Dopaminergic neurons similarly admit of tonic and phasic activity that allow the organism to predict, respectively, negative or positive reward (Aston-Jones and Cohen 2005; Montague, Hyman, and Cohen, 2004).

Regardless of whether we agree with their Shannon-Weaver model, Figueredo and his colleagues provide several helpful illustrations of mutual constraint across different levels of biological and psychological organization.

Conclusion

In this introduction, we have tried to organize our authors' contributions in order to show that they can complement, rather than simply conflict with, one another. As we have seen, their contributions are presented at different levels of biological organization and embody different, albeit overlapping, scientific and philosophical approaches to information-related issues within their respective levels. Some, like Baddeley and his colleagues, show how the Shannon-Weaver model actually informs their neuroscientific research, though in a way that also shows the model's limits. Others, like Marcos, Boeckx and Uriagereka, and Figueredo and his colleagues, use the Shannon-Weaver model as a point of departure for developing interpretations within their disciplines that go significantly beyond its original intent. In the remaining discussions, however, the objective was to develop functional, semantic, or in some sense intentional interpretations of information independently of the Shannon-Weaver model. At lower organizational levels, these functional interpretations were principally informed by biological, including genetic, evolutionary, and epigenetic considerations (Shanks and Pyles; Boi), but also sometimes interwoven with philosophical ones, such as a Peircean theory of signs (Queiroz and colleagues) or an Aristotelian account of formal causation (Ruiz-Mirazo and Moreno; López-Moratalla and Cerezo). Finally, at the level of emotion and cognition, the previous biological disciplines were combined with additional (and sometimes newer) disciplines, such as neuroeconomics and optimal foraging theory, to generate a value-based account of information (Hardy-Vallée), or epistemology, to provide a semantic account of perceptual information.

Our hope, then, is that these contributions will bring greater attention to issues surrounding the informational nature of biological organization and that they will inspire students and researchers to offer their own contributions and improvements.

References

Abbott, Alison. 1999. A post-genomic challenge: Learning to read patterns of protein synthesis. *Nature* 402 (6763): 715–720.

Adams, Frederick. 2003. The informational turn in philosophy. *Minds and Machines* 13 (4): 471–501.

Almouzni, Geneviéve, Saadi Khochbin, Stefan Dimitrov, and Adolfus Wolffe. 1994. Histone acetylation influences both gene expression and development of *Xenopus laevis. Developmental Biology* 165 (2): 654–669.

Aristotle. 1995. *Physics* and *Metaphysics*. In *The complete works of Aristotle*, 2 vols., ed. Jonathan Barnes. Princeton: Princeton University Press.

Arp, Robert. 2008. *Scenario visualization: An evolutionary account of creative problem solving*. Cambridge, Mass.: MIT Press.

Aston-Jones, Gary, and Jonathan D. Cohen. 2005. An integrative theory of locus coeruleus-norepinephrine function: Adaptive gain and optimal performance. *Annual Review of Neuroscience* 28: 403–450.

Barwise, Jon. 1986. Information and circumstance. *Notre Dame Journal of Formal Logic* 27 (3): 324–338.

Barwise, Jon, and Jerry Seligman. 1997. *Information flow: The logic of distributed systems*. Cambridge: Cambridge University Press.

Benner, Steven A. 1999. How small can a microorganism be? In *Size limits of very small microorganisms*, ed. National Research Council, 126–135. Washington, DC: National Academy Press.

Benner, Steven A., and A. Michael Sismour. 2005. Synthetic biology. *Nature Reviews: Genetics* 6 (7): 533–543.

Bickhard, Mark H. 2004. The dynamic emergence of representation. In *Representation in mind: New approaches to mental representation*, ed. Hugh Clapin, Phillip Staines, and Peter Slezak, 71–90. Amsterdam: Elsevier.

Boeckx, Cedric. Forthcoming. Some reflections on Darwin's problem in the context of Cartesian biolinguistics. In *The Biolinguistic enterprise: New perspectives on the evolution and nature of the human language faculty*, ed. Anna Maria Di Sciullo and Cedric Boeckx. Oxford: Oxford University Press.

Broadbent, Donald E. 1958. *Perception and communication*. London: Pergamon Press.

Brooks, Daniel R., and E. O. Wiley. 1988. *Evolution as entropy: Toward a unified theory of biology*. Chicago: University of Chicago Press.

Bruni, Luis E. 2003. A sign-theoretic approach to biotechnology. Unpublished Ph.D. dissertation, Institute of Molecular Biology, University of Copenhagen.

Cartwright, Nancy. 1989. *Nature's capacities and their measurement*. Oxford: Oxford University Press.

Carruthers, Peter. 2002. The cognitive functions of language. *Behavioral and Brain Sciences* 25 (6): 657–673.

Carruthers, Peter. 2006. *The architecture of the mind: Massive modularity and the flexibility of thought*. Oxford: Oxford University Press.

Cherry, E. Colin. 1953. Some experiments on the recognition of speech with one and two ears. *Journal of the Acoustical Society of America* 25 (5): 975–979.

Chomsky, Noam. 1956. Three models for the description of language. *IRE Transactions on Information Theory: Proceedings of the Symposium on Information Theory* IT-2 (3): 113–124.

Chomsky, Noam. 1957. *Syntactic structures*. The Hague: Mouton.

Chomsky, Noam. 1968. *Language and mind*. New York: Harrcourt, Brace, and World.

Chomsky, Noam. 1995. *The minimalist program*. Cambridge, Mass.: MIT Press.

Chomsky, Noam. 2005. Three factors in language design. *Linguistic Inquiry* 36 (1): 1–22.

Coen, Enrico. 1999. *The art of genes: How organisms make themselves*. Oxford: Oxford University Press.

Corning, Peter A., and Stephen J. Kline. 1998. Thermodynamics, information and life revisited, part II: "Thermoeconomics" and "control information." *Systems Research and Behavioral Science* 15 (6): 453–482.

Cornish-Bowden, Andreas, and Miguel Cárdenas. 2001. Complex networks of interactions connect genes to phenotypes. *Trends in Biochemical Sciences* 26 (8): 463–465.

Damasio, Antonio R. 1994. *Descartes' error: Emotion, reason, and the human brain*. New York: Putnam.

Darwin, Charles. 1859/2003. *On the origin of species by means of natural selection*. New York: Fine Creative Media.

Dennett, Daniel C. 1987. *The intentional stance*. Cambridge, Mass.: MIT Press.

Dretske, Fred I. 1981. *Knowledge and the flow of information*. Oxford: Blackwell.

Dretske, Fred I. 1994. The explanatory role of information. *Philosophical Transactions of the Royal Society A: Mathematical, Physical, & Engineering Sciences* 349 (1689): 59–69.

El-Hani, Charbel Niño, João Queiroz, and Claus Emmeche. 2006. A semiotic analysis of the genetic information system. *Semiotica* 160 (1/4): 1–68.

El-Hani, Charbel Niño, Argyris Arnellos, and João Queiroz. 2007. Modeling a semiotic process in the immune system: Signal transduction in B-cells activation. *TripleC: Cognition, Communication, Co-operation* 5 (2): 24–36.

Fodor, Jerry A. 1983. *The modularity of mind: An essay on faculty psychology*. Cambridge, Mass.: MIT Press.

Foster, David H., and Patrick A. Ward. 1991. Horizontal-vertical filters in early vision: Predict anomalous line-orientation identification frequencies. *Proceedings of the Royal Society B: Biological Sciences* 243 (1306): 83–86.

Fraga, Mario, Esteban Ballestar, Maria Paz, Santiago Ropero, Fernando Setien, Maria L. Ballestar, Damia Heine-Suñer, et al. 2005. Epigenetic differences arise during the lifetime of monozygotic twins. *Proceedings of the National Academy of Sciences of the United States of America* 102 (30): 10604–10609.

Gallistel, Charles R., and Audrey E. Cramer. 1996. Computations on metric maps in mammals: Getting oriented and choosing a multi-destination route. *Journal of Experimental Biology* 199 (1): 211–217.

Gánti, Tibor. 1975. Organization of chemical reactions into dividing and metabolizing units: The chemotons. *BioSystems* 7 (1): 15–21.

Glimcher, Paul W. 2003. *Decisions, uncertainty, and the brain: The science of neuroeconomics.* Cambridge, Mass.: MIT Press.

Gould, Stephen Jay. 2002. *The structure of evolutionary theory.* Cambridge, Mass.: Belknap.

Griffiths, Paul E. 1997. *What emotions really are: The problem of psychological categories.* Chicago: University of Chicago Press.

Grünwald, Peter D., and Paul M. B. Vitányi. 2003. Kolmogorov complexity and information theory. *Journal of Logic Language and Information* 12 (4): 497–529.

Hubel, David H., and Torsten N. Wiesel. 1962. Receptive fields, binocular interaction and functional architecture in the cat's visual cortex. *Journal of Physiology* 160 (1): 106–154.

Jaenisch, Rudolf, and Adrian Bird. 2003. Epigenetic regulation of gene expression: How the genome integrates intrinsic and environmental signals. *Nature Genetics* 33 (March): 245–254.

Jenuwein, Thomas, and C. David Allis. 2001. Translating the histone code. *Science* 293 (5532): 1074–1080.

Jones, Peter A., and Daiya Takai. 2001. The role of DNA methylation in mammalian epigenetics. *Science* 293 (5532): 1068–1070.

Joyce, Gerald F. 1994. Foreword. In *Origins of life: The central concepts,* ed. David W. Deamer and Gail R. Fleischaker, xi–xii. Boston: Jones and Bartlett Publishers.

Kauffman, Stuart. 2000. *Investigations.* Oxford: Oxford University Press.

Kauffman, Stuart. 2003. Molecular autonomous agents. *Philosophical Transactions of the Royal Society: Mathematical, Physical and Engineering Sciences* 361 (1807): 1089–1099.

Keller, Evelyn Fox. 2007. A clash of two cultures. *Nature* 445 (7128): 603.

Knutson, Keith L., Mary L. Disis, and Lupe G. Salazar. 2007. CD4 regulatory T cells in human pathogenesis. *Cancer Immunology, Immunotherapy* 56 (3): 271–285.

Kolmogorov, Andrey N. 1965. Three approaches to the quantitative definition of information. *Problems of Information Transmission* 1 (1): 1–7.

Küppers, Bernd-Olaf. 1990. *Information and the origin of life.* Cambridge, Mass.: MIT Press.

Li, Ming, and Paul Vitányi. 1997. *An introduction to Kolmogorov complexity and its applications.* New York: Springer-Verlag.

Liebermann, Philip. 2007. The evolution of human speech. *Current Anthropology* 48 (1): 39–66.

Lutter, Leonard, Luann Judis, and Robert Paretti. 1992. Effects of histone acetylation on chromatin topology in vivo. *Molecular and Cellular Biology* 12 (11): 5004–5014.

MacArthur, Robert H., and Eric R. Pianka. 1966. On optimal use of a patchy environment. *American Naturalist* 100 (916): 603–609.

MacKay, Donald M. 1969. *Information, mechanism and meaning.* Cambridge, Mass.: MIT Press.

Mayr, Ernst. 1982. *The growth of biological thought: Diversity, evolution, and inheritance.* Cambridge, Mass.: Harvard University Press.

Maturana, Humberto R., and Francisco J. Varela. 1973. *De máquinas y seres Vivos: Una teoría sobre la organización biológica.* Santiago de Chile: Editorial Universitaria S.A.

Maynard Smith, John. 2000. The concept of information in biology. *Philosophy of Science* 67 (2): 177–194.

McIntosh, Andy. 2006. Functional information and entropy in living systems. *WIT Transactions on Ecology and the Environment* 87:115–126.

Miller, George A. 1956. The magical number seven, plus or minus two: Some limits on our capacity for processing information. *Psychological Review* 63 (2): 81–97.

Misteli, Tom. 2001. Protein dynamics: Implications for nuclear architecture and gene expression. *Science* 291 (5505): 843–847.

Mithen, Steven. 1996. *The prehistory of the mind: The cognitive origins of art, religion, and science.* London: Thames & Hudson.

Montague, P. Read, Steven E. Hyman, and Jonathan D. Cohen. 2004. Computational roles for dopamine in behavioral control. *Nature* 431 (7010): 760–767.

Montague, P. Read. 2006. *Why choose this book? How we make decisions.* New York: Penguin.

Moray, Neville. 1969. *Attention: Selective processes in vision and hearing.* London: Hutchinson Educational.

Moreno, Alvaro, and Julio Fernández. 1990. Structural limits for evolutive capacities in molecular complex systems. *Biology Forum* 83 (2–3): 335–347.

Neumann, John von. 1949. *Theory of self-reproducing automata.* Urbana: University of Illinois Press.

Niv, Yael, Michael O. Duff, and Peter Dayan. 2005. Dopamine, uncertainty and TD learning. *Behavioral and Brain Functions* 1:6.

Oyama, Susan. 1985. *The ontogeny of information: Developmental systems and evolution.* Cambridge: Cambridge University Press.

Oyama, Susan, Paul E. Griffiths, and Russell D. Gray, eds. 2003. *Cycles of contingency: Developmental systems and evolution*. Cambridge, Mass.: MIT Press.

Parham, Peter. 2000. *The immune system*. New York: Garland Science.

Pattee, Howard H. 1977. Dynamic and linguistic modes of complex systems. *International Journal of General Systems* 3 (4): 259–266.

Pattee, Howard H. 1982. Cell psychology: An evolutionary approach to the symbol-matter problem. *Cognition and Brain Theory* 5 (4): 325–341.

Peirce, Charles S. 1931–1966. *Collected papers of Charles Sanders Peirce*, ed. Charles Hartshorne, Paul Weiss, and A. W. Burks. Cambridge, Mass.: Harvard University Press.

Peirce, Charles S. 1932–1935. *Collected papers of Charles Sanders Peirce*, 6 vols., ed. Charles Hartshorne and Paul Weiss. Cambridge, Mass.: Harvard University Press.

Pietroski, Paul M. 2005. Meaning before truth. In *Contextualism in philosophy: Knowledge, meaning, and truth*, ed. Gerhard Preyer and Georg Peter, 253–300. Oxford: Oxford University Press,

Popper, Karl. 1990. Towards an evolutionary theory of knowledge. In *A world of propensities*, ed. Karl Popper, 27–51. Bristol: Thoemmes.

Posner, Michael I. 1978. *Chronometric exploration of mind*. Hillsdale, N.J.: Lawrence Erlbaum Associates.

Posner, Michael I. 1980. Orienting of attention. *Quarterly Journal of Experimental Psychology* 32 (1): 3–25.

Posner, Michael I., Michel Nissen, and Michael Ogden. 1978. Attended and unattended processing modes: The role of set for spatial location. In *Modes of perceiving and processing information*, ed. Herbert L. Pick and Elliot Saltzman, 137–158. Hillsdale, N.J.: Lawrence Erlbaum.

Posner, Michael I., Charles R. R. Snyder, and Brian J. Davidson. 1980. Attention and the detection of signals. *Journal of Experimental Psychology: General* 109 (2): 160–174.

Pylyshyn, Zenon W. 1984. *Computation and cognition: Toward a foundation for cognitive science*. Cambridge, Mass.: MIT Press.

Pylyshyn, Zenon W. 1999. Is vision continuous with cognition? The case for cognitive impenetrability of visual perception. *Behavioral and Brain Sciences* 22 (3): 341–423.

Queiroz, João, and Charbel Niño El-Hani. 2006a. Semiosis as an emergent process. *Transactions of the Charles S. Peirce Society. A Quarterly Journal in American Philosophy* 42 (1): 78–116.

Queiroz, João, and Charbel Niño El-Hani. 2006b. Towards a multi-level approach to the emergence of meaning processes in living systems. *Acta Biotheoretica* 54 (3): 174–206.

Queiroz, João, Claus Emmeche, and Charbel Niño El-Hani. 2005. Information and semiosis in living systems: A semiotic approach. *SEED* 5 (1): 60–90. Available at http://www.library.utoronto.ca/see/SEED/Vol5-1/Queiroz_Emmeche_El-Hani_abstract.htm.

Reth, Michael, and Jürgen Wienands. 1997. Initiation and processing of signals from the B cell antigen receptor. *Annual Review of Immunology* 15: 453–479.

Rolfe, David F. S., and Guy C. Brown. 1997. Cellular energy utilization and molecular origin of standard metabolic rate in mammals. *Physiological Reviews* 77 (3): 731–758.

Sagan, Carl. 1970. Life. In *The Encyclopaedia Britannica*. London: William Benton.

Schrödinger, Erwin. 1945. *What is life? The physical aspect of the living cell.* New York: Macmillan.

Shanks, Niall. 2004. *God, the devil and Darwin: A critique of intelligent design theory.* Oxford: Oxford University Press.

Shannon, Claude E. 1948. A mathematical theory of communication. *Bell System Technical Journal* 27: 379–423, 623–656.

Shannon, Claude E., and Warren Weaver. 1949. *The mathematical theory of communication.* Urbana: University of Illinois Press.

Sober, Elliott. 2000. *Philosophy of biology.* Boulder, Colo.: Westview Press.

Solomonoff, Ray J. 2003. The universal distribution and machine learning. *Computer Journal* 46 (6): 598–601.

Spelke, Elizabeth S. 2003. What makes us smart? Core knowledge and natural language. In *Language in mind: Advances in the study of language and thought*, ed. Dedre Gentner and Susan Goldin-Meadow, 277–311. Cambridge, Mass.: MIT Press.

Strahl, Brian D., and C. David Allis. 2000. The language of covalent histone modifications. *Nature* 403 (6765): 41–45.

Torres-Padilla, Maria-Elena, David-Emlyn Parfitt, Tony Kouzarides, and Magdalena Zernicka-Goetz. 2007. Histone arginine methylation regulates pluripotency in the early mouse embryo. *Nature* 445 (7124): 214–218.

Turner, Bryan M. 2000. Histone acetylation and an epigenetic code. *BioEssays* 22 (9): 836–845.

Varela, Francisco. 1994. On defining life. In *Self-production of supramolecular structures*, ed. Gail R. Fleischaker, Stefano Colonna, and Pier Luigi Luisi, 23–31. Dordrecht: Kluwer Academic Publishers.

Vincent, Benjamin T., and Roland Baddeley. 2003. Synaptic energy efficiency in retinal processing. *Vision Research* 43 (11): 1283–1290.

Vincent, Benjamin T., Roland Baddeley, Tom Troscianko, and Iain Gilchrist. 2005. Is the early visual system optimised to be energy efficient? *Network* 16 (2–3): 175–190.

Vogelauer, Maria, Jiansheng Wu, Noriyuki Suka, and Michael Grunstein. 2000. Global histone acetylation and deacetylation in yeast. *Nature* 408 (6811): 495–498.

West-Eberhard, Mary Jane. 2003. *Developmental plasticity and evolution.* New York: Oxford University Press.

Westerhoff, Hans V., and Bernhard O. Palsson. 2004. The evolution of molecular biology into systems biology. *Nature Biotechnology* 22 (10): 1249–1252.

White, Donald W., Lawrence M. Dill, and Charles B. Crawford. 2007. A common, conceptual framework for behavioral ecology and evolutionary psychology. *Evolutionary Psychology* 5 (2): 275–288.

Wicken, Jeffrey S. 1987. *Evolution, thermodynamics and information: Extending the Darwinian program.* New York: Oxford University Press.

Wicken, Jeffrey S. 1988. Thermodynamics, evolution and emergence: Ingredients for a new synthesis. In *Entropy, information, and evolution: New perspectives on physical and biological evolution,* ed. Bruce H. Weber, David J. Depew, and James D. Smith, 139–188. Cambridge, Mass.: MIT Press.

Zhang, Ke, Jian Zeng Guo, Yueqing Peng, Wang Xi, and Aike Guo. 2007. Dopamine-mushroom body circuit regulates saliency-based decision-making in Drosophila. *Science* 316 (5833): 1901–1904.

Zuckerman, Marvin. 1991. *Psychobiology of personality.* Cambridge: Cambridge University Press.

Zuckerman, Marvin. 1995. Good and bad humors: Biochemical bases of personality and its disorders. *Psychological Science* 6 (6): 325–332.

I
The Definition of Life

1
The Need for a Universal Definition of Life in Twenty-first-century Biology

Kepa Ruiz-Mirazo and Alvaro Moreno

Twentieth-century biology was based on and shaped by the concept of the gene, thus yielding extraordinary results, like the development of molecular biology and the modern synthesis of evolutionary theory (Keller 2000). The idea that processes in biological systems could be explained solely in terms of the properties of their molecular components was so prominent that it overshadowed the idea that there is something else relevant for biology to do, namely, to explain the complex dynamic *integration* of those components into functional units—that is, living cells. This reductionist approach concealed the organicist currents of thought that, having emerged from the fields of theoretical biology and cybernetics, sought to overcome the classical confrontation between vitalism and mechanicism (von Bertalanffy 1952; Elsasser 1966; Maturana and Varela 1973; Rashevsky 1938; Rosen 1991).

However, since the dawn of the twenty-first century—and particularly after the expectations about the human and nonhuman genome projects were not met—there has seemed to be an ever-increasing awareness of the limitations of a biology that is exclusively centered on the gene concept (Westerhoff and Palsson 2004; *Science* 2002). This awareness has provoked a revival of the idea that living systems constitute a particular type of *organization*, conceived in systemic, fundamental, and universal terms. This organization is *systemic* because it highlights the notion of an organism, understood as an integrated functional unit, as a crucial concept in biology, irreducible to (bio)molecular properties. The organization is *fundamental* because it searches for the origins of biological phenomena in physics and chemistry; that is, it does not try to detach biology from its deep material roots and constraints, but rather seeks to establish a framework in which these are naturally included. Finally, such biological organization is *universal* because it attempts to face the challenge, posed at the end of the 1980s by proponents of artificial life

(Langton 1989) and most recently by both astrobiology and synthetic biology, of considering life not only "as we know it" but as "it could be."

In the context of this change, in which biology is moving away from analytic and descriptive approaches toward more integrated and synthetic ones (Benner and Sismour 2005), it is important to work on the problem of the definition of life as a way to clarify and organize the achievements made thus far, as well as to indicate the challenges that remain. Some authors (e.g., Cleland and Chyba 2002) maintain that the current situation is not ripe enough and that it is necessary to wait for a general theory of biology before a proper definition of life can be successfully articulated (in the same way that a proper definition of water was not possible until the molecular theory of chemistry was developed). Nevertheless, the work of synthesis needed to generate and defend such a definition could actually help form the basis for a general theory of biological systems.

How, then, should we tackle the problem of defining life, not only as we know it, but as it could exist or might exist on other planets, or even as it might at some future time be synthesized in a terrestrial lab? Our method will be to elaborate and defend a concrete proposal, which derives from the analysis of the conditions and mechanisms that underlie minimal forms of organization, including the minimal forms of life found on Earth so far. Although we are aware that our particular proposal will not be a final definition, we hope that it will bring us closer to one than previous attempts have. That is the only way to advance in this kind of project and to respond to the criticism of the always-present skeptical minds (e.g., Machery 2006): make a provisional claim and then, after discussion, be ready to modify and improve it.

Today, we know that life is a more intricate and multifarious phenomenon than it was believed to be a century ago, and even at that time, our knowledge of it revealed unexpected levels of complexity. So our primary aim is to gather and clarify the presently available knowledge of this complex phenomenon and its possible origins, convinced that a conceptual synthesis contributes to the production of new knowledge, which later on will bring about new syntheses, and so on, until an overall consensus can be achieved.

Biology itself has obviously developed as a science in a dialectic manner, with moving boundaries. A descriptive demarcation of a given set of phenomena is initially achieved, followed by closer investigation into the mechanisms or principles underlying such phenomena. As a result of this investigation, new discoveries are made, which imply a redemarcation of the limits of the original set of phenomena (providing a new characterization of them, including certain cases, excluding others, and so forth). In this way, biology has always put forth a more or less established, though typically implicit, definition of life. It is a basic task of *philosophers* of biology to make that definition as explicit as possible, collecting and summarizing all knowledge achieved in the field to date, and through conceptual clarity, offering insights for future research avenues.

Specifying the Type of Definition Required

Definitions, in general, can be constructed with two main purposes in mind: (1) to demarcate or classify a certain type of phenomenon, and (2) to grasp or express the fundamental nature of that type of phenomenon. The first purpose normally leads to *descriptive* definitions that consist in a set of properties—typically, a checklist—containing all that is required to determine whether a phenomenon belongs to a particular kind, whereas the second, *essentialist* definitions characterize a given phenomenon in terms of its most basic functional mechanisms and organization.

In the case of life, definitions that consist of a list of properties (e.g., self-organization, growth, development, functionality, metabolism, adaptability, agency, reproduction, inheritance, susceptibility to death) have not been very helpful. If one reviews different proposals (Mayr 1982; de Duve 1991; Farmer and Belin 1992; Koshland 2002), it is hard to tell whether any of them includes just the necessary and sufficient ones (so that there are no redundant properties and no additional ones that ought to be included). The reason is that these catalogs do not provide a hierarchy or an account through which the chosen properties can be related to each other. In addition, lists do not offer any hint to clarify the source or process of integration of the phenomenon under analysis. This is crucial, because a definition of life that is expected to be truly universal must be built from general principles, some of which should stem from our knowledge in physics and chemistry. In other words, the definition has to include primitive concepts that help to bridge the gap between the physicochemical and the biological domains. This has already been highlighted by Oparin (1961), who claimed that the problem of defining life is tightly intertwined with the problem of its origin.

Therefore, the set of requirements that a definition of life should meet can be condensed in the following points (see Emmeche 1998; also Ruiz-Mirazo, Peretó, and Moreno 2004). The definition should

(A) be fully coherent with current knowledge in biology, chemistry, and physics;
(B) avoid redundancies and be self-consistent;
(C) possess conceptual elegance and deep explanatory power—that is, it must provide a better understanding of the nature of life, guiding our search into its origins and its subsequent maintenance and development;
(D) be universal, in the sense that it must discriminate the necessary from the contingent features of life, selecting just the former;
(E) be minimal, but specific enough—that is, it should include just those elements that are common to all forms of life (not being, in principle, restricted to life on Earth), and at the same time, it must put forward a clear operational criterion to tell the living from the inert, clarifying border-line cases, contributing to determine biomarkers, and so on.

According to these criteria, it is wiser and more useful to opt for an essentialist type of definition that at the same time sheds some light on the process that leads to the constitution of living phenomena and offers a natural framework for generalization, that is, for specifying the universal features of all possible life. Thus, in the next section we examine several essentialist definitions of life proposed to date.

Why Current Essentialist Definitions Are Not Satisfactory

There are several examples in the literature that belong to the class of essentialist definitions and fulfill, by and large, the previous requisites. However, we will briefly explain why they are not fully satisfactory. For instance, Shapiro and Feinberg's definition states that life is the activity of a "highly ordered system of matter and energy characterized by complex cycles that maintain or gradually increase the order of the system through an exchange of energy with its environment" (1990, 147). This proposal, which can be applied to the whole biosphere, is formulated in very broad terms and with a well-intended purpose of bringing biology closer to physics.

However, the apparent advantage of this formulation—particularly in relation to requirements (A) and (D)—turns out to be a shortcoming: for it lacks the necessary level of specificity. In this scheme, where can we look for properties like self-production, reproduction, adaptation, heredity, selective evolution, or alternatively, individual agency (i.e., the capacity to act in the world as a distinctive unit), which are so characteristic of biological organisms? The motivation to establish a link between the inert and the living domains should not lead us to forget all of what we actually know about biology (or chemistry!). Thus, the previous definition is not suitable, as it should at least give an indication of where to search in this direction—in other words, requirement (E) is also not adequately fulfilled by Shapiro and Feinberg's proposed definition.

Emmeche's *biosemiotic* definition, according to which "life is the functional interpretation of signs in self-organized material code-systems making their own umwelts" (1998, 11) presents a different problem. It is certainly self-consistent and conceptually elegant, but has an important drawback: it assumes "signs" (i.e., information) as a primitive natural kind, when physics and chemistry do not. Therefore, although it does not directly contradict physicochemical knowledge, it does not come to terms with it, and thus requirement (A) is not, strictly speaking, fulfilled.

According to the autopoietic school, the minimal living organization should be defined as a "network of processes of production (synthesis and destruction) of components such that these components: (i) continuously regenerate and realize the network that produces them, and (ii) constitute the system as a distinguishable unit in the domain in which they exist" (Maturana and Varela 1973; Varela 1994, 26). Apart from being excessively abstract (Ruiz-Mirazo, Peretó, and Moreno 2004), which does not facilitate a natural bridge with physics and chemistry, the main problem of this definition is that the characteristic evolutionary capacity of living beings is not taken into account. For these reasons, the account in question fails to meet requirements (A) and (C).[1] Although, in principle, an autopoietic system should be capable of reproduction (through autocatalytic growth and division), adapting to external perturbations (organizational homeostasis), and even modifying its type of organization (through the accumulation of structural changes), it is not necessarily part of a process of Darwinian evolution, because the molecular mechanisms that make this possible are not included (rather, they are deliberately excluded) from its specification.

This leads us to one of the main issues we would like to address in this article: the tension between geneticist approaches to the definition of life, which explicitly take into account evolutionary aspects, and those that prefer to simply not consider them, concentrating instead on the operational logic or metabolic organization of biological organisms. For example, recall Sagan's definition that "life is a system capable of evolution by natural selection" (1970), or the more elaborate and precise version of Joyce (1994, xi): "life is a *self-sustained chemical* system capable of undergoing Darwinian evolution" (which is, in fact, NASA's working definition—see http://www.nasa.gov/vision/universe/starsgalaxies/life's_working_definition.html). It is clear that the underlying conception is very close to the standard view that supports the neo-Darwinian paradigm in biology, where the stress is placed on the evolutionary dynamics of biological systems. According to this view, the key properties that a system must show in order to evolve through natural selection are "reproduction, variability, and inheritance" (Maynard Smith 1986). It follows, then, that any system with those properties, including a population of replicating molecules, could readily fulfill the definition.

On the other hand, Luisi (1998) and other authors are critical of this perspective, arguing that a proper definition of "living being" must primarily take into account the characteristic way in which the components of such a system get organized as a coherent whole. The second version of the "standard definition" given earlier might seem less vulnerable to this criticism, as it introduces the idea of *self-maintenance*. However, the core idea stays the same. In Joyce's own terms, while the "notion of Darwinian evolution subsumes the processes of self-reproduction, material continuity over an historical lineage, genetic variation, and natural selection, . . . [self-maintenance] refers to the fact that living systems contain all the genetic information necessary for their own constant production (i.e., metabolism)" (1994, xi). The geneticist bias is rather obvious: here, metabolism seems to be the result of acquiring a complete enough pool of genes so as to achieve the constant production of a system.

As Luisi points out, this definition was created from a conception of life that is fundamentally molecular, in tune with a general research program that looks into the roots of Darwinian evolution in the context of populations of replicating molecules undergoing some selective dynamics—such as models of quasi-species, hypercycles, and so on (Luisi 1998; see also Eigen and Schuster 1979; Eigen and Winkler-Oswatitsch 1992). But the basic problem with this type of definition is that it does not afford a proper characterization of the type of material organization that could actually lead to the beginning of a process of Darwinian evolution: precisely, some sort of pregenetic metabolic organization (Ruiz-Mirazo, Umerez, and Moreno 2008).

Integrating Individual-Collective Dimensions: Living Being versus Life?

Life reveals itself, as Maynard Smith (1986) emphasizes, both as a collective population of self-reproducing hereditary systems (life as evolution), and as individual self-maintaining dissipative units (life as metabolism). The main problem with definitions of life up to now is that they tend to focus on just one of those two dimensions: they do not work out properly, in a well-balanced way, the tension between the individual-metabolic and the collective-historic-ecological spheres of the phenomenon. However, a good definition of life should both highlight and help explain the link between those two spheres.[2] Otherwise, we will run into problems with our present conception of life, or we will need to distinguish "the life we know on planet Earth" from other types of life.

Imagine, for the sake of the argument, that some sort of metabolic system without reproductive and evolutionary capacity is found on some other planet, or synthesized in the lab. That, of course, would be a revolutionary discovery. But it is reasonable to think that the characteristics of that system would be very different from the metabolisms of which we are aware. In particular, even if it is a relatively robust metabolism in relatively stable external conditions, it would lack the capacity for long-term sustainability: it would be a phenomenon prone to disappear, sooner or later, as a consequence of perturbations from an environment that unavoidably (and sometimes radically) changes over time. That is why, to rephrase the famous claim by Dobzhansky (1973), little in the biology that we know on earth makes sense, except in the light of evolution.

But the issue is quite tricky. It seems that one is somehow forced to choose between the analysis of a single system (a living being) or of a population of systems (life in a global sense). Could a definition convey both dimensions at the same time? Gánti (1975, 1987, 2003), who was keenly aware of this problem, proposed the distinction between "absolute" or "actual" life criteria and "potential" life criteria. The former apply to "every living being at every moment of its life without exception," and the latter, "not being necessary criteria for the living state of the individual organisms, are indispensable with regard to the surviving of the living world" (Gánti 1987, 68–69). The absolute life criteria include inherent unity, stability, metabolism, information-carrying subsystem, and program control. The potential criteria include growth, reproduction, capability of hereditary change, evolution, and mortality. This distinction also allows Gánti to overcome the problems posed by certain limiting cases, like resting seeds, a frozen tissue culture, sterile animals, and so forth. However, Gánti does not define life, as such, in a statement; instead, he elaborates a whole theory of soft automata and proposes the *chemoton* as the minimal operational unit that fulfills the previous criteria. Hence, the principle of life is the coupling of the three autocatalytic subsystems (the membrane, metabolic, and informational networks) required to put together that minimal unit.

For our purposes, the main contribution of Gánti is not the specific set of life criteria he proposes—which shows some of the inherent shortcomings of lists (as explained previously)—but the actual development of a theory of minimal individual organization for life that explicitly includes evolutionary potential, in contrast to the hypotheses of Maturana and Varela (1973), or Rosen (1991). We believe that the type of evolutionary dynamics that characterizes biological systems (Darwinian, *open-ended*, evolution) requires a type of individual organization that goes beyond autopoiesis. And the chemoton is a possibility worth exploring on these lines, even if it probably falls short in various respects: (1) as a minimal model, it is too far removed from minimal life as we know—or can extrapolate—it; that is, it does not capture the complexity of the organization of a hypothetical prokaryote common ancestor cell; (2) it does not incorporate a definite theory (general scheme of transitions) for its own emergence; so it is not particularly enlightening for the origins of life problem.[3] In short, it constitutes an attractive and well-founded theory that enables us to study intermediate, protobiological systems that could play a central role at some stage of prebiotic evolution (like an RNA or one-polymer-world stage).

Our own proposal (see following discussion) involves an alternative framework from which it will be possible to draw more demanding material and organizational life requirements, as well as to shed light on a wider range of critical transitions in the process of the origins of life. But before we are ready to do that, it is convenient to offer a preliminary synthesis of the knowledge we have of the material mechanisms and constraints underlying all biological systems on Earth. After this analysis, we will propose a definition of life as a global phenomenon, and then we will show what that definition really implies at the level of individual systems. A living being, after all, cannot exist but in the context of a global network of similar systems. This is reflected in the fact that, in order to be functional, genetic components (which specify the metabolic machinery and organization of single biological entities) must be shaped through a process that involves a great number of individual systems, and also very many consecutive generations, or reproductive steps. And, as we have already mentioned and will argue more extensively, such a collective process—which has both a synchronic-ecological side and a diachronic-evolutionary one—is actually crucial for the sustainability of the living phenomenon as a whole.

Searching for Life's Organizational Requirements

The fact that life can be conceived only within an evolutionary framework poses an additional challenge to the attempts to define it. Because biological systems are actually the result of a long process of tinkering with and (re)construction of what has accumulated over the course of an evolutionary history, rather than an inexorable consequence of the action of physical and chemical laws, it could be that they do not really fit into any general definition, as suggested by Keller (2007a). To some extent, that difficulty is indeed inescapable. The evolutionary perspective has put forward a conception of the biological world as a domain constituted by systems historically linked through processes of improvised construction and natural selection, in which a large amount of features may be simply *frozen accidents*. Nevertheless, as we shall argue, there are necessary organizational requirements that allow for an open-ended evolutionary history through cumulative tinkering and the building up of increasing levels of complexity.

But first, let us clarify what we mean by the concept of *organization*. Machines and living things are *organized* because their parts play distinctive, systematic, and essential roles in the whole. Now, where does this order come from? As Polanyi (1968) pointed out, machines are not the result of physical or chemical laws, but of the *constraints* imposed on these laws by human intelligent action. Natural organizations also derive from a set of material constraints, but these are spontaneous, external-agent-free constraints, which actually hold the key to make the project of a universal definition of life possible. Indeed, in order to define the phenomenon of life in the most general or universal sense, what matters is determining the set of *enabling constraints* on the physical and chemical laws that are indispensable for the emergence of a biological world. The search for a universal definition of life will thus converge with the investigation of the constraints required to generate increasingly complex forms of organization.

Current lines of research in fields such as the origins of life, astrobiology, complex systems, and artificial life are beginning to provide us with new experimental and theoretical keys to determine the general mechanisms and principles of organization underlying biological phenomena. Although there were surely many local, contingent events in terrestrial biogenesis, which would be different in the (hypothetical) processes of the origins of life in other conditions, it seems quite reasonable to assume that some of those events were necessary steps.[4] Otherwise, it would be impossible to apply, in the first place, standard scientific methods to solve the problem (see de Duve 1991; Morowitz 1992; Fry 1995).

Even if the complete empirical reproduction of the process of terrestrial biogenesis is impracticable, and even if different forms of life on other planets remain inaccessible to us, the continuous advances in techniques or platforms of simulation in silico, together with the rise of systemic approaches in vitro, will allow researchers to keep exploring general and stable forms of organization and eventually enable them to discern the contingent from the universally necessary aspects. Computational models/simulations constitute, in this sense, a new way to carry out conceptual experiments (Dennett 1994), making the precise analysis of different theoretical assumptions, reproducible results, and their implications possible. In addition, those models/simulations are including more and more ingredients and parameters related directly to experimental conditions (Solé et al. 2007). At the same time, as the wet experiments in the lab also get closer to holistic approaches to the problem, there will be an increasing amount of available data to match up to by the theoretical models.

Starting from Self-organization

As for already established knowledge on this issue, we want briefly to highlight and justify the relevance of insights from research carried out on *self-organizing phenomena* (in both physicochemical and biological systems), *minimal self-constructing (protometabolic) systems*, and *cells with minimal genetic machinery*, because they will provide the conceptual basis of our proposal.

It is clear that the one-shot, order-for-free, kind of self-organization found in some physical or chemical systems—the *dissipative structures* of Prigogine and Stengers (1979)—is too simple: life requires a cumulative-iterative, step-by-step form of self-organization that incorporates new elements that allow for the internal distinction of functional parts (Keller 2007b). But here is where problems arise: nobody understands well how this more complex type of self-organization could emerge. However, this does not mean we have to get rid of this idea; rather, we should continue working on it, if only because it is the most solid bridge from physics and chemistry toward biology.

Some authors—for example, Eigen and Winkler-Oswatitsch (1992) and Szathmáry (2006)—contend that, instead of self-organization, the main driving force toward life is the combination of self-reproduction and natural selection. In their view, what really can explain the natural origin of complex organizations (and eventually that required for minimal life) is the dynamics of certain self-replicating modular structures (templates) that can (1) produce diverse offspring and (2) eliminate the less-fit variants though a selective process. Doubtless, self-reproduction with (some form of) heredity and selection mechanism is crucial to understand how the aforementioned complex type of self-organization may appear and develop. However, things are far more intricate, because it turns out that evolution by natural selection itself requires systems with an organizational apparatus that allows, in the first place, a wide enough space for functional diversity (Wicken 1987; Moreno 2007; Ruiz-Mirazo, Umerez, and Moreno 2008).[5] So some sort of self-organization mechanisms or constraints are necessary, though quite obviously not sufficient.

This is where research on minimal protometabolic networks and protocells may help, because one should search specifically for *chemical* systems that have a self-producing organization, so that they can actively contribute to and regulate the maintenance of that organization. It is only in this kind of recursively self-maintaining system that proper functional mechanisms can be grounded and develop (Bickhard 2000). In turn, this naturally leads to systems that are capable of constraining, for their own needs, the flow of matter and energy between the network of internal reactive processes and its environment, or what we have called and analyzed elsewhere as *basic autonomy* (Ruiz-Mirazo, Peretó, and Moreno 2004). To achieve this fundamental property—which is similar to metabolism in its minimal sense—chemical systems must be able to build their own boundaries. Furthermore, these boundaries must be cellular, rather than closed, structural compartments to overcome problems like component diffusion and to manage the interaction with the external medium, controlling the input/output of compounds and energy into/out of the system (like real biological membranes do). So a combination of self-assembled global constraints (e.g., a closed lipidic bilayer) with local, specific ones (e.g., rudimentary channels, proto-enzymatic catalysts) is required.

According to Kauffman, chemically autonomous systems would go beyond "collective catalytic closure" and be able to "act on their own behalf" (2000, 114) thanks to an organization that generates a set of constraints (namely, the components of the system) that drive the flow of energy so that it becomes useful for the reproduction of those very constraints. In Kauffman's words, "work [useful energy] begets constraints begets work," which is an elegant way to argue for the relevance of concepts like *work* (rooted in thermodynamics) or *functional constraint*, stemming from the thorough analysis of a certain type of coupled cyclic chemical network happening in biological systems, where both adequately and mutually fit in.

Genetically Instructed Metabolisms

However, as we argued previously, this basic autonomy is still not enough for life: the capacity of such highly distributed component-production networks to maintain their specificity and level of complexity depends on the fidelity of many local interactions. And this would be a rather fragile, and clearly not long-term sustainable, form of metabolic organization, unless specific and molecularly reliable regulatory control mechanisms are created. In fact, the creation of these control mechanisms, insofar as they should also be transmitted reliably to subsequent generations (apart from contributing to the metabolic efficiency of individual autonomous systems), should open up their evolutionary potential as a population. In this kind of system, what is the simplest form of regulatory control that makes possible the robust maintenance—and, thus, also the potential increase over time—of their complexity?

As Pattee (1982) has argued, following von Neumann's (1949) earlier ideas, open-ended evolution requires an almost inert, modular type of "memory," whose units allow an indefinite sequential variety (free compositionality), decoupled from the dynamics actually regulated by this memory. These conditions imply the formation of a system organized in two different, but complementary, levels: on the one hand, the quasi-inert strings of the memory, and on the other hand, the rate-dependent catalytic dynamics of metabolism. The first level acts as a regulatory control of the second, but only because the latter converts the former inert sequences into dynamically active building blocks, which thereupon assemble into complex catalytic components. If there was not a separate (in a dynamic sense, partially decoupled) memory, the construction and preservation of complex, sequence-specific catalysts would be impossible; if a suitable fine-tuned metabolic machinery did not exist, the preservation and expression of the chemical memory (i.e., of its content) would not be possible.

Thus, each level implies the other, so a new form of organizational closure (called *semantic closure* by Pattee [1982]) is achieved. It is important to highlight that these two levels, being dynamically decoupled, need also to be *hierarchically* coupled by means of a translation apparatus (genetic code). Here is where a careful analysis in terms of *information* needs to be done (Ruiz-Mirazo and Moreno 2006). Information cannot simply be associated with the sequence of a polymer, because it is, by definition, a relational concept. Otherwise, all semantics disappears: semantics not in the sense that *we* interpret the content of those sequences as more or less functional for the system, but in the sense that the *system itself* is continuously doing that interpretive task (e.g., protein synthesis). Furthermore, this is a task whose functionality will be evaluated through the long-term time scale of natural selection.

In fact, the construction of this complex type of metabolism must have taken a very long time (i.e., many generations of cellular protometabolisms). Of course, it must have been preceded by other major stages—the so-called RNA or one-polymer worlds—as it is now generally claimed. But, perhaps less evident, the innovations taking place at the biochemical-physiological level of individual protocells must have been accompanied by innovations at the collective-historical level (e.g., by the generation of proto-ecosystems, global recycling of material resources, cooperation/competition strategies, reproduction/replication reliability over generations, and the like). So information is not only a relational concept within the metabolic organization of each organism, but also exists in a more encompassing, evolutionary sense.

Any genetically instructed metabolism—that is, any living being we know of—depends on material records, the production of which cannot be explained solely in terms of its activity as an individual organism. Leaving aside other, also relevant epigenetic mechanisms, genes play a crucial double role: (1) as instructions to be read out for the synthesis of the most complex components in catalytically very efficient metabolisms, and (2) as reliably transmitted records that allow for a process of evolution (of *collections* of metabolisms) over a much longer time scale. So information is a central concept *as a principle of organization* in biology. Although further insights will surely come from top-down research that is being carried out on cells with a minimal genetic apparatus (Gil et al. 2004; Castellanos, Wilson, and Shuler 2004; Glass et al. 2006; Gabaldón et al. 2007), or even from more radical semisynthetic approaches (Szostak, Bartel, and Luisi 2001; Luisi, Ferri, and Stano 2006), we consider this absolutely relevant for the present discussion, as it does not seem to be a contingent feature of terrestrial life, but a universal requirement for any form of self-maintaining, self-reproducing population of systems capable of open-ended evolution and long-term sustainability.

A Definition of Life in Universal Terms: Our Proposal

We are now in a position to present the following tentative definition of life (see also Ruiz-Mirazo, Pereto, and Moreno 2004): "Life is a complex network of self-reproducing *autonomous* agents whose basic organization is instructed by material records generated through the *open-ended*, historical process in which that collective network evolves." Thus, (1) at the individual level, living beings should be conceived as a special kind of autonomous system, whose metabolic, adaptive, and reproductive properties are genetically instructed/controlled; and (2) at the collective level, life should be conceived as a metanetwork of those autonomous agents that generate indefinitely new informational records that, once embedded/expressed in the individual organizations, allow for new, open-ended, functional components and interactions. These components and interactions are a result of both ontogenic-synchronic processes of cooperation and competition for limited resources in the ecological sphere, and philogenic-diachronic connections in the long-term evolutionary sphere.

There is such a deep interconnection between these two levels that both the individual and the collective organization of life are cause and consequence of each other. Nevertheless, it is important to underscore the fact that there is also a basic asymmetry between the individual (metabolic) network and the collective (ecological-evolutionary) one: both are self-maintaining and self-producing organizations, but only individual living beings (organisms) are autonomous *agents* with an active physical border, which allows for an extremely high degree of *functional integration* among components, including a machinery for *hereditary reproduction*.

In this framework, the relationship between autonomy and genetic information appears as the key to understanding the intricate connection between the functional dynamics of individual organisms, and the evolutionary dynamics of the global biological network. It should also be highlighted that an open-ended process of evolution (which does not mean an evolutionary arrow pointing toward higher and higher complexity, but only toward the mere *possibility* of such complexity) is precisely what allows the continuous renewal of particular individual organisms, and particular types of organism, for the sake of keeping the overall process running.

This open-ended historical process takes place on a much longer time scale than the typical lifetimes of individual systems, and it begins only when those systems are endowed with genetic machinery that instructs their metabolism and is transmitted reliably through generations (Ruiz-Mirazo, Umerez, and Moreno 2008). Nevertheless, the evolutionary dynamics of ecosystems do not just suddenly appear: they must be progressively articulated during stages that precede the origins of genetics (i.e., prior to the actual origins of life). So according to this scheme, the fundamental pillars of *ecopoiesis* (i.e., the process of constitution of an ecosystem [Haynes 1990]) and thus the key individual and collective mechanisms required to generate a whole sustainable biosphere, would be established along with the development of protobiological systems.[6]

Because the material resources of any real physical environment are limited, life must learn how to make the best use of what is available, and also of what it continuously produces. In the individual sphere, this means that living, autonomous agents have to compete for organic compounds, and under selective pressure, depredation strategies (food webs) are bound to appear. In the collective sphere, the ecosystem eventually has to deal with the problem of recycling bioelements on a global scale. Otherwise, there would be a major crisis that would threaten the continued survival of the entire biological world. The solution is to couple geophysical and geochemical processes that take place on the planet and to establish global bio-geo-chemical cycles. The regulation of environmental conditions, such as the composition of the atmosphere (Lovelock and Margulis 1974) by the biological dwellers of the planet, would obviously contribute to their long-term maintenance. Therefore, the problems of the origin of life and of the constitution of a sustainable biosphere in constant, open-ended evolution would be tightly linked.

Final Remarks: Some Implications of the Definition

Obviously, nobody can expect that all biologically relevant items are explicitly addressed in a brief statement like a definition. Nevertheless, as we mentioned at the beginning of this work, a good definition should be well-balanced, containing the key conceptual tools to develop a theory or scheme around it that is coherent and provides enough "hints" to establish a natural connection between the physicochemical and biological realms. We believe that the present version contains the fundamental (necessary and sufficient) theoretical ingredients to capture the minimal and universal features of living phenomena, as well as to reconstruct the basic steps for its origins, from self-organizing physicochemical phenomena up to the constitution of systems with a level of complexity equivalent to that of the last universal common ancestor (LUCA) of terrestrial forms of life. An autopoietic type of definition, by itself, would not lead us all the way up to that point (as an open-ended type of evolution requires the development of nontrivial mechanisms, like those supporting hereditary reproduction). On the other hand, the standard geneticist definitions are not satisfactory, because they do not provide the key conceptual tools to characterize the basic material organization required to get there.

In that sense, our proposal is more restrictive and more demanding, yet at the same time has a higher explanatory potential than the standard definition: autonomy requires, and provides, much more than self-maintenance of a chemical network. Equally important, our proposed definition leads to specific requirements for candidate systems that go beyond those that emerge from autopoietic or chemoton criteria. In particular, as we have argued and explained elsewhere (Ruiz-Mirazo, Pereto, and Moreno 2004), an *autonomous system with open-ended evolutionary capacities* should comprise: a semipermeable boundary (i.e., a *membrane*), carrying out active transport processes and controlling the flow of matter-energy through the system; an energy transduction/conversion apparatus (i.e., a set of *energy currencies*, including both chemical and chemosemiotic ones) to distribute those energy resources efficiently; and at least two types of interdependent macromolecular components, some directly performing and coordinating self-construction processes (*catalysts*), others storing and transmitting information necessary to maintain those processes in the course of subsequent generations (*records*).

Finally, our definition, articulated as a pair of basic concepts that characterize living systems (autonomy and open-ended evolution), also involves a particular hypothesis of life's origins. Between chemical autocatalytic networks and full-fledged biological cells, a series of complex transitions had to take place, both at the level of individual autonomous systems (from "minimal" to "hereditary" autonomous systems) and at the level of their collective interactions (ecopoiesis). The highly sophisticated molecular and organizational constraints required to implement genetic mechanisms (necessary for a fully open-ended evolution) would make these more plausible in the latest stages of the origins of life, whereas autonomous cellular systems (which would control matter-energy resources necessary to keep any far-from-equilibrium component-production system running) would play the bridging role.

Acknowledgments

Kepa Ruiz-Mirazo holds a *Ramon y Cajal* research fellowship. Juli Peretó, coauthor of one of our previous publications on the definition of life, contributed to the development of the ideas present in this article. Funding for this work was provided by grants FFI2008-06348-C02-01/FISO (KRM and AM) and FFU2009-12895-C02-02 (AM), from the Spanish MICINN, together with IT 505-10 (KRM and AM), from the Basque Government.

Notes

1. Similar difficulties are faced by Rosen, who analyzes the living organization wholly from an organism/individual point of view, eventually defining life as a system that is "closed to efficient cause" (1991).

2. This can be achieved even if the definition is formulated from the point of view of individual organisms, that is, in terms of what a living being is and how it organizes and behaves (like authors of the autopoietic school do). The crucial point (which those authors, however, do not consider relevant) is to include in the definition the main feature(s) that will reveal, at the individual-metabolic level of analysis, the important molecular and organizational implications of being inserted in a collective-evolutionary dynamics.

3. In fact, Gánti does not really bet on which of the three autocatalytic subsystems (membrane, metabolic, and informational subsystems) came first, or on which two of them got coupled first.

4. For instance, the particular genetic code that all living systems on Earth share and continuously use for protein synthesis could be a frozen accident, or something that has been constructed in the course of very primitive evolutionary stages (Wong 1975). But the actual existence of a genetic code may still be a general requirement for life.

5. By functional diversity, we mean that distinguishable parts of the system perform actions that contribute in a distinctive, though sometimes hard-to-track, way (due to nonlinearities, intermediates, and so on) to the overall maintenance of the system as an integrated operational unit.

6. The metabolic potential and diversity of those protobiological systems, in turn, would be linked to their capacity to establish the ecological networks necessary for their long-term sustainability.

References

Benner, Steven A., and A. Michael Sismour. 2005. Synthetic biology. *Nature Reviews: Genetics* 6 (7): 533–543.

Bickhard, Mark H. 2000. Autonomy, function, and representation. *Communication and Cognition—Artificial Intelligence* 17 (3–4): 111–131.

Castellanos, Mariajose, David B. Wilson, and Michael L. Shuler. 2004. A modular minimal cell model: Purine and pyrimidine transport and metabolism. *Proceedings of the National Academy of Sciences of the United States of America* 101 (17): 6681–6686.

Cleland, Carol, and Christopher F. Chyba. 2002. Defining "life." *Origins of Life and Evolution of the Biosphere* 32 (4): 387–393.

Dennett, Daniel C. 1994. Artificial life as philosophy. *Artificial Life* 1 (3): 291–292.

Dobzhansky, Theodosius. 1973. Nothing in biology makes sense except in the light of evolution. *American Biology Teacher* 35 (3): 125–129.

de Duve, Christian. 1991. *Blueprint for a cell: The nature and origin of life.* Burlington, N.C.: Neil Patterson Publishers.

Eigen, Manfred, and Peter Schuster. 1979. *The hypercycle: A principle of natural self-organization.* New York: Springer.

Eigen, Manfred, and Ruthild Winkler-Oswatitsch. 1992. *Steps towards life: A perspective on evolution.* New York: Oxford University Press.

Elsasser, Walter M. 1966. *Atom and organism: A new approach to theoretical biology.* Princeton, N.J.: Princeton University Press.

Emmeche, Claus. 1998. Defining life as a semiotic phenomenon. *Cybernetics & Human Knowing* 5 (1): 3–17.

Farmer, J. Doyne, and Alletta d'Andelot Belin. 1992. Artificial life: The coming evolution. In *Artificial II*, ed. Christopher G. Langton, Charles Taylor, J. Doyne Farmer, and Steen Rasmussen, 815–838. Redwood City, Calif.: Addison-Wesley.

Fry, Iris. 1995. Are the different hypotheses on the emergence of life as different as they seem? *Biology and Philosophy* 10 (4): 389–417.

Gabaldón, Toni, Juli Pereto, Francisco Montero, Rosario Gil, Amparo Latorre, and Andrés Moya. 2007. Structural analyses of a hypothetical minimal metabolism. *Philosophical Transactions of the Royal Society B: Biological Sciences* 362 (1486): 1751–1762.

Gánti, Tibor. 1975. Organization of chemical reactions into dividing and metabolizing units: The chemotons. *BioSystems* 7 (1): 15–21.

Gánti, Tibor. 1987. *The principle of life.* Budapest: OMIKK.

Gánti, Tibor. 2003. *The principles of life.* Oxford: Oxford University Press.

Gil, Rosario, Francesco Silva, Juli Pereto, and Andrés Moya. 2004. Determination of the core of a minimal bacteria gene set. *Microbiology and Molecular Biology Reviews* 68: 518–537.

Glass, John I., Nacyra Assad-Garcia, Nina Alperovich, Shibu Yooseph, Matthew R. Lewis, Mahir Maruf, Clyde A. Hutchison III, Hamilton O. Smith, and J. Craig Venter. 2006. Essentials genes of a minimal bacterium. *Proceedings of the National Academy of Science of the United States of America* 103: 425–430.

Haynes, Robert H. 1990. Ecce ecopoiesis: Playing god on Mars. In *Moral expertise: Studies in practical and professional ethics*, ed. Don MacNiven, 161–183. London: Routledge.

Joyce, Gerald F. 1994. Foreword. In *Origins of life: The central concepts*, ed. David W. Deamer and Gail R. Fleischaker, xi–xii. Boston: Jones and Bartlett Publishers.

Kauffman, Stuart. 2000. *Investigations.* Oxford: Oxford University Press.

Keller, Evelyn Fox. 2000. *The century of the gene.* Cambridge, Mass.: Harvard University Press.

Keller, Evelyn Fox. 2007a. A clash of two cultures. *Nature* 445 (7128): 603.

Keller, Evelyn Fox. 2007b. The disappearance of function from "self-organizing systems." In *Systems biology: Philosophical foundations*, ed. Fred C. Boogerd, Frank J. Bruggeman, Jan-Hendrik S. Hofmeyr, and Hans V. Westerhoff, 303–317. Dordrecht: Elsevier.

Koshland, Daniel E., Jr. 2002. The seven pillars of life. *Science* 295 (5563): 2215–2216.

Langton, Christopher G. 1989. Artificial life. In *Artificial life*, ed. Christopher G. Langton, 1–47. Santa Fe Institute Studies in the Sciences of Complexity, vol. 6. Reading, Mass.: Addison-Wesley.

Lovelock, James E., and Lynn Margulis. 1974. Homeostatic tendencies of the Earth's atmosphere. *Origins of Life and Evolution of the Biosphere* 5 (1–2): 93–103.

Luisi, Pier Luigi. 1998. About various definitions of life. *Origins of Life and Evolution of the Biosphere* 28 (4–6): 613–622.

Luisi, Pier Luigi, Francesca Ferri, and Pasquale Stano. 2006. Approaches to semi-synthetic minimal cells: A review. *Naturwissenschaften* 93 (1): 1–13.

Machery, Édouard. 2006. Why I stopped worrying about the definition of life . . . And why you should as well. Unpublished manuscript.

Maturana, Humberto R., and Francisco J. Varela. 1973. *De máquinas y seres Vivos: Una teoría sobre la organización biológica*. Santiago, Chile: Editorial Universitaria S.A.

Maynard Smith, John. 1986. *The problems of biology*. Oxford: Oxford University Press.

Mayr, Ernst. 1982. *The growth of biological thought: Diversity, evolution, and inheritance*. Cambridge, Mass.: Harvard University Press.

Moreno, Alvaro. 2007. A systemic approach to the origin of biological organization. In *Systems biology: Philosophical foundations*, ed. Fred C. Boogerd, Frank J. Bruggeman, Jan-Hendrik S. Hofmeyr, and Hans V. Westerhoff, 243–268. Dordrecht: Elsevier.

Morowitz, Harold J. 1992. *Beginnings of cellular life: Metabolism recapitulates biogenesis*. New Haven, Conn.: Yale University Press.

Oparin, Aleksandr Ivanovich. 1961. *Life: Its nature, origin and development*, trans. Ann Synge. Edinburgh: Oliver and Boyd.

Pattee, Howard H. 1982. Cell psychology: An evolutionary approach to the symbol-matter problem. *Cognition and Brain Theory* 5 (4): 325–341.

Polanyi, Michael. 1968. Life's irreducible structure. *Science* 160 (3834): 1308–1312.

Prigogine, Ilya, and Isabelle Stengers. 1979. *La nouvelle alliance: Métamorphose de la science*. Paris: Gallimard.

Rashevsky, Nicolas. 1938. *Mathematical biophysics: Physico-mathematical foundations of biology*. New York: Dover Publications.

Rosen, Robert. 1991. *Life itself: A comprehensive inquiry into the nature, origin, and fabrication of life*. New York: Columbia University Press.

Ruiz-Mirazo, Kepa, and Alvaro Moreno. 2006. The maintainance and open-ended growth of complexity in nature: Information as a decoupling mechanism in the origins of life. In *Rethinking complexity*, ed. Fritjof Capra, Pedro Sotolongo, Alicia Juarrero, and Jacco van Uden, 55–72. Mansfield, Mass.: ISCE Publishing.

Ruiz-Mirazo, Kepa, Juli Peretó, and Alvaro Moreno. 2004. A universal definition of life: Autonomy and open-ended evolution. *Origins of Life and Evolution of Biospheres* 34 (3): 323–346.

Ruiz-Mirazo, Kepa, Jon Umerez, and Alvaro Moreno. 2008. Enabling conditions for "open-ended evolution." *Biology and Philosophy* 23 (1): 67–85.

Sagan, Carl. 1970. Life. In *The Encyclopaedia Britannica*, 985–1002. London: William Benton.

Science. 2002. Special issue—Systems Biology. *Science* 295 (5560): 1661–1682.

Shapiro, Robert, and Gerald Feinberg. 1990. Possible forms of life in environments very different from the Earth. In *Physical cosmology and philosophy*, ed. John Leslie, 248–255. New York: Macmillan.

Solé, Ricard V., Andreea Munteanu, Carlos Rodriguez-Caso, and Javier Macía. 2007. Synthetic protocell biology: From reproduction to computation. *Philosophical Transactions of the Royal Society B: Biological Sciences* 362 (1486): 1727–1739.

Szathmáry, Eörs. 2006. The origins of replicators and reproducers. *Philosophical Transactions of the Royal Society B: Biological Sciences* 361 (1474): 1761–1776.

Szostak, Jack W., David P. Bartel, and Pier Luigi Luisi. 2001. Synthesizing life. *Nature* 409 (6818): 387–390.

Varela, Francisco. 1994. On defining life. In *Self-production of supramolecular structures*, ed. Gail R. Fleischaker, Stefano Colonna, and Pier Luigi Luisi, 23–31. Dordrecht: Kluwer Academic Publishers.

von Bertalanffy, Ludwig. 1952. *Problems of life: An evaluation of modern biological thought*. London: Watts & Co.

von Neumann, John. 1949. *Theory of self-reproducing automata*. Urbana: University of Illinois Press.

Westerhoff, Hans V., and Bernhard O. Palsson. 2004. The evolution of molecular biology into systems biology. *Nature Biotechnology* 22 (10): 1249–1252.

Wicken, Jeffrey S. 1987. *Evolution, thermodynamics and information: Extending the Darwinian program*. Oxford: Oxford University Press.

Wong, J. Tze-Fei. 1975. A co-evolution theory of the genetic code. *Proceedings of the National Academy of Sciences of the United States of America* 72 (5): 1909–1912.

2
Energy Coupling

Yaşar Demirel

As opposed to nonliving systems, living systems can reproduce as well as collect, process, and exchange information in order to control and direct energy and matter they receive from their environments. Living systems extract free energy from the sun, store it, and use it for sorting and motile activities. Chloroplasts use energy to initiate electron transfer cycles and proton gradients to produce adenosine triphosphate (ATP) and carbohydrates in photosynthesis. Mitochondria use carbohydrates as food molecules to create electron transfer cycles and proton gradients to produce ATP and carbon dioxide in oxidative phosphorylation. Feeding carbon dioxide back to photosynthesis completes the carbon cycle. The production and utilization of ATP require energy coupling processes. Photosynthesis, oxidative phosphorylation, and hydrolysis of ATP represent some of the energy coupling systems of the carbon cycle. Here, coupling refers to a *flow*, such as heat flow, mass flow, or chemical reaction flow, occurring without its primary thermodynamic driving force, or opposite to the direction imposed by its primary driving force. The principles of thermodynamics allow the progress of a process without, or against, its primary driving force only if it is coupled with another spontaneous process. This is consistent with the Second Law of Thermodynamics, which states that a finite amount of organization may be obtained at the expense of a greater amount of disorganization in a series of coupled, spontaneous processes.

In this chapter, I describe and elucidate the energy coupling processes of living systems. First, I discuss equilibrium and nonequilibrium systems, and introduce living systems as open, nonequilibrium, and dissipative structures continuously interacting with their surroundings. Then, the roles of thermodynamics and Gibbs free energy as they apply to energy coupling phenomena are summarized. Because protein structures play a crucial role in information processes and energy couplings, their structures are discussed next. The "well-informed" character of living systems and the control of free energy (exergy) by information are also briefly discussed. Finally, using the linear nonequilibrium thermodynamic approach, energy couplings in ATP production through oxidative phosphorylation and active transport of ions by chemical pumps are discussed as a part of bioenergetics.

Equilibrium and Nonequilibrium Systems

A nonliving system reaches a state of thermodynamic equilibrium when it is left for a long time with no external interactions, and the internal properties are fully determined by the external properties. For example, if the system's temperature is not uniform, heat is exchanged with the immediate surroundings until the system reaches a thermal equilibrium. Therefore, flows and driving forces, such as temperature gradient or chemical potential gradient, are zero in equilibrium systems (see figure 2.1).

Here, μ is the chemical potential, T is the temperature, and P is the pressure; J denotes a flow, and A and B represent subsystems. Equilib-

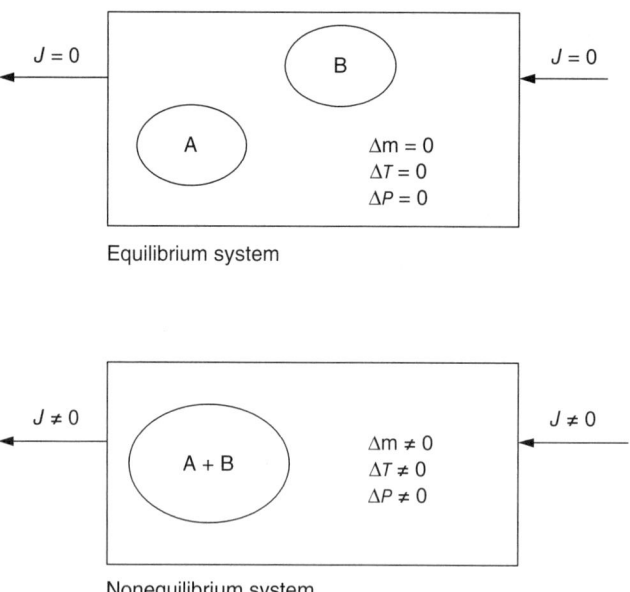

Figure 2.1
Equilibrium and nonequilibrium systems

rium states are stable, and for small disturbances the stability can be recovered.

Many natural systems, however, consist of flows caused by unbalanced driving forces, and the description of such systems requires a larger number of properties in space and time. In nonequilibrium systems, various driving forces exist, causing flows of matter and energy. These forces may interact and lead to induced flows, such as a mass flow caused by temperature gradient. Therefore, in a system with many driving forces, all of the flows are caused by all of the driving forces. The magnitude of these driving forces measures the distance from equilibrium, and control the coupling phenomena and the evolution of systems in time and space. After a sufficiently long time, these nonequilibrium and open systems may reach stationary states and establish steady flows.

Thermodynamic Branches

Thermodynamics can assess the behavior of systems based on their distance from equilibrium in three different states: (1) systems at, or near, thermodynamic equilibrium; (2) systems that are some distance from equilibrium and can return to equilibrium; and (3) systems that are far from global equilibrium and constrained by gradients (thermodynamic forces) to occur at some critical distance from an equilibrium state. Figure 2.2 designates these three different states as *thermodynamic branches*. States that are a continuous extension of global equilibrium belong to a linear region, in which linear extrapolation is possible. However, when the distance from global equilibrium reaches a critical point ΔX_c, the system reaches the nonlinear region, in which it bifurcates into multiple states. The degree to which a system has been moved from equilibrium is measured by the gradients imposed on the system. These states may be stable, metastable, or unstable. In nonlinear regions, systems will utilize all means available to counter the applied gradient(s) effectively and move to self-organized *dissipative structures* (Prigogine 1967), if dynamics and/or kinetic conditions are favorable. Therefore, a critical distance from equilibrium may reveal the diverse behavior hidden in the nonlinearities, and act as a controlling parameter in the behavior and evolution of a system. Dissipative structures are stable for a finite range of parameters and are sensitive to the supply of matter, energy, and information from outside.

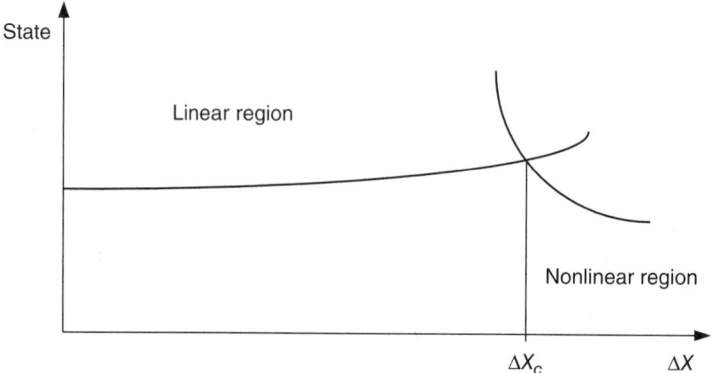

Figure 2.2
Schematic of thermodynamic branches displaying behavior of a system away from an equilibrium state

Thermodynamics and Living Systems

The First Law of Thermodynamics describes energy conversion and conservation. *Exergy* is a measure of the maximum capacity of an energy system to perform useful work while the system proceeds to equilibrium with its surroundings. The Second Law of Thermodynamics requires that if there are irreversible processes occurring in the system, the exergy in that system will degrade. The Second Law states that the entropy of the universe increases after any irreversible process, and hence describes the direction of processes. The Gibbs free energy equation combines these two principles and describes the energy conversion and utilization in isothermal processes. Gibbs free energy is $\Delta G = \Delta H - T\Delta S$, where $\Delta G = G_{final} - G_{initial}$ represents the Gibbs free energy difference between the final and initial states of a system, $\Delta H = H_{final} - H_{initial}$ is the difference in enthalpy, $\Delta S = S_{final} - S_{initial}$ is the difference in entropy, and T is the absolute temperature. A process occurs spontaneously if its free energy decreases: $\Delta G < 0$. The system is in equilibrium when $\Delta G = 0$. The concepts of entropy, enthalpy, and free energy help us understand biochemical reaction cycles and the transport systems with energy coupling.

Ludwig Boltzmann ([1886] 1974) related entropy to the number of ways in which the elements of a system can be arranged. For example, different structures of proteins represent different levels of information, as each structure corresponds to different arrangements of elements and molecules. Therefore, entropy measures the level of organization and randomized energy.

Thermodynamics views nature as decaying toward randomness in accordance with the Second Law of Thermodynamics. However, some natural systems, such as convection cells, autocatalytic chemical reactions, and living systems, move from disorder and equilibrium into highly organized structures (Schneider and Kay 1995). As living systems grow (by adding more of the same types of pathways) and develop (by adding new types of pathways), they increase their dissipation of energy, matter, and information. As noted by Schrödinger (1945), this organization may be obtained at the expense of a greater amount of disorganization in a series of coupled processes. For instance, plants are synthesized from disordered atoms and molecules found in atmospheric gases and soils. Schrödinger (1945) proposed that the study of living systems from a nonequilibrium perspective would reconcile biological self-organization and thermodynamics. Wicken (1987) advanced the notion that living organisms are "informed thermodynamic systems" and, that without "information," the inflow of energy (exergy) alone would not lead to self-organization. Living systems also represent certain patterns of coupled chemical, transport, and information processes. Fluctuations created by the nonlinear chemical cycles continuously send innovating signals to the system, while transport processes capture, relay, and stabilize them; order appears to be a compromise between them. The dynamical system generating chaos acts as an efficient selector rejecting the vast majority of random sequences and keeping only those compatible with the underlying transport and rate laws. Incorporating irreversibility and entropy production in these laws could lead to a preferred direction of reading and create attractors for asymptotic stability and hence reproducibility.

Linear Nonequilibrium Thermodynamics

Although a system, as a whole, may not be in equilibrium, the local volumes containing a sufficient number of molecules may be in local thermodynamic equilibrium. Under local equilibrium, the temperature and the internal energy are well defined, and thermodynamic properties are related to the state variables in the same manner as in equilibrium (Kondepudi and Prigogine 1999). In nonequilibrium systems, energy and entropy can be defined in terms of energy and entropy densities. The postulate of local thermodynamic equilibrium is valid if the gradients of intensive thermodynamic functions (thermodynamic forces) are small, and their local values vary slowly in comparison with the relaxation time of the local state of the system.

In a nonequilibrium system, the gradients spontaneously decay toward an equilibrium state, and internal entropy production (d_iS/dt) and entropy exchanged with the surrounding (d_eS/dt) together determine the entropy change of the system (dS/dt): $dS/dt = d_eS/dt + d_iS/dt$. Entropy production is always positive for irreversible processes and zero for reversible processes; entropy exchanged with the surroundings may be positive or negative. At a stationary state, $dS/dt = 0$, and hence $d_eS/dt = - d_iS/dt$, which implies that internal entropy production is exported to the surroundings. Living systems progress toward increasing size and complexity, sustain finite thermodynamic forces, and hence export entropy. In a stationary state, entropy production reaches its minimum value (Prigogine 1967).

Based on the local equilibrium, nonequilibrium thermodynamics determines the rate of local entropy production (d_iS/dt) from the general balance equations, including the entropy balance and Gibbs relation (Prigogine 1967; Demirel 2007). The rate of entropy production is directly proportional to the dissipated Gibbs free energy or power (Ψ): $\Psi = T(d_iS/dt)$, and can be estimated by multiplying the flows (J) with the thermodynamic forces (X). The sum of the rates of change of free energy due to each process in a system yields the total dissipation, $\Psi = \sum J_k X_k \geq 0$, which is always positive for irreversible processes. Under steady state conditions, the setting of the forces will lead to unique values for the flows. Even if one force or flow is set at zero, the system will evolve in a characteristic way. For example, if the input force is finite and the output force is zero, the system exhibits a flow level similar to that observed in many biological transport systems.

Linear Phenomenological Equations

When a system is in the vicinity of global equilibrium, a linear relation exists between flow J_i and thermodynamic driving force X_k: $J_i = L_{ik} X_k$, where the parameters L_{ik} are called the phenomenological coefficients. For example, Fick's law represents a linear flow-force relation between the flow of a substance and its concentration gradient (thermodynamic force). Choice of a force X_i conjugate to a flow J_i requires that their product, ($J_i X_i$), should have the dimension of power. If a nonequilibrium system consists of several flows caused by various forces, we have $J_i = \sum L_{ik} X_k$. These equations are called the *phenomenological equations*, which are capable of describing the multiflow systems and the induced effects of the nonconjugate forces on a flow. For example, in a two-flow system, the linear phenomenological equations become $J_1 = L_{11}X_1 + L_{12}X_2$ and $J_2 = L_{21}X_1 + L_{22}X_2$. These equations show that a flow occurs due to all the forces present within the system.

In living systems, many parts of oxidative phosphorylation and active transport of ions exhibit linear relationships between stationary flows and their conjugate thermodynamic forces outside of equilibrium (Rothschild et al. 1980). Under sufficient conditions, an enzyme mechanism exhibits a multidimensional inflection point around which a set of linear flow-force equations will be valid over an extended range outside of equilibrium. Enzyme catalyzed reactions obey, as an approximation, the Michaelis-Menten rate equation, which can show a high degree of linearity in the chemical affinity for a certain range of substrate concentrations.

Onsager's Reciprocal Rules

When the flows and forces in the linear phenomenological equations are properly chosen, the matrix of phenomenological coefficients is symmetrical ($L_{ik} = L_{ki}$ with $i \neq k$) according to Onsager's reciprocal rules. These relations are proved to be an implication of the property of microscopic reversibility; for a simple reversible reaction, for example, if one of the paths is preferred for the forward reaction, the same path must also be preferred for the reverse reaction. Phenomenological coefficients are not the function of the thermodynamic forces and flows; on the other hand, they can be functions of the parameters of both the local state and the nature of a substance. The values of L_{ik} must satisfy the conditions $L_{ii} > 0$ and $L_{ii}L_{kk} > (1/4)(L_{ik} + L_{ki})^2$ because entropy production is positive for all irreversible processes.

Coupled Phenomena

When two or more processes occur simultaneously in a system, they may couple, that is, interact, and induce new effects. Coupling implies an interrelation between flow i and flow j, so a flow (e.g., heat, or mass flow, or chemical reaction) occurs without its own thermodynamic driving force, or in opposition to the direction imposed by its own driving force. For example, in regard to the sodium pump, ions can flow against the direction imposed by their electrochemical potential gradients only by coupling to energy releasing ATP hydrolysis.

As there is no restriction of the sign of individual contributions to total dissipation ($\Psi = \sum J_k X_k \geq 0$), a negative sign for the dissipation of a given process implies that this process must be coupled to at least one other spontaneous process. For example, for a reaction-diffusion system, the rate of total dissipation of Gibbs free energy may be represented by $\Psi = J_1 X_1 - J_r A_r > 0$, where J_1 is the flow of an ion, X_1 is the electrochemical potential gradient of the ion, J_r is the flow (velocity) of chemical reaction and A_r is the affinity that drives that reaction. The affinity is equivalent to Gibbs free energy with an opposite sign ($-\Delta G$). Here, the dissipation due to the chemical reaction process is negative ($J_r A_r < 0$), and thus it must be coupled to the flow process with the much larger positive dissipation ($J_1 X_1 > 0$). The total dissipation becomes positive only by coupling, and the energy of the flow process drives the chemical reaction.

Degree of Coupling

For a two-flow system, the degree of coupling q is obtained from: $q = L_{12}/(L_{11}L_{22})^{1/2}$ (Kedem and Caplan 1965). Because the total dissipation must be positive, the value of q changes only between -1 and $+1$. Coupling does not occur if the cross coefficient L_{12} is zero, and $q = 1$ represents a complete coupling. Chemical reactions are scalar and transport processes are vectorial processes, and the coupling between them occurs only in an anisotropic medium, according to the Curie-Prigogine principle (Prigogine 1967). Membranes display anisotropic character, as they consist of compartmental structures. For example, during the active transport of sodium ions, the direction of flow is determined by the property of the membrane in mitochondria. The medium may be locally isotropic, although it is not spatially homogenous. In this case, the coupling coefficients are associated with the whole system (Caplan and Essig 1983). In the case of the calcium pump, hydrolysis of ATP creates a directional flow (Waldeck et al. 1998). Linear nonequilibrium thermodynamics can describe the macroscopic energy coupling without the need for detailed information on microscopic mechanisms of coupling (Rottenberg 1979; Caplan and Essig 1983; Kondepudi and Prigogine 1999; Demirel and Sandler 2004).

Protein Structures

Proteins are polymers, and each type of protein structure consists of a precise sequence of amino acids that allows it to fold up into a particular three-dimensional shape or conformation. The lowest level of the hierarchy in protein structure is the polypeptide chain. Under the proper conditions, the polypeptide chain can fold back and form secondary structures known as *helices* and *sheets*. These helices and sheets pack against each other and form tertiary, or fully folded, structures. Strong chemical bonds strengthen the polypeptide backbone and lead to stable protein structures. The highest level of the hierarchy occurs when proteins interact with each other and balance internal tensions between the counteracting tendencies for order (Schutt and Lindberg 2000). The disorder may be represented in the entropy of the backbone chain and the surrounding solvent molecules; the order is represented by hydrogen bonds, salt bridges, and disulfide links. Folding may be a mechanism of transporting heat (entropy) away from the protein structure as it moves to a more ordered form. The most stable state of the polypeptide chain in solution occurs at minimum Gibbs free energy of the complete system that is the protein structure plus solvent. The larger number of arrangements exists within the surrounding solvent (water) molecules. This is known as the *hydrophobic effect* or *entropy stabilization* and may be due to the set of hydrophobic amino acids in proteins.

The folding of polymers is a coupled process between conformational transitions in the two single strands and their binding for duplex formation (Cao and Lai 1999). The substance that is bound by the protein is called a *ligand*. Separate regions of the protein surface generally provide binding sites for different ligands, allowing the protein's activity to be specific and regulated. Proteins reversibly change their shapes when ligands bind to their surfaces. One ligand may affect the binding of another ligand, and metabolic pathways are controlled by feedback regulation, in which some ligands inhibit and others activate enzymes early in a pathway (Wang et al. 2007). The binding affinities, enthalpies, and Gibbs free energy of formations may change with temperature, which may be affected by the biochemical reaction cycles that either release or require energy. The ability to bind to other molecules enables proteins to act as catalysts, signal receptors, information processors, switches, motors, or pumps. For example, proteins walk in one direction by coupling one of the conformational changes with the hydrolysis of an ATP molecule bound to the protein.

Information and Living Systems

Definition and quantification of information (like energy) have created broad discussions. The field of information systems, with its role in living systems, is constantly evolving (Corning and Kline 1998). Shannon and Weaver (1949) defined *information* as the capacity to reduce statistical uncertainty in the communication of messages between a sender and a receiver; however, this definition bears no relationship to natural systems, such as living organisms, that are "informed thermodynamic systems" (Wicken 1987). Later information theorists introduced *structural* or *functional information* to account for the self-organizing capabilities of living systems, and *instructional information*, which is a physical array (Brooks and Wiley 1988). However, linkages with the field of semiotics established a much more compatible approach to biological information (Salthe 1998). Within this trend, *control information* is defined as the capacity to control the acquisition, disposition, and utilization of matter, energy, and information flows in purposive (cybernetic) processes (Corning and Kline 1988; McIntosh 2006. According to Layzer (1988), information is the difference between the observed entropy state of any system and the maximum possible entropy.

A string of digits can represent digital information, and basic biological information is digital. The nucleic acids, the "molecules of genetic information," are digitally organized; they consist of chains of just four different units (nucleotides). Nucleic acids consist of a sugar phosphate backbone with purine or pyrimidine bases attached to the sugar molecules. The basic element of nucleic acids consists of a phosphate, a sugar, and a base, and is called a *nucleotide*. The bases of deoxyribonucleic acid (DNA) are guanine, adenine (both purines), cytosine, and thymine (both pyrimidines). Information is stored in DNA as a sequence of base pairs. Ribonucleic acid (RNA) is a nucleic acid polymer consisting of nucleotide monomers that translate genetic information from DNA into protein products; three types of RNA molecules are involved in the translation and cooperation of different functions. RNA is very similar to DNA, and serves as the template for translating genes into proteins, transferring amino acids to ribosomes to form proteins, and also translating the transcript into proteins. RNA is transcribed (synthesized) from DNA by enzymes called *RNA polymerases* and further processed by other enzymes (Emmeche 1991).

As Pierce (2002) suggested, biological systems diversify at bifurcation points as the information within the system becomes too complex and

Table 2.1
Information systems and thermodynamics

Information systems (sources)	Thermodynamic potential
Information	Free energy (exergy)
Information inflow (replication)	Exergy inflow
Information outflow (death)	Waste exergy
Information exchange (sorting)	$d_e S$
Internal information processing (growth)	$d_i S$

random. The bifurcation points are stimulated by intrinsic mechanisms or informational entropy, and are sensitive to controlling parameters. Living systems consist of organized structures and processes of informed, self-replicating, and dissipative autocatalytic cycles. They are capable of funneling energy, mass, and information flows into their own production and reproduction, and contribute to pathways of autocatalytic processes.

The Second Law of Thermodynamics operates on information in the form of DNA and RNA to generate biological variety and order (Brooks and Wiley 1988). More developed dissipative structures are capable of degrading more energy and of processing more complex information through developmental and environmental constraints; this establishes mechanisms for energy coupling in the pathways of chemical cycles and transport systems, and interactions between the dissipative structures. Table 2.1 illustrates that through replication, the source of information may keep the information system away from equilibrium.

Miller (1978) used the word "information" in the same sense that the word "message" is used. The value of the message is implied before receipt. A message becomes information only after the receiving system or subsystem assigns value to it by calculating an adjustment of its behavior in state space. Information is used either as a measure of the uncertainty, or surprise in the communication event as a whole, and as a measure of the degree of order (or complexity) in the system.

Energy Coupling

Cells are classified as either *prokaryotic* or *eukaryotic*. Prokaryotic cells have relatively simple structures and utilize three principal energy-coupling processes: glycolysis, respiration, and photosynthesis. Eukaryotic cells are the basic organizing units of higher life forms. The cytoplasm of eukaryotic cells contains many membrane-bounded organelles, including mitochondria. All membranes in a cell are made of lipids, proteins, cholesterol, and glycoproteins, and are involved in selectively transporting substances in and out of the cell and organelles, as well as communicating via signal transduction proteins. Therefore, membrane contents facilitate the flow of information and coupling to various degrees. Eukaryotic cells are capable of many motile activities, such as cell divisions. Neurons and immune cells are especially rich in motile activity. Membranes show structural anisotropy, and the spatial arrangement of enzymes within the membrane and vectorial character of energy coupling is now widely recognized (Mannella 2000; Frey, Renken, and Perkins 2002).

The flow of free energy originates from the sun. Chloroplasts in plants and algae, and mitochondria in plants and in animals, house the free energy conversion to ATP. Cells are capable of coupling the energy released by the hydrolysis of ATP to drive most of the energy-demanding processes. ATP production and utilization involve various driving and driven processes through the following energy couplings: (a) during photophosphorylation in chloroplasts, the absorption of photons is converted into the flow of electrons, which creates a transmembrane proton electrochemical potential gradient, and the energy of this gradient is coupled to ATP production; (b) during oxidative phosphorylation in mitochondria, nutrients from food entering the respiration system initiate electron transfer cycles, which create a transmembrane proton electrochemical potential gradient, and the energy of this gradient is coupled to ATP production; (c) all the energy-demanding biological processes, such as muscle work, chemical synthesis, cell division, and the transport of substrates are coupled to both the energy released by the hydrolysis of ATP and the energy of electrochemical gradients of ions in highly regulated and controlled cyclic systems.

Mitochondria utilize carbohydrates as nutrients, and oxygen to produce carbon dioxide and ATP; chloroplasts utilize light, carbon dioxide, and water to produce oxygen, carbohydrates, and ATP. Feeding the carbon dioxide back to the food chain completes the carbon cycle. ATP production and utilization are not isolated energy-coupling phenomena; they are synchronized and regulated, and hence represent overall energy coupling processes of the carbon cycle.

Mitochondria contain inner and outer membranes made of bilayers, which may influence the coupling and local gradients of ions, molecules, and macromolecules in the regulation of energy metabolism. The inner membrane houses the respiratory chain and ATP synthesis. The energy metabolism has to optimize the rate of energy production and energy demand, such as increasing thermogenesis at low temperatures, information processing, and the efficient use of nutrients. This optimization incorporates the use of information as well as several control mechanisms, including slips of proton pumps and proton leaks across the bilayer (Kadenbach et al. 2000; Stucki, Compiani, and Caplan 1983).

Oxidative Phosphorylation

Oxidative phosphorylation uses the input force of redox potential of the oxidization of nutrients to synthesize ATP and the output force of the phosphate potential to drive many energy-utilizing processes in the cell. The synthesis of ATP is matched and synchronized to the utilization of cellular ATP according to the chemiosmotic theory: $ADP + P_i + nH^+_{in} = ATP + H_2O + nH^+_{out}$, where "in" "and" out denote two phases separated by a membrane, and n is the ratio H^+/ATP, showing the level of transmembrane proton transport for each ATP to be synthesized (Turina, Samoray, and Gräber 2003). The enzyme-catalyzed reactions, the electron transport chain, and proton translocation are composed of a series of elementary reactions in both forward and backward directions.

One way of describing oxidative phosphorylation utilizes the linear nonequilibrium thermodynamics model, which does not require the detailed microscopic mechanisms of the coupling phenomena. Experiments and empirical analysis show that linear relations exist between the rate of respiration (flow of oxygen) and growth of many microbial systems (flow of ATP), with mechanisms to optimize the efficiency of energy couplings (Stucki 1980a, 1991; Fontaine et al. 1997). This might be the result of a special kinetics of enzyme reaction, such as the Michaelis-Menten equation, which shows that the enzyme reactions in certain regions can be approximated as linear. Also, the existence of a multiple inflection point in the force-flow space of a system of enzyme-mediated reactions suggests that a steady state in the vicinity of the linear region occurs over a considerable range, as the local asymptotic stability is guaranteed (Rothschild et al. 1980; Soboll and Stucki 1985; Stucki 1991). Intrinsic linearity would have an energetic advantage as a consequence of evolution (Stucki 1984). Stucki (1984, 1991) demonstrated that variation of the phosphate potential at constant oxidation potential yields linear flow-force relationships in the mitochondria. Extensive ranges of linearity are also reported for the active transport of sodium in epithelial membranes, where sodium transport occurs close to a stationary state, and the dissipation of power reaches a minimum value (Caplan and Essig 1983).

A representative and simplified overall dissipation function for the oxidative phosphorylation may be $\Psi = J_p X_p + J_o X_o = $ −output power + input power ≥ 0, where the input force X_o is the redox potential of oxidizable substrates, and X_P is the output force representing the affinity A of the chemical reaction. The associated input flow J_o is the net oxygen consumption, and the outflow J_P is the net rate of ATP production. For the dissipation function to be positive, both contributions may be positive: $J_o X_o > 0$, $J_p X_p > 0$, or the contribution of flow process may be large and positive, while the contribution of chemical system may be negative: $J_o X_o > 0$, $J_p X_p < 0$. The oxidative phosphorylation may be represented by $J_p X_p < 0$ (driven process), hence must be coupled to the respiration $J_o X_o > 0$ (driving process) to comply with the condition of positive overall dissipation of energy: $\Psi > 0$. This dissipation may also be due to the passive permeation of protons and ions (leaks) or to the decrease of the efficiency of chemical pumps (slips). In uncoupled respiration, no gradients are formed, and respiration is not controlled by energy consumption in the cell. This regulated and optimized energy coupling phenomena may result through the sequence of couplings controlled at switch points where the mobility and catalytic activity of the coupling protein structures can be altered (Soboll 1995; Krupka 1998).

It is customary to use the following representative linear phenomenological equations for the overall oxidative phosphorylation: $J_p = L_p X_p + L_{po} X_o$ and $J_o = L_{op} X_p + L_o X_o$. Here, L_o shows the influence of substrate availability on oxygen flow, and L_p is the feedback of the phosphate potential on ATP production flow. The cross-phenomenological coefficient L_{op} represents the macroscopic mechanism of coupling, and according to Onsager's reciprocal rules, $L_{po} = L_{op}$. The cross coefficient L_{op} shows the phosphate influence on oxygen flow, and L_{po} shows the substrate dependency of ATP production. These equations suggest that each flow results from the forces of X_p and X_o when L_{po} does not vanish. The matrix of the phenomenological coefficients must be positive for entropy production to be positive.

Molecular Motors and Pumps

A protein structure called *ATP synthase*, or F_oF_1, couples the energy of the proton electrochemical potential gradient to ATP synthesis, or the energy of ATP hydrolysis in F_1 to the proton translocation through the subunit F_o rotation. The direction of rotation for ATP synthesis may be different from that of ATP hydrolysis. Changes in protein shape can be driven in a unidirectional manner by the expenditure of chemical energy. By coupling allosteric shape changes to ATP hydrolysis, for example, proteins can do useful work and drive chemical pumps. Many motor proteins can generate directional movement, including the muscle motor protein myosin and the kinesin proteins of microtubules. For example, the redox proton pump utilizes the energy released by a redox reaction cycle to the transport of protons across the membrane, and establishes an electrochemical potential gradient (protonmotive force). The reversible ATPase-proton pump then utilizes the protonmotive force to synthesize and hydrolyze ATP (Sambongi et al. 2000; Tomashek and Brusilow 2000).

Muscle fibers lengthen and contract in small volume changes to perform mechanical work as they utilize chemical energy released by the hydrolysis of ATP. In general, the cyclic changes in the structure of muscle proteins enable them to convert the chemical free energy into work. There must be constant interchange of chemical bond energy and energy dissipation within structures that can undergo order-disorder transitions. Energy is released when a bond forms. For example, the bond energy of salvation can change the flexibility of some part of the polypeptide chain and the heat capacity of the structure, which exchanges energy with its surroundings to maintain a uniform temperature.

The Energy Coupling Efficiency

The efficiency of energy coupling η is defined as the ratio of output and input powers. When there is no heat effect, we have the dissipation equation $\Psi = J_p X_p + J_o X_o = $ −output power + input power ≥ 0, and the efficiency becomes $\eta = -J_p X_p /(J_o X_o)$. In terms of the normalized flow ratio (j) and the normalized force ratio (x), the energy coupling efficiency becomes $\eta = jx = -(x+q)/(q+1/x)$, where $j = -J_p/(J_o Z)$, $x = X_p Z / X_o$, and Z is called the *phenomenological stoichiometry*, defined by $Z = \sqrt{L_p / L_o}$ (Caplan and Essig 1983; Fontaine et al. 1997; Demirel 2007). Thus, the efficiency

Table 2.2
Representative linear nonequilibrium thermodynamics formulations for oxidative phosphorylation

Flow	Force	Linear relations	Dissipation of power Ψ/ Coupling efficiency η	Coefficients	Energy coupling (q)
J_i	X_i	$J_i = \sum L_{ik} X_k$	$\Psi = \sum J_k X_k \geq 0$	L_{ii}, L_{ik}	$q = L_{12}/(L_{11}L_{22})^{1/2}$
J_o	X_o	$J_o = L_o X_o + L_{op} X_p$	$\Psi = J_o X_o + J_p X_p \geq 0$	L_o, L_p, L_{op}, L_{po}	$q = L_{op}/(L_p L_o)^{1/2}$
J_p	X_p	$J_p = L_{po} X_o + L_p X_p$	$\Psi = (L_o X_o + L_{op} X_p) X_o$ $+ (L_{po} X_o + L_p X_p) X_p)$	$L_o L_p > (L_{op})^2$	$Q = 0$ if $L_{op} = L_{po} = 0$
			$\eta = -J_p X_p/(J_o X_o)$	$L_{op} = L_{po}$	$0 < q < |1|$

depends on the force ratio x and the degree of coupling q. The ratio J_p/J_o is the conventional phosphate-to-oxygen consumption ratio P/O. The energy coupling efficiency is zero when either J_p or X_p is zero. Therefore, at intermediate values of J_p and X_p, the efficiency passes through an optimum (maximum) defined by $\eta_{opt} = \left(q/\left(1+\sqrt{1-q^2}\right)\right)^2$. Here, q represents a lump-sum quantity for the various individual degrees of coupling of different processes of oxidative phosphorylation. This equation shows that optimal efficiency depends only on the degree of coupling and increases with increasing values of q.

When oxidative phosphorylation progresses with a load J_L, such as the active transport of ions, then the total dissipation becomes $\Psi_c = J_p X_p + J_o X_o + J_L X_p$. Here, J_L is the net rate of ATP utilized, and it is assumed that the phosphate potential X_p is the driving force: $J_L = L X_p$. L is called the conductance matching of oxidative phosphorylation, and is defined by $L = L_p \sqrt{1-q^2}$ (Stucki 1980a). Table 2.2 lists the linear nonequilibrium thermodynamics description of energy coupling in oxidative phosphorylation.

Stucki (1980a, 1980b) and Cairns et al. (1998) reported the various output modes shown in table 2.3. These experimental results show that the degree of coupling is not a constant but a variable subject to metabolic requirements and physiological organ roles. For example, fatty acids decrease the degree of coupling, and act as uncouplers (Wojtczak and Schönfeld 1993).

Mitochondria could regulate the degree of coupling of oxidative phosphorylation depending on the energy demand of the cell (Stucki 1980a,

Table 2.3
Output required from energy coupling

Outputs	Loci of the optimal efficiency states	Degree of coupling	Energy cost
Maximal net rate of ATP production at optimal efficiency: $(J_p)_{opt}$	From the plot of $(J_p)_{opt}$ versus x	$q_f = 0.786$	$\eta = \eta_{opt}$
Maximal output power at optimal efficiency: $(J_p X_p)_{opt}$	From the plot of $(J_p X_p)_{opt}$ versus x	$q_p = 0.910$	$\eta = \eta_{opt}$
Economic net output power: $(J_p X_p \eta)_{opt}$	From the plot of $(J_p X_p \eta)_{opt}$ versus x	$q_p^{ec} = 0.972$	Energy cost min.

1991; Soboll and Stucki 1985; Soboll 1995; Cairns et al. 1998; Jencks 1989; Demirel and Sandler 2002). Some uncoupling activities are favorable for the performance of metabolism and even for the energy-conserving functions of cellular respiration. For example, oxidative phosphorylation in fed rats operates very close to conductance matching, that is, at the economic degree of coupling permitting a maximal net flow of ATP synthesis at low energy costs with optimal efficiency (Stucki 1980a). The overall energy coupling is synchronized by various feedback regulations (Stucki, Lehmann, and Mani 1984). Figure 2.3 shows the change of the efficiencies in terms of flow ratio x between -1 and 0, and for the particular degrees of coupling: q_f, q_p, q_f^{ec}, and q_p^{ec} (Stucki 1984; Cairns et al. 1998; Soboll 1995).

There are three steady states of linear energy converters: (1) static head in which there is no net output flow: $J_p = 0$, as the two forces balance each other; (2) level flow, which results from vanishing force: $X_p = 0$; and (3) the state of optimal efficiency at which a symmetric

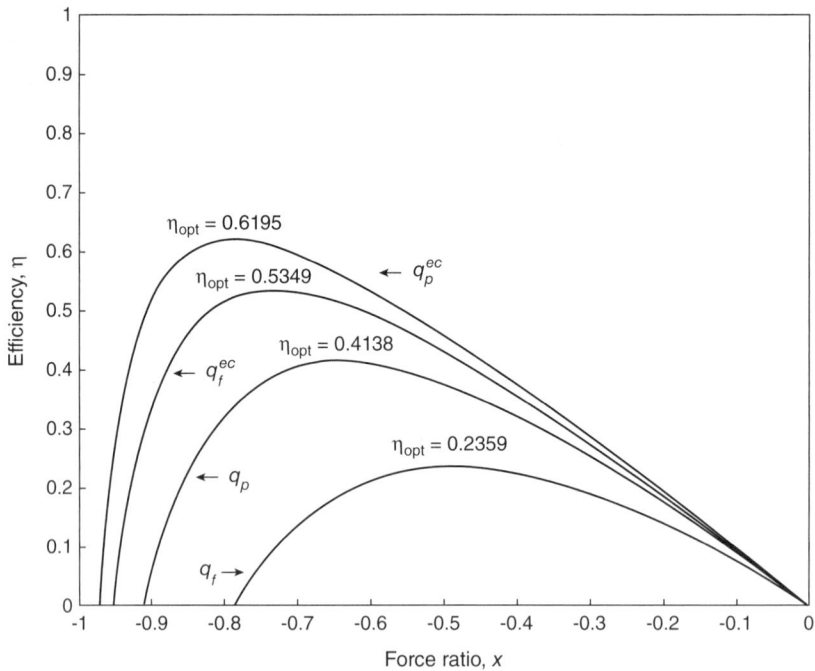

Figure 2.3
Change of efficiencies in energy coupling for various outputs in terms of force ratio x

control of the reactions for the input and output process exists (Caplan and Essig 1983).

Determination of the Degree of Energy Coupling

At static head, where $J_p = 0$, the coupling between the respiratory chain and oxidative phosphorylation maintains a phosphate potential X_p, which can be obtained from $(X_p)_{sh} = -qX_o/Z$, and the static head force ratio x_{sh} becomes $x_{sh} = -q$. The oxygen flow J_o at static head is obtained as $(J_o)_{sh} = L_o X_o (1-q^2)$. If an uncoupling agent, such as dinitrophenol, is used the ATP production vanishes, and we obtain $(J_o)_{unc} = L_o X_o$. Combining these two equations leads to $(J_o)_{sh} = (J_o)_{unc}(1-q^2)$. Using the experimentally attainable static head condition (state 4 in mitochondria) and the uncoupled oxygen flow $(J_o)_{unc}$, we can determine the degree of coupling by $q = [1-(J_o)_{sh}/(J_o)_{unc}]^{1/2}$.

Information and Biological Systems

Information flow, storage, and utilization are critical in the dynamics of biological systems. The unified theory of evolution attempts to explain the origin of biological order as a manifestation of the flow of energy and the flow of information on various spatial and temporal scales. For example, in a low-level biological information system like *Escherichia coli*, the main task of data processing is the production of polypeptides at the time they are needed for the metabolism of the cell. The environment of *Escherichia coli* determines the selection of what is transcribed. The transcription of the enzymes for the cleavage of sugar lactose into galactose and glucose is controlled; if the environment has no lactose, a repressor molecule blocks transcription. Genes, therefore, act as specific informational units with addresses and regulatory support systems to form the required enzymes and regulatory and structural proteins. However, the size and complexity of multicellular organisms, especially in animal life, require more specialized information signaling and processing procedures, including defending themselves against microorganisms and viruses.

Every living system functions as a whole and is made up of subsystems maintained with a constant flow through them of energy, substances, and information. A subsystem is highly specialized to perform needed functions of coordination; it can receive messages, process them, and prepare the required responses. Information theory names these processes with various terms such as "messenger RNA," "genetic code," "genetic information," "transcription," "translation," "cell signaling," and "chemical signals." For example, our bodies are made up of a respiratory system, a digestive system, and a reproductive system. The cooperation among these systems and their larger whole guides the relationships. Hence, living systems respond to change and survive within constantly changing environmental conditions by maintaining their form in a kind of fluctuating, dynamic balance. For example, warm-blooded animals rely on homeostasis to maintain their body temperature within a certain range.

The basic genetic information system collects signals and information from the environment through sense organs. Nerves transmit signals to the central nervous system where they are processed, and necessary and appropriate responses are generated. To maintain their integrity, systems tend to respond to the flow of information in ways that counteract any deviation from their established patterns of interaction. Such responses change the system's relationship to the environment so as to restore conditions to a tolerable range (Schneider and Kay 1995). Memory stores previous information and models or patterns of how the information has interacted, and of what decisions were made based on these models. This constitutes a library of strategies. The quality of this library constitutes a measure of the information inherent in the order and complexity of the system. On the other hand, consciousness imposes divisions between the realms of subjective and objective; thus living systems gain a complimentary dynamic between self and not-self in their essential exchanges of information with their environments. Furthermore, the survival and optimal outcome of living systems demand a degree of flexibility during the course of adaptation to the intrinsic uncertainty in their interactions with random processes (Dunne and Jahn 2005).

Information storage in DNA, and transforming this information into protein clusters, plays the central role in maintaining thermodynamic stability, information processing, increased organization, and the transport of matter and heat. The genome is the source of cellular information, and any cellular structure, such as lipids and polysaccharides, may store and transmit information according to information theory (Gatenby and Frieden 2005, 2007). In addition, thermodynamic forces, in the form of the transmembrane gradients of H^+, Na^+, K^+, Ca^{2+} and their consequent electric potential, cause significant displacements from equilibrium (randomness), and are therefore potential sources of information. Gatenby and Frieden (2005, 2007) suggest that the genome-protein system may be a component of a large ensemble of cellular structures, which store, encode, and transmit this information.

In biological applications, information theory initially used Shannon's methods (Shannon and Weaver 1949) and measured the information content in genes, RNA, and proteins. However, recent information theory uses bioinformatic and dynamical systems to determine the topology and dynamics of cellular and intracellular information networks. Such modern applications help us understand the mechanisms by which genetic information and information structures—including ion gradients and molecular flows across cell membranes—are translated into energy, control of energy, and self-organization (Gatenby and Frieden 2007).

Information for Directing and Controlling Exergy

Under steady state conditions, the natural setting of thermodynamic forces would lead to unique values of flows, such as *level flow* and *static head*. During oxidative phosphorylation, both level flow and static head occur, and the rate of phosphorylation depends on the mechanisms for changing the transmembrane potential, while the mitochondrial respiration level is controlled by mechanisms for reducing the protonmotive force (Caplan and Essig 1983; Lemeshko and Anishkin 1998). Also, studies on the initial electron transfer dynamics during photosynthesis in *Rhodobacter sphaeroides* show that the driving force and the kinetics of charge separation vary over a broad range; it is possible to fit the complex electron transfer kinetics of each mutant quantitatively, varying only the driving force (Wang et al. 2007). As McIntosh (2006) suggested, flows and forces must be directed, varied, and controlled by the information required for a particular organization. Therefore, some specific information may create and control the mechanisms for changing the flows and forces, hence directing and controlling the exergy for required outputs from coupled biological systems. For example, the information within a living plant enables photosynthesis to take place by combining energy from the sun with carbon dioxide and water. Molecular network motifs, generated by genes and proteins, enable the cell to process information and respond to stimuli. The network motifs consist of transcription-regulation and protein-protein interactions (Yeger-Lotem et al. 2004).

Active Transport-Chemical Pumps

Allosteric proteins can utilize energy released from the hydrolysis of ATP, the ion gradients, or the electron transport processes to pump specific ions or small molecules across a membrane. For example, the calcium transport protein (Ca^{2+} pump or the Ca^{2+} ATPase) maintains the low cytoplasmic calcium concentration of resting muscle by pumping calcium out of the cytosol into the membrane-enclosed sarcoplasmic reticulum; then, in response to a nerve impulse, Ca^{2+} is rapidly released (through other channels) back into the cytosol to trigger muscle contraction. The Ca^{2+} pump is homologous to the Na^+-K^+ pump. These pumps maintain nonequilibrium ion concentrations across the membrane. The Ca^{2+} pump is autophosphorylated during its reaction cycle when a large, cytoplasmic head region binds and hydrolyzes ATP. The protein thereby shifts to a high-energy, phosphorylated state that is tightly bound to two Ca^{2+} ions that were picked up from the cytosol. This form of the protein then decays to a low-energy, phosphorylated state, and releases the Ca^{2+} ions into the lumen. Dephosphorylation of the enzyme resets the protein for its next round of ion pumping. A preferred directional transport of an ion requires the cycle to use energy through the motor protein's conformational changes (Waldeck et al. 1998; Jencks 1989).

Compartmental structure of mitochondria may influence the level of local ion gradients generated by respiration, the internal diffusion of adenine nucleotides, and other substances. However, the symmetry properties alone are not sufficient for identifying physical coupling; the actual physics of the processes and the specific medium structure, such as anisotropy, are necessary. For example, some degree of imperfection, due to parallel pathways of reaction or intrinsic uncoupling within the pathway itself, may lead to leaks and slips in mitochondria (Stucki 1980a).

Linear and Nonlinear Energy Coupling

Comparison of energy couplings in linear and nonlinear regions requires a combination of thermodynamic and kinetic considerations of nonlinear η_{nl} and linear η_l modes. Donating γ as a measure for the relative binding affinity of a substrate (H^+) on either side of the membrane, the following inequalities are obtained for $\gamma \geq 1$: $\eta_{nl} > \eta_l$ for $\zeta = 1/q$ and $\eta_{nl} < \eta_l$ for $\zeta = q$, where ζ is the reduced stoichiometry (Z/n), which is subjected to kinetic limitation $(1/q) \geq \zeta \geq q$, and n is the mechanistic stoichiometry.

These inequalities suggest that at a certain value of ζ, the efficiencies of nonlinear and linear modes become equal to each other. This means that there exist values for ζ where the energy coupling operates more efficiently (Rottenberg 1979; Stucki 1991). A gain ratio η_r of efficiencies at linear and nonlinear modes η_l/η_{nl} is a measure for the efficiency gain in linear mode operation. The efficiency in linear mode depends only on q, and the efficiency in nonlinear mode depends, in addition to q, on input force X_o. In the nonlinear region, the efficiency decreases at high values of input force, and the force ratio at optimum operation $x_{opt,nl}$ is shifted toward level flow where $x = 0$. In oxidative phosphorylation, the input force is the redox potential of the oxidizable substrates and the output force is the phosphate potential. If these two forces are balanced, the system operates close to reversible equilibrium. This means that the system would be completely coupled, and two processes would merge into a single process governed by a single force. However, experiments show that coupling in the cyclic processes of mitochondria are incomplete ($q < 1$), and input force is well above $50RT$, which leads to a dissipative structure.

The gain ratio η_r (η_l/η_{nl}) can be calculated at a reference force ratio, such as x_{opt}, which is a natural steady-state force ratio of oxidative phosphorylation. This is seen as the result of adaptation of oxidative phosphorylation to various metabolic conditions and also results from the thermodynamic buffering of reactions catalyzed by enzymes (Stucki, Lehmann, and Mani 1984). Experimentally observed linearity of the several energy converters operating far from equilibrium may be due to enzymatic feedback regulation with an evolutionary drive toward higher efficiency. Within the framework of the dissipative structures, the thermodynamic buffering represents a new bioenergetics regulatory principle for the maintenance of a far-from-equilibrium regime. Due to the ATP production in oxidative phosphorylation, the phosphate potential is shifted far from equilibrium. Since ATP drives many processes in the cell, the shift in X_p to far from equilibrium results in a shift of all the other potentials into a far-from-equilibrium regime.

Conclusions

Living systems consist of organized structures and processes of informed, self-replicating, and dissipative autocatalytic cycles. Self-organizing structures resist and dissipate externally applied gradients, which would move the system away from equilibrium, and hence dissipative structures with order emerge. Living systems survive as they are capable of funneling material, energy, and information into their own production and reproduction and contribute to pathways of autocatalytic dissipative processes. Production and utilization of free energy require energy-coupling process and cycles, such as photosynthesis, oxidative phosphorylation, and active transport of ions by chemical pumps. This requirement links biology with the fundamental principles of nonequilibrium thermodynamics, which may describe the coupled systems without the detailed coupling mechanisms. Protein structures play a crucial role in information transferring/processing, directing the flow of energy, regulating the energy coupling, and thus lead to the "well-informed" character of living systems. This concerns the basic definition of how information is defined and connected to the fundamental principles of thermodynamics. Incorporating self-organization and information into physical sciences may give rise to important advances in such areas as biological evolution, regulation of energy coupling, and molecular mechanisms of energy coupling.

References

Boltzmann, Ludwig. [1886] 1974. The second law of thermodynamics. In *Ludwig Boltzmann, theoretical physics and philosophical problems: Selected writings*, ed. Brian McGuinness, 13–32. New York: D. Reidel.

Brooks, Daniel R., and E. O. Wiley. 1988. *Evolution as entropy: Toward a unified theory of biology*. Chicago: University of Chicago Press.

Cairns, Charles B., James Walther, Alden H. Harken, and Anirban Banerjee. 1998. Mitochondrial oxidative phosphorylation thermodynamic efficiencies reflect physiological organ roles. *American Journal of Physiology. Regulatory, Integrative and Comparative Physiology* 274 (5): R1376–R1383.

Cao, Wei, and Luhua Lai. 1999. A thermodynamic study on the formation and stability of DNA duplex at transcription site for DNA binding proteins GCN4. *Biophysical Chemistry* 80 (3): 217–226.

Caplan, S. Roy, and Alvin Essig. 1983. *Bioenergetics and linear nonequilibrium thermodynamics: The steady state*. Cambridge, Mass.: Harvard University Press.

Corning, Peter A., and Stephen J. Kline. 1998. Thermodynamics, information and life revisited, part II: "Thermoeconomics" and "control information." *Systems Research and Behavioral Science* 15 (6): 453–482.

Demirel, Yaşar. 2007. *Nonequilibrium thermodynamics: Transport and rate processes in physical, chemical and biological systems*. Amsterdam: Elsevier.

Demirel, Yaşar, and Stanley I. Sandler. 2002. Thermodynamics and bioenergetics. *Biophysical Chemistry* 97 (2–3): 87–111.

Demirel, Yaşar, and Stanley I. Sandler. 2004. Nonequilibrium thermodynamics in engineering and science. *Journal of Physical Chemistry B* 108 (1): 31–43.

Dunne, Brenda J., and Robert G. Jahn. 2005. Consciousness, information, and living systems. *Cellular and Molecular Biology* 51 (7): 703–714.

Emmeche, Claus. 1991. A semiotical reflection on biology, living signs and artificial life. *Biology and Philosophy* 6 (3): 325–340.

Fontaine, Eric M., Anne Devin, Michel Rigoulet, and Xavier M. Leverve. 1997. The yield of oxidative phosphorylation is controlled both by force and flux. *Biochemical and Biophysical Research Communications* 232 (2): 532–535.

Frey, Terrence, Christian W. Renken, and Guy A. Perkins. 2002. Insight into mitochondrial structure and function from electron tomography. *Biochimica et Biophysica Acta* 1555 (1–3): 196–203.

Gatenby, Robert A., and B. Roy Frieden. 2005. The role of non-genomic information in maintaining thermodynamic stability in living systems. *Mathematical Biosciences and Engineering* 2 (1): 43–51.

Gatenby, Robert A., and B. Roy Frieden. 2007. Information theory in living systems, methods, applications, and challenges. *Bulletin of Mathematical Biology* 69 (2): 635–657.

Jencks, William P. 1989. Utilization of binding energy and coupling rules for active transport and other coupled vectorial processes. *Methods in Enzymology* 171: 145–164.

Kadenbach, Bernhard, Maik Hüttemann, Susanne Arnold, Icksoo Lee, and Elisabeth Bender. 2000. Mitochondrial energy metabolism is regulated via nuclear-coded subunits of cytochrome *c* oxidase. *Free Radical Biology & Medicine* 29 (3–4): 211–221.

Kedem, Ora, and S. Roy Caplan. 1965. Degree of coupling and its relation to efficiency in energy conversion. *Transactions of the Faraday Society* 61: 1897–1911.

Kondepudi, Dilip, and Ilya Prigogine. 1999. *Modern thermodynamics: From heat engines to dissipative structures.* New York: Wiley.

Krupka, Richard M. 1998. Channeling free energy into work in biological processes. *Experimental Physiology* 83 (2): 243–251.

Layzer, David. 1988. Growth of order in the universe. In *Entropy, information, and evolution: New perspectives on physical and biological evolution*, ed. Bruce H. Weber, David J. Depew, and James D. Smith, 23–40. Cambridge, Mass.: MIT Press.

Lemeshko, Viktor V., and Andriy G. Anishkin. 1998. Mathematical simulation of energy coupling in mitochondria within the framework of the proton-chemical hypothesis. *Biophysics* 43 (2): 285–291.

Mannella, Carmen A. 2000. Introduction: Our changing views of mitochondria. *Journal of Bioenergetics and Biomembranes* 32 (1): 1–4.

McIntosh, Andy. 2006. Functional information and entropy in living systems. *WIT Transactions on Ecology and the Environment* 87: 115–126.

Miller, James Grier. 1978. *Living systems*. New York: McGraw-Hill.

Pierce, Stephanie E. 2002. Non-equilibrium thermodynamics: An alternate evolutionary hypothesis. *Crossing Boundaries* 1 (2): 49–59.

Prigogine, Ilya. 1967. *Introduction to thermodynamics of irreversible processes*. New York: Wiley.

Rothschild, Kenneth J., Samuel A. Ellias, Alvin Essig, and H. Eugene Stanley. 1980. Nonequilibrium linear behavior of biological systems: Existence of enzyme-mediated multidimensional inflection points. *Biophysical Journal* 30 (2): 209–230.

Rottenberg, Hagai. 1979. Nonequilibrium thermodynamics in energy conversion in bioenergetics. *Biochimica et Biophysica Acta* 549 (3–4): 225–253.

Salthe, Stanley N. 1998. The role of natural selection in understanding evolutionary systems. In *Evolutionary systems: Biological and epistemological perspectives on selection and self-organization*, ed. Gertrudis van de Vijver, Stanley N. Salthe, and Manuela Delpos, 13–20. Dordrecht: Kluwer Academic Publishers.

Sambongi, Yoshihiro, Ikuo Ueda, Yoh Wada, and Masamitsu Futai. 2000. A biological molecular motor, proton-translocating ATP synthase: Multidisciplinary approach for a unique membrane enzyme. *Journal of Bioenergetics and Biomembranes* 32 (5): 441–448.

Schneider, Eric D., and James J. Kay. 1995. Order from disorder: The thermodynamics of complexity in biology. In *What is life? The next fifty years: Reflections on the future of biology*, ed. Michael P. Murphy and Luke A. J. O'Neill, 161–172. New York: Cambridge University Press.

Schrödinger, Erwin. 1945. *What is life? The physical aspect of the living cell*. New York: Macmillan.

Schutt, Clarence E., and Uno Lindberg. 2000. The new architectonics: An invitation to structural biology. *Anatomical Record* 261 (5): 198–215.

Shannon, Claude E., and Warren Weaver. 1949. *The mathematical theory of communication*. Urbana: University of Illinois Press.

Soboll, Sibylle. 1995. Regulation of energy metabolism in liver. *Journal of Bioenergetics and Biomembranes* 27 (6): 571–582.

Soboll, Sibylle, and Jörg W. Stucki. 1985. Regulation of the degree of coupling of oxidative phosphorylation in intact rat liver. *Biochimica et Biophysica Acta* 807 (3): 245–254.

Stucki, Jörg W. 1980a. The optimal efficiency and the economic degrees of coupling of oxidative phosphorylation. *European Journal of Biochemistry* 109 (1): 269–283.

Stucki, Jörg W. 1980b. The thermodynamic-buffer enzymes. *European Journal of Biochemistry* 109 (1): 257–267.

Stucki, Jörg W. 1984. Optimization of mitochondrial energy conversions. *Advances in Chemical Physics* 55: 141–167.

Stucki, Jörg W. 1991. Nonequilibrium thermodynamic sensitivity of oxidative phosphorylation. *Proceedings of the Royal Society B: Biological Sciences* 244 (1311): 197–202.

Stucki, Jörg W., Mario Compiani, and S. Roy Caplan. 1983. Efficiency of energy-conversion in model biological pumps optimization by linear nonequilibrium thermodynamics relation. *Biophysical Chemistry* 18 (2): 101–109.

Stucki, Jörg W., Lilly H. Lehmann, and Peter Mani. 1984. Transient kinetics of thermodynamic buffering. *Biophysical Chemistry* 19 (2): 131–145.

Tomashek, John J., and William S. A. Brusilow. 2000. Stoichiometry of energy coupling by proton-translocating ATPases: A history of variability. *Journal of Bioenergetics and Biomembranes* 32 (5): 493–500.

Turina, Paola, Dietrich Samoray, and Peter Gräber. 2003. H+/ATP ratio of proton transport-coupled ATP synthesis and hydrolysis catalysed by CF0F1±liposomes. *European Molecular Biology Organization Journal* 22 (3): 418–426.

Waldeck, A. Reginald, Karel van Dam, Jan Berden, and Philip W. Kuchel. 1998. A non-equilibrium thermodynamics model of reconstituted Ca2+-ATPase. *European Biophysics Journal with Biophysics Letters* 27 (3): 255–262.

Wang, Haiyu, Su Lin, James P. Allen, JoAnn C. Williams, Sean Blankert, Christa Laser, and Neal W. Woodbury. 2007. Protein dynamics control the kinetics of initial electron transfer in photosynthesis. *Science* 316 (5825): 747–750.

Wicken, Jeffrey S. 1987. *Evolution, thermodynamics and information: Extending the Darwinian program.* New York: Oxford University Press.

Wojtczak, Lech, and Peter Schönfeld. 1993. Effect of fatty acids on energy coupling processes in mitochondria. *Biochimica et Biophysica Acta* 1183 (1): 41–57.

Yeger-Lotem, Esti, Shmuel Sattath, Nadav Kashtan, Shalev Itzkovitz, Ron Milo, Ron Y. Pinter, Uri Alon, and Hannah Margalit. 2004. Network motifs in integrated cellular networks of transcription-regulation and protein-protein interaction. *Proceedings of the National Academy of Sciences of the United States of America* 101 (16): 5934–5939.

II
Information and Biological Organization

3

Bioinformation as a Triadic Relation

Alfredo Marcos

Information is seen today not as a unitary concept, but as a family of concepts whose members lack any clear interconnection. The relationships of this family's members with neighboring notions, such as knowledge, form, entropy, correlation, probability, meaning, order, organization, and complexity, also require clarification. Furthermore, measurements of information are normally interpreted as gauges of structural complexity, correlation, thermodynamic order, or potential information, and as such, are unable to discriminate biological functionality. The clarification of the concept of information and its relationships with other surrounding notions, along with the development of a measure of functional information are, therefore, important tasks for philosophers of biology, especially as formulating biological theories in informational terms has become increasingly common. I hope that the following discussion will help fulfill these objectives.

In this chapter, I first show how the concept of information gradually became a key notion in the biological sciences, and I describe some of the controversies that resulted from this development. Then, I go on to defend a concept of information as a triadic relationship. This concept contributes to producing a general measure of information, as well as aids in integrating the measure and specific uses of the concept of information into a single framework. Next, I present a general measure of information that is based on this concept. Finally, I revisit the role of information in biology and try to apply this concept of information to the case of biology. The combined effect of the several parts of this discussion will be to clarify the relationships between information and surrounding concepts important for biology, and to form a sound approach to certain biological questions concerning bioinformation itself, such as the problem of its location.

Biology and the Rise of Information

The term "information" derives from the Latin *informatio*, meaning "explanation," "idea," or "representation." The verb *informo* can mean "to give form or shape," "to sketch," "to draw," "to instruct," "to represent," or "to form an idea" of something. In ancient times, the term was used in both everyday and learned discourse, appearing, for example, in the works of Virgil, Cicero, Tertullian, and Augustine of Hippo (see Capurro 1978). It was used in different contexts, including ontological ("to shape something"), epistemological ("to become acquainted through the sensorial or intellectual reception of a form"), pedagogical, and moral ("to instruct") ones, but was not the object of special philosophical reflection.

During the Middle Ages, the verb *informo* and its derivatives were incorporated into philosophical language from scholastic discourse. It is frequently used, for example, in the writings of Thomas Aquinas (1225–1274 CE). Throughout this period, the verb retained its ontological, epistemological, didactic, and moral aspects as well as its active sense, whereby *informatio* was an action rather than a thing, namely, that of shaping and its result. Before going on to the modern uses of the term "information," let me mention two cases of its application, which could even be seen as precedents of its use in genetics. First, Varro (116–27 BCE) describes the development of a fetus as a process of information, whereby it is "shaped" or "informed" (*informatur*). Second, Thomas Aquinas distinguishes between the biological process that brings a new life into the world *per modum informationis* and a nonbiological process that brings something to life *per modum creationis* (see Capurro and Hjørland 2003).

When, around the fourteenth and fifteenth centuries, the use of the term "information" spread into European languages from French, "investigation," "education," and "intelligence" were added to its traditional connotations. Perhaps because of the rejection of scholastic terminology, from then on "information" ceased to be a philosophical term, and others, such as "impression," "idea," and "representation," were preferred. Descartes, Locke, Hume, Berkeley, Bacon, Kant, and others did not think of their epistemologies in terms of information. And in the few places where we find a word that derives from the term "information," it had lost its ontological meaning and had become mainly an epistemological term. As a result, the term came to be understood as an idea or a representation that enters the subject's mind. Modern idea-ism is obviously related to this shift from the view of information as an action to an idea (see Musgrave 1993).

The term's growth in importance, to the point of acquiring a central place in contemporary culture, began in the nineteenth century. It was bound up with the development of communication technologies—especially the telegraph—and with the use given to it in military intelligence service (Adriaans 2007). Thus, information acquired economic and political value. Since then, theories of communication have been developed that seek to facilitate the transmission of the greatest possible amount of this new commodity at the lowest possible cost, in the shortest possible time, and with the maximum security. Since the Second World War, these developments have accelerated, as have those concerning computation. The linking of these two technologies, and their omnipresence, has done the rest. The term "information" now occupies a central place in everyday speech and in almost all sciences and disciplines, from computer science, biology, and librarianship to journalism, sociology, and philosophy.

The central reference for the theory of information is the classical book by Claude E. Shannon and Warren Weaver (1949). However, the term "information" does not even appear in its title, *The Mathematical Theory of Communication*. The expression "theory of information" probably comes from an article by Hartley (1928) entitled *Theory of Information Transmission*. Although Shannon focuses attention on the engineering problems of communication, we should understand that his theory deals specifically with the communication *of information*. In their work, Weaver clarifies the basic concept of information:

> Information must not be confused with meaning. . . . To be sure, this word information in communication theory relates not so much to what you *do* say, as to what you *could* say. That is, information is a measure of one's freedom of choice when one selects a message. If one is confronted with a very elementary situation where he has to choose one of two alternative messages, then it is arbitrarily said that the information, associated with this situation, is unity. . . . The amount of information is defined, in the simplest cases, to be measured with the logarithm of the number of available choices. (Shannon and Weaver 1949, 8–9)

Shannon identifies the elements that comprise the communication of information processes. He represents them by means of the diagram in figure 3.1 (34).

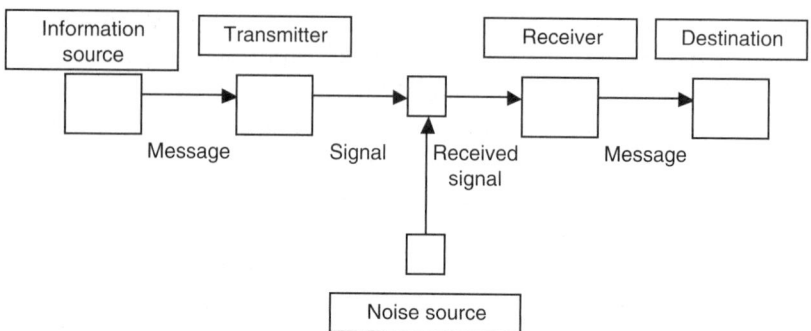

Figure 3.1
Shannon's mathematical theory of communication (Shannon and Weaver 1949, 34)

Shannon's main objective was to apply his theory to the technical systems of communication. This is the reason why his diagram includes a transmitter, the function of which is to transform the original message—for instance, a sequence of letters—into a signal suitable for transmission over the channel. Shannon thinks of the channel as a "pair of wires, a coaxial cable, a band of radio frequencies, a beam of light, etc." (Shannon and Weaver 1949, 34). Near the other end of the channel is a receiver that performs the inverse operation of that performed by the transmitter. But we could design even more detailed diagrams, with more *boxes*, depending on the nature and precision of the problems to which we are applying the theory (see, e.g., Escarpit 1976; Guéroult 1965; Moles 1972). In Shannon's diagram, the functions of encoding and decoding the message are performed by the transmitter and the receiver, respectively, but we could envisage new *boxes* for an encoder and a decoder. In fact, Shannon himself introduced another, more complex diagram, with a secondary feedback channel for data correction (Shannon and Weaver 1949, 68).

On the other hand, researchers, like Singh (1966), have constructed simpler diagrams with no more than three elements: a source or emitter, a channel, and a receiver. We can even adopt a more abstract interpretation of Shannon's theory, a nondimensional interpretation, free from spatiotemporal connotations. For example, Abramson (1963) interprets an information channel from a mathematical point of view as a simple relationship between the probabilities of two sets of symbols. An information channel is determined by only an incoming alphabet, an outgoing alphabet, and a set of conditional probabilities. For instance, $P(b_j|a_i)$ is the probability of receiving the symbol b_j if a_i were emitted. Here, a source of information is no more than a set of symbols and their corresponding probabilities.

As Shannon himself warns (Shannon and Weaver 1949, 31), his theory does not anticipate all problems regarding the concept of information. In order to understand which types of problems are at stake here, the distinction established by Weaver (Shannon and Weaver 1949) is still a very useful guide. He notes three types of problems concerning information:

1. There are *technical problems* concerning the maximum amount of information that a message can convey. These concern the statistical regularities of the source, like the internal structure and constraints of the messages, together with the conditions of noise and equivocation of the channel. Given these conditions, we ask, "What is the best possible configuration of the message?" Thus, we have problems at a *syntactic* level, of the type dealt with by Claude Shannon's mathematical theory of communication. The measure of complexity proposed by Andrey Nikolaevich Kolmogorov is also at the syntactic level (Kolmogorov 1965; Solomonoff 2003; Li and Vitányi 1997; Grünwald and Vitányi 2003).

2. There are *semantic problems* concerning the meaning and truth of the messages, and the correlation between the message and some other thing. Weaver makes it clear that Shannon's theory does not seek to explain problems at this level or at the next one. In the last few decades, several theories have appeared that do deal with semantic aspects of information (Barwise and Seligman 1997).

3. Finally, there are *pragmatic problems* concerning the efficiency of the message in regard to altering the receiver's behavior. Weaver says that "the *effectiveness problems* are concerned with the success with which the meaning conveyed to the receiver leads to the desired conduct on his part" (Shannon and Weaver 1949, 5). In biological terms, we find here the functional aspects of information, its ability to affect the receiver's behavior in a functional or adaptive sense.

Recently, Luciano Floridi distinguished among "information *as* reality," "information *about* reality," and "information *for* reality" (2007). It is tempting to correlate these categories with Weaver's levels. On the syntactic level, what we study is "information as reality," that is, the properties of the message itself. On the semantic level, we deal with "information about reality," or what a message tells us about another part of reality. On the pragmatic level, we observe the capacity of a message to alter reality. This is like saying that we observe the message as "information for (making or modifying) reality." A variety of approaches have arisen to address the levels of information. Floridi (2007) identifies as many as seven, but as he notes, all of them are concerned with syntactic or semantic problems. However, because our interest in this discussion focuses on living systems, our concern is mainly with pragmatic or functional problems.

Since the 1950s, the notion of information—either as a metaphor, analogy, or real entity—has become increasingly important in most fields of biology (see Paton 1992). It has even been used to define life (Tipler 1995, 124–127). The biological sciences have adopted a theoretical perspective derived from information theory, together with developments in modern genetics and evolutionary science. This perspective holds that all biological processes involve the transfer of information, and has been called *bioinformational equivalence* (Stuart 1985; see also Burian and Grene 1992, 6).

A glance at the current bibliography (see Queiroz, Emmeche, and El-Hani 2005; Jablonka 2002) will suffice to show that, since Stuart's 1985 paper, the use of the concept of information in biology has become more widespread (for a historical perspective, see Kull 1999; Collier 2007). In molecular biology, biomolecules are considered to contain information and are the result of informational processes (Holzmüller 1984). In genetics, especially, biological thinking is shaped by the idea of information transfer (Brandt 2005; Kjosavik 2007); in developmental biology, much is said about the expression of information and phenotypic information (Waddington 1968; Oyama 2000). In cell biology (Albrecht-Buehler 1990; Marijuán 1989, 1991), tissue biology, zoology, and botany, we study different ways of communicating information with chemical, neuronal, or linguistic bases (Stegmann 2005; Pfeifer 2006). In ecology, the concepts of complexity and biodiversity are closely bound up with information through notions of entropy and order (Margalef 1968).

In neurophysiology, the study of communication, storage, and processing of information is central, as are the various electric and chemical codes (Baddeley, Hancock, and Földiák 2000). The immune system is also researched in terms of knowledge, both acquired and accumulated (Forrest and Hofmeyr 2000). Evolution, from the origin of life onward, is thought of as the accumulation of information in macromolecules (Elsasser 1975; Küppers 1990; MacLaurin 1998; Moreno and Ruiz-

Mirazo 2002). The latest research into the human genome, and the genomes of other organisms, has required the application of powerful methods of computation, and this coming together of disciplines has given rise to what is known as *bioinformatics* (see Arp, Smith, and Spear 2011; Nishikawa 2002; and the journal *Bioinformatics*).

The concept of information, however, is also central to epistemology and the cognitive sciences, and as several research programs are attempting to link the cognitive phenomenon with its biological basis, it would be desirable to have one general concept of information that could be applicable to both cognitive and biological contexts. We might remember, as classical examples of such programs, evolutionary epistemology along the lines of Lorenz and Wuketits (1983) or Popper (1990), Piagetian epistemology (Piaget 1970), psychobiology (Bond and Siddle 1989), evolutionary psychology (Horan 1992), cognitive ethology (Allen 1992), neural Darwinism (Edelman 1987), and in general, a widespread current tendency to naturalize epistemology (Giere 1988). An analogy could be drawn among the programs of artificial life, computational science, and the social sciences, where the confluence with biology is evident and the need for a common concept of information is urgent.

Despite its applicability to a broad range of disciplines, the informational perspective in biology is not without polemics, which in recent years have arisen over its need and usefulness. Some authors consider information a distinctively human phenomenon, so that its application in other fields is purely metaphorical. For others, the use of information concepts is redundant in fields like biology, which are subject to general laws of matter and energy. Such researchers think that biological phenomena should be explained in mechanical, electromagnetic, chemical, and thermodynamic terms, thus rendering informational considerations superfluous. According to this last perspective, to speak of information in biology would just be an odd way of speaking of correlation and causation (Stuart 1985; Griffiths 2001; Sarkar 1996, 2000; Janich 1992; Kitcher 2001).

Many researchers, however, think that the informational perspective sheds considerable light on biological phenomena, allowing us to understand them in a way otherwise impossible (Maynard Smith 2000a; Queiroz, Emmeche, and El-Hani 2005). Here, I do not seek to argue in favor of this position, which I consider more reasonable, as it is already defended elsewhere in this book (see also *Philosophy of Science* 67 [2]; Maynard Smith 2000b; Godfrey-Smith 2000; Griffiths 2001; Roederer

2005; Avery 2003; Yockey 2005). Thus, assuming the informational perspective to be useful in biology, I now propose to present the kernel of a theory (concept and measure) that should contribute to clarifying the biological applications of information.

Information as a Triadic Relationship

Some authors have viewed information as a thing, third substance, or primitive element. Wiener thinks that information "is information, not matter or energy" (1961, 132). Also, information has been seen as a property of a thing in terms of form, order, organization, negative entropy (Brillouin 1962), complexity (Kolmogorov1965), or diversity (Margalef 1980). Information as a property raises the problem of its location, which is a recurrent difficulty, and as such, one of the major arguments against the bioinformational paradigm. Actually, the problem of information location would be unsolvable unless we abandon this view of information as a simple property. Third, we find information as a dyadic (semantic) and a triadic (pragmatic or functional) relation. As Barwise notes:

> But is information relational? Surely so. The basic intuition about the information content C_s of a situation s is that it is information *about* something besides s. . . . The account of the information content C_s of a situation s given by Dretske and that given by Perry and me differ on many points, but they do agree on the relational nature of information. (1986, 326; see also Dennett 1987; MacKay 1969; Küppers 1990; Queiroz, Emmeche and El-Hani 2005)

On the other hand, information as a thing or basic substance should be the last hypothesis to explore, for the principle of ontological economy implies that, other things being equal, if some other hypothesis works, it is clearly preferable. The other three possibilities could be equated with the three parts of Weaver's classical distinction (Shannon and Weaver 1949). The *technical problems*, which Weaver places at level A, are studied by considering the formal and statistical properties of messages. At this level, we are dealing with information as a property. The semantic problems, or level B problems, are concerned with the dyadic relationship between the message and its meaning. The *effectiveness problems*, or problems of level C, imply three elements. Weaver suggests that they are the message, its meaning, and a change in the receiver's behavior caused by the reception of the message (5). Therefore, problems of level C have a pragmatic aspect, which in biological contexts could be construed as a function. For instance, the change in a cell's behavior caused by the reception of a genetic message may consist of the accomplishment of a given function, such as the synthesis of a determined protein.

I shall argue that information should be conceived of as a relationship, specifically demonstrating the need for a *triadic* relationship. On my account, pragmatic or functional information is viewed as the basic and more general concept of information, and the others could be derived by abstraction, ellipsis, or addition. Even Shannon implies the functional and relative aspects of information, stating that "the fundamental problem of communication is that of reproducing at one point either exactly or approximately a message selected at another point" (1949, 31). One must suppose that the mere material transfer of what has been produced would not be information, and that reproduction itself is worthless unless it refers to what has been produced. The receiver of information can be so called only if it can relate what is received to what was emitted.

I also draw inspiration from Charles S. Peirce's ideas, but using neither his terminology nor his technical niceties. Nor, unlike Queiroz, Emmeche, and El-Hani, do I attempt a thorough application of his semiotics to the concept of information. On the other hand, I do share with these thinkers an essentially Peircean conception of information as a triadic relation.

Another point in favor of information as a triadic relation is that it enables us to defend a general measure of information as well as to integrate the different measures and notions of information. This pragmatic or functional concept of information is also the concept that best adapts to biological contexts, where functional explanations are very common. We consider an explanation of an organ or a molecule satisfactory only if it includes reference not only to its structure and material composition, but also to its function in the organism. For Peirce, precedents include the following:

All dynamical action, or action of brute force, physical or psychical, either takes place between two subjects . . . or at any rate is a resultant of such actions between pairs. But by semiosis I mean, on the contrary, an action or influence which is or involves a cooperation of three subjects, such as a sign, its object and its interpretant, this three-relative influence not being in any way resolvable into actions between pairs. (1931–1935, 5:484)

From these precedents, let us now construe the triadic informational relation in a slightly different terminology. Information implies a relationship between (1) a message, m, which may be any event, linguistic or otherwise; (2) a system of reference, S, which the message informs the receiver about; and (3) a receiver, R. The receiver is a formal scheme resident in a concrete subject (a human being, another living system, a part of a living system, an ecosystem, a cell, a computer, etc.). A concrete subject could, of course, use more than one receiver and use them alternately (playing with different "hypotheses") or successively (owing to an evolutionary or individual process of learning). Peirce could be quoted again, as he clearly differentiates the interpreter (the concrete subject) from the interpretant (the abstract scheme connecting sign and object). We can also see the receiver as an internal (that is, resident in a concrete subject) predictive model of S, along the lines suggested by Rosen, who characterizes living beings as "anticipatory systems" (1985).

Some elements entering into one informational relationship could participate in another by playing a different role: the element playing the role of receiver in one informational relation could be a message in another. For example, a scientific theory can be viewed as a receiver offering us expectations about some domain. At this level, empirical data are messages to the theory. But a scientist could opt for a certain theory, considering it better confirmed than others, in which case the theory becomes a message to a receiver (in the scientist) dealing with theoretical alternatives. A system of alternative messages in one relation can, in another relation, be a system of reference, and vice versa—and the process could be iterated ad infinitum. A segment of DNA can be a message informing the appropriate part of the cell about the mRNA to be synthesized. The same mRNA, hitherto part of a system of reference, may later become a message informing the cytoplasm about the synthesis of a certain protein, and so on. As Queiroz, Emmeche, and El-Hani state using Peircean vocabulary, "semiosis entails the installation of chains of triads" (2005, 60). This is why a metaphor like "the flow of information" is sometimes useful.

Comparing this triadic scheme to the classical Shannon-Weaver one (see figure 3.1), it may seem surprising that the emitter or source is not even mentioned, but that is because it should be considered a system of reference when the information that R receives through a message is about the emitter itself. On the other hand, in determining intended meaning, the emitter acts as a virtual receiver, and could be mathematically construed as such. Likewise, Millikan states that we should "focus on representation *consumption*, rather than representation production. Devices that *use* representations determine them to be representations" (1989, 283–284). Furthermore, there is often no specific emitter in nonlinguistic contexts, like some biological ones, so a general theory of information should not demand the presence of an emitter.

The case of channel is more complex, because usually we have a dimensional image of it. However, it is possible to construe a channel in a more abstract way, as a set of conditional probabilities, along the lines suggested by Abramson (1963). In the same spirit, Barwise and Seligman (1997) suggest that a channel could be understood, basically, as an objective correlation of any degree between two domains.

Also relevant is the fact that a message gives information on a system, that is, on its possible states, not on only one of them. "The concept of information," as Weaver states, "applies not to the individual messages . . . but rather to the situation as a whole" (Shannon and Weaver 1949, 9). If a message increases the estimated probability of a state in the system, that of the others obviously decreases. This is one of the reasons why I prefer to talk about a *system of reference* rather than an *object*, as in Peircean terminology.

Most of the conceptual problems concerning information stem from ellipsis, even the opinion that there are many different unrelated concepts of information. We often speak about the information of a message with no reference to a receiver or a referential system, although both of them exist implicitly. Information is always, as it were, functional, transitive, and pragmatic. The message is always referred to something by a receiver; otherwise it is not a message, just an event (Millikan 1989, 286). If messages were not referred to something by a receiver, Griffiths would be perfectly right to say that "most information talk in biology is a picturesque way to talk about correlation and causation" (2001, 400).

However, factors conditioning information are often mistaken for information itself. Such is the case regarding the formal characteristics of the system of reference, and those of either the message or the system to which it belongs. The correlation between the messages and the system the information is about also affects the amount of information involved, but neither this correlation nor this form constitutes the information itself. For instance, we could obtain information on the hour represented by a clock by observing another one. This is possible when a close correlation exists between both of them. Nevertheless, correlation by itself does not equal information but is only a factor that conditions the amount of information.

In order to make my proposal clear, let us say that our rendering of the notion of information differs from the classical Shannonian diagram, but is, at the same time, clearly related to it. In the first place, I propose to begin from level C, that of the pragmatic or functional problems. Then, I shall try to give a reinterpretation of other theories dealing with problems on levels A and B, as restricted or ideal pragmatic theories. If we agree with MacKay that "information is what information does" (1969, 41; see also van Benthem 2007), then a general theory of information should start from the pragmatic level and only then move on to a reconstruction of the rest. Second, I propose taking the receiver as a pivotal point for the information relation, following Millikan's advice (1989). As we will see later, from a mathematical point of view, the distributions of probabilities defining the receiver will be sufficient for the accomplishment of the functions traditionally assigned to the source and the channel. This possibility is already suggested by the abstract interpretation of Shannon's theory given by Abramson (1963).

The relationship among the three above-mentioned elements (m, R, and S) is informative when it changes the receiver's knowledge of the system of reference. By "knowledge," I mean the distributions of probabilities of the possible states of the system of reference in the receiver. Knowledge, therefore, should be understood here along the lines suggested by Karl Popper in a very general way: "Can only animals know? Why not plants? Obviously, in the biological and evolutionary sense in which I speak of knowledge, not only animals and men have expectations and therefore (unconscious) knowledge, but also plants; and, indeed, all organisms. . . . Flowering plants know that warmer days are about to arrive . . . according to sensed changes in radiation" (1990, 9, 10, 35). In a remarkably parallel way, Rosen states, "I cast about for possible biological instances of control of behavior through the utilization of predictive models. To my astonishment I found them everywhere . . . the tree possesses a model, which *anticipates* low temperature on the basis of shortening days" (1985, 7). This understanding of "knowledge" does not necessarily imply consciousness, so the notion is applicable to human and nonhuman living systems, and even to a computer.

We can describe information (I) as a relationship between a message (m), a receiver (R), and a system of reference (S). To this relationship, there belongs the triad formed by a message, receiver, and system of reference in which the message alters the receiver's previous knowledge of the system of reference (Dretske 1981, 2007). Moreover, the more probable an alternative is to a receiver, the more information will be received when a message says that a different one has occurred, unless it is a simple contradiction. So, for example, the introduction of a certain genetic message into the cytoplasm increases the probability of the cell carrying out a certain function, for the probabilities of alternative behavior decrease. Now, we can say that the receiver knows—or knows better—how to do something.

I am aware that by linking information with knowledge I also introduce epistemological problems like that of truth. As Dretske says, "we must carefully distinguish meaning, something that need not be true, from information which must be true" (2007, 2). How can the concept of truth be applied to the case of living systems or computers? I cannot go into such a complex problem here, so I shall just make a suggestion. Perhaps it would be correct in these cases to talk about a kind of *practical truth*, a concept derived from Aristotle (1995, NE, IV, 4:2). My answer would follow Popperian lines: truth goes hand in hand with functionality; *misinformation* exists when the message contributes negatively to the survival or functionality of the system. In continuity with this idea, at the conscious level, we can say that only nonfalsified theories prosper, that is, expectations that are not gainsaid. This position allows us to retain a belief in the possibility of a unified concept of information.

The informational relation may be perfectly objective (see Barwise 1986; Fodor 1986; Denbigh and Denbigh 1985). For example, it is quite objective that a genetic message informs the cytoplasm about synthesizing proteins, or that a statement informs one about the weather. It is clear that even if one's expectations about the weather are subjective, and the probabilities of a cell behaving in different ways are objective, both are equally objective phenomena to an observer. The observer does not directly take into account the probabilities in one's mind, but those that he or she reckons one holds; neither does the observer directly bear in mind the probabilities of a given cell behavior, but those he or she calculates. This makes it easier to reconcile the concepts of information used in biology with those used in classically epistemological contexts.

This does not mean that the information has been in the world since the beginning, preceding any subject capable of using it, as Dretske (1981) says. Without cellular machinery, there is no connection between DNA and protein, just as without a hearer there is no connection between statement and weather. As Moreno and Ruiz-Mirazo state, the genetic message is, in principle, "decoupled from the dynamical organization of the system" (2002, 73).

Information can be measured from the magnitude of its effects, that is, by the changes to the receiver's knowledge of the system of reference. This is a traditional and standard way of measuring different physical magnitudes. Measuring information—like measuring anything else—requires a subject to do it, and this subject acts according to theoretical grounds. To assess the quantity of information given by a genetic message to the cytoplasm, we need extensive biochemical knowledge. Our results, however, could be right or wrong, and we must remain aware that any attribution of a receiver to a subject is hypothetical, having at best the conjectural truthfulness of a scientific theory, but this does not make the informational relation any less objective.

Finally, it should be remembered that in measuring information, our knowledge of a given reality (the informational relation measured) changes, so the measurement of information is also an information relation (capable of being measured). But this does not cause any confusion of the two informational relations, because it is a normal recursive phenomenon (Hofstadter 1979).

Measuring Information as a Relation

Our ways of measuring information do not do justice to the concept of information as described in biological literature. For example, genetic variation is understood to increase the capacity for information, whereas selection determines which variations are really functionally informative (Collier 1988, 2007; Mayr 1982; Wicken 1987; Queiroz, Emmeche, and El-Hani 2005). Conversely, the standard measures of information are normally interpreted as gauges of potential information, structural complexity, or thermodynamic order; they cannot discriminate biological functionality.

But the proposed measures of specifically biological information are also problematic. For example, the key to the measure of information proposed by Gatlin (1972) is a deviation from the most random distribution. In theory, the absence of selective forces acting on the formation of nucleic acids and proteins is thought to bring about a highly random configuration, any deviation from which will respond to a selective bias. Several difficulties arise here.

First, the measure does not allow for any distinction between deviations produced by the effects of natural selection and those derived from prebiotic conditioners (Wicken 1987). There is also the conceptual problem whereby, according to Gatlin's formulae, information increases along with redundancy, which ultimately leads to the absurd situation of maximum information with maximum redundancy. Gatlin keeps the functional meaning of information while restricting biological functions to the production of copies. Increased redundancy favors this function, but if this were the only tendency in evolution, the complexity of living things would not have increased. The limit imposed on the growth of redundancy is based on the need to perform (with competitive success) a series of bodily functions that are not strictly reproductive. Therefore, biological information cannot generally be identified with redundancy. The measure of information suggested by Brooks and Wiley (1986) has the same problem as Gatlin's.

In accordance with the previously discussed concept of information, I shall now establish a measure of information as a function of the magnitude of its effects, that is, a change in knowledge. This measure is inspired by the ideas of Peirce and Popper, as well as by those of MacKay (1969), Dretske (1981), and Wicken (1987, 1988). The basic requirements of such a measure are that it agree with our intuitive notion of information and be coherent with the best theory of syntactical information available, namely, Shannon's. Concerning the first condition, Dretske writes (2007, 2), "In formulating a theory of information we respect ordinary intuitions about what information is—and why else would one call it a theory of information?" The importance of the second requirement may be demonstrated by quoting Weaver: "It seems clear that an important contribution has been made to any possible general theory of communication. . . . It is almost certainly true that a consideration of communication on levels B and C will require additions . . . but it seems equally likely that what is required are minor additions and not real revision" (Shannon and Weaver 1949, 26).

Here, the main thesis is that information can be considered as a (functional or pragmatic) relationship between a receiver, R, a message, M, and a system of reference, S:

1. a message, m_i, is an element of a set of alternative messages, M. So M = $\{m_1, m_2, ..., m_n\}$.
2. S can be any system and $\sigma = \{s_1, s_2, ..., s_q\}$ is a set of alternatively accessible states of S.
3. R is characterized by
3a. a set of (a priori) probabilities associated with the different alternative states of the referential system: $P(s_1), ..., P(s_q)$, where $\Sigma_k P(s_k) = 1$.
3b. a function assigning an (a posteriori) probability, $P(s_k|m_i)$, to each pair $<m_i,s_k>$; where $\Sigma_k P(s_k|m_i) = 1$.

Information of m_i-to-R-about-S can therefore be measured by taking into account the difference, D, between the probabilities before the reception of the message, $P(s_1), ..., P(s_q)$ and afterward, $P(s_1|m_i), ..., P(s_q|m_i)$:

$D_{(m_i, R, S)} = \Sigma_k | P(s_k) - P(s_k|m_i) |$.

Accordingly, we propose a measure of information in the function of the binary logarithm of D (see figure 3.2):

$I_{(mi, R, S)} = -\log(1 - (D/2))$.

We can find the average of information from M-to-R-about-S by weighting the information that each m_i carries with its frequency:

$I_{(M, R, S)} = \Sigma_i P(m_i) \cdot I_{(mi, R, S)}$

It is possible and trivial to prove that

$0 \leq D \leq 2$.

Therefore:

if D = 0, then I = 0

if D = 2, then there is no real value to I

if 0 < D < 2, then the amount of information, I, approaches ∞ when D approaches 2.

D = 0 means that there is no change in R's knowledge of S despite R's receiving the message, m_i, in which case, information, I, logically equals zero.

D = 2 happens only if the message m_i informs of something happening that R previously considered impossible. In this case, our measure, I, has no real value, a situation where a radical restructuring of the subject's expectations is seen to be required. The receiver used so far by the subject has been surpassed, and an alternative one, if possible, must be found. We could, therefore, now assess the quantity of information in relation to a (meta)receiver dealing with second-order alternatives. For example, a statement considered literally impossible, such as "man is a wolf to men," would invite a metaphorical interpretation. The information rendered by this statement in relation to our *literal receiver* has no real value; instead, it has a positive value in relation to our *metaphorical receiver*. In addition, we have obtained some positive amount of information in relation to a receiver dealing with second-order alternatives, such as literal/metaphorical interpretation, as we now realize that the statement in question requires a metaphorical and not a literal interpretation.

This case is important, because all learning processes (biological and cultural evolution, Piagetian development of cognitive structures, Kuhnian dynamics of scientific theories, etc.) seem to involve two kinds of change: (1) accumulative or gradual (assumed within the limits

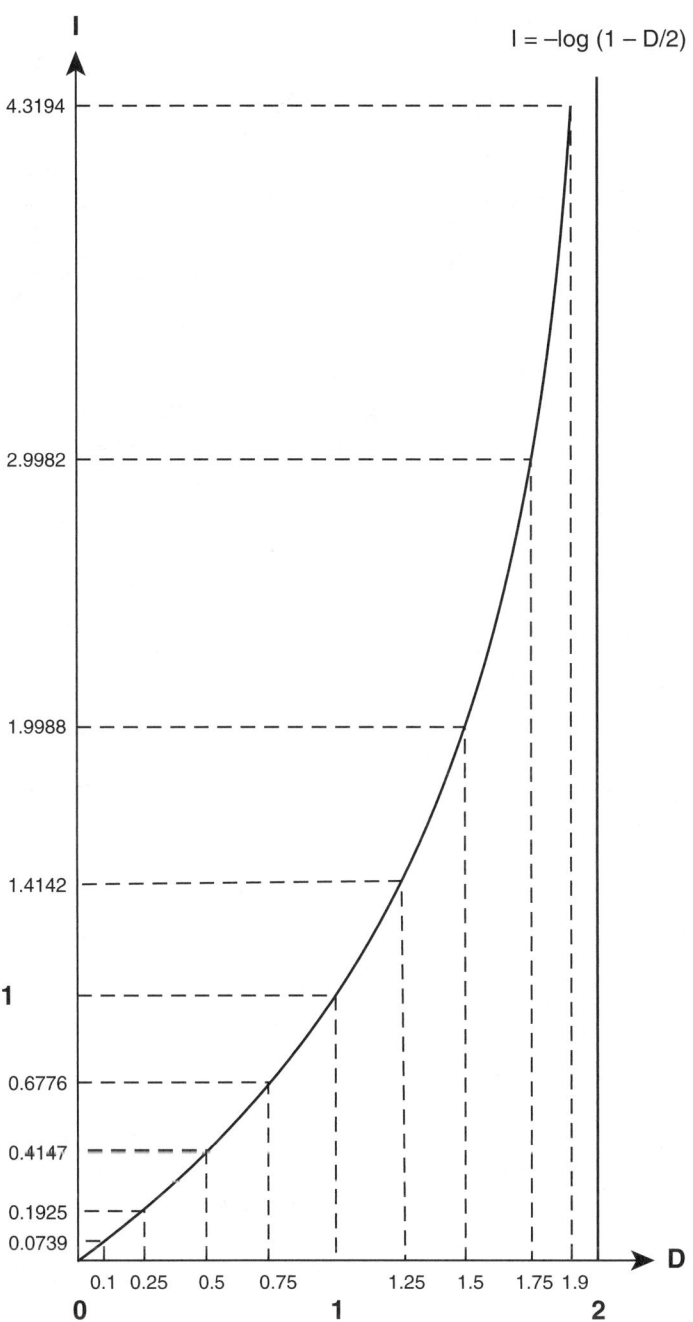

Figure 3.2
Information measure in the function of a binary logarithm

of a given receiver and rendering a positive amount of information), and (2) reorganizational or saltational (when our measure yields no real value, indicating that a radical change—a change to a new receiver—is required).

As Piaget observed, a little child believes that two containers of the same height can contain the same amount of liquid. The child uses a receiver in which the height of the liquid in the container is strictly correlated with the total amount of liquid. Nevertheless, disappointment about these expectations arrives sooner or later. Then, if the child's intelligence follows a normal development, he or she suffers a change in worldview and shifts to another receiver. For this new receiver, the total amount of liquid in a container depends on the container's volume, and not simply on its height. This step in learning development is an abrupt one, preceded by a radical frustration regarding previous expectations.

According to Kuhn's interpretation of the history of science (1996), we can distinguish periods in which new data are accommodated within an existing paradigm—in our terms, they render information in relation to a certain receiver—followed by others in which traditional expectations are broken and a shift to a new paradigm is required. We can say, then, that a change of receiver is felt as necessary by some part of the scientific community.

Evolutionary phenomena, like species extinction, are also understood as the raising of new environmental circumstances that defy the expectations associated with a given kind of organism. In informational terms, we can say that new messages seem to frustrate the expectations of the receiver currently used by an organism. If no organism of a species has the ability to use a different receiver, extinction inevitably results. However, if part of the species population, by chance, had a suitable alternative receiver, then this part could survive. I believe that our measure of information helps us to formally capture these intuitions.

In the other cases, our measure, I, approaches ∞ if D approaches 2. This means that the greater the number of possible states of the system and the greater the disagreement with R's previous knowledge (without reaching D = 2), then the greater the amount of information (see figure 3.2). Such results are obviously coherent with our intuitive notion of information.

Under certain restricted conditions, however, our formulæ can be demonstrated to yield the same outcomes as the standard Shannonian ones. In the classical theory of information, the amount of information produced by an event is measured by its probability. But can it be reduced to a three-element framework? What does m_i give information about? We may consider that the classical theory of information deals with systems giving information about themselves to receivers of this type:

	s_1	...	s_k	...	s_q
m_1	1	0	0	0	0
...	0	1	0	0	0
m_i	0	0	1	0	0
...	0	0	0	1	0
m_n	0	0	0	0	1

$q = n$
$s_k \equiv m_i$ iff $k = i$
$P(s_k) = P(m_i)$ if $k = i$
$P(s_k|m_i) = 1$ iff $k = i$
$P(s_k|m_i) = 0$ iff $k \neq i$

where $D_{(mi, R, S)} = 2 - 2P(s_k)$.
Therefore:

$P(s_k) = 1 - (D/2)$

whence

$I_{(m_i, R, S)} = -\log P(s_k)$

$I_{(M, R, S)} = -\Sigma_k P(s_k) \cdot \log P(s_k)$.

That is, under certain restricted conditions (M≡σ), our formulae for $I_{(m_i, R, S)}$ and $I_{(m_i, R, S)}$ yield the same outcomes as the standard Shannonian ones:

$I(m_i) = -\log P(m_i)$

$H(M) = -\Sigma_i P(m_i) \cdot \log P(m_i)$

In Shannon's terms, $I(m_i)$ is the information attributed to a message (m_i), and $H(M)$ is the entropy of a source. So, for Shannonian sources of information—that is, sources informing about themselves to a receiver of the type described previously—the equations:

$I(m_i, R, S) = I(m_i)$

and

$I_{(M, R, S)} = H(M)$

are fulfilled.

Our measure of information meets the two basic requirements previously mentioned, as we have demonstrated that it is in accordance with our intuitive notion of information and clearly connected with Shannon's theory. It is also of some interesting use in biological contexts, as I shall attempt to show.

In order to clarify how our measure works, I offer an (oversimplified) biological example:

1. Let the message, m_i, be a codon in a mature mRNA sequence being processed by a given ribosome, for example, the codon UCG. Let M be the set of all possible codons, so M = {UUU, UUC, . . . , GGA, GGG}.

2. Let S be the next possible step in the peptide chain being synthesized by a given ribosome. In this case, a state of the system, s_k, should be considered the ribosome adding a given amino acid to the chain or stopping the synthesis. So σ = {phenylalanine, leucine, serine, tyrosine, . . . , stop, . . . , glutamic acid, glycine}.

3. Let R be a functional cytoplasm with a random supply of amino acids. R is characterized by

3a. a set of (a priori) probabilities associated with the next possible step in the peptide chain:

P(phe) = P(leu) = P(ser) = P(tyr) =, ..., = P(stop) =, ..., = P(glu) = P(gly) = 1/21 = 0.047619

3b. a function assigning an (a posteriori) probability. For the sake of argument, let us assume the following values:

P(phe|UCG) = P(leu|UCG) = P(tyr|UCG) =, ..., = P(stop|UCG) =, ..., = P(glu|UCG) = P(gly|UCG) = 0.001

and

P(ser|UCG) = 0.999

whence

$D_{(mi, R, S)} = \Sigma_k | P(s_k) - P(s_k|m_i) | = 1.883761$.

Accordingly, the information that the codon UGC gives to the cytoplasm's receiver on the next step that a given ribosome should take during the current process of synthesis is

$I_{(mi, R, S)} = -\log(1 - (D/2)) = 4.1$ units of functional information (UFIs).

This quantity could vary, obviously, if we change any of the *relata* of the informational relationship, and there is no sense in attributing the amount to any particular *relatum*. So the same genetic message could yield a different quantity of information when related to another kind of cytoplasm, as is well known in the fields of transgenic engineering and developmental biology (see Alberts et al. 2002). For instance, a genetic fragment coding for hemoglobin in an erythrocyte becomes mute in a neuron; the same protein in the same organism could perform a different function in different kinds of cells (Gilbert and Sarkar 2000), and a signal pathway effector may lead to the induction of different gene products in different gene lineages (Brisken et al. 2002). Moreover, the same genetic message could render more or less information to the same cytoplasm if we ask for any other biological function as the referential system. One could ask for the information that a codon gives to a certain cytoplasm in the synthesis of a functional protein. It is often the case that one amino acid can be replaced by another in a protein without any loss of functionality (see Salemi and Vandamme 2003; Nelson and Cox 2004). This is one of the reasons why we find a considerable degree of genetic variability in natural populations. So the a priori probability of synthesizing a functional amino acid is higher than the probability of synthesizing a given amino acid, like serine. Consequently, the amount of information, on average, produced by *the same* codon to *the same* cytoplasm is higher regarding the synthesis of a given amino acid than regarding the synthesis of any functional amino acid.

Bioinformation Revisited

Bioinformation and Thermodynamic Entropy

Information first appears in biology in connection with the concept of physical entropy and its different measures (thermodynamic or statistic). This well known history begins with Clausius (1822–1888) and Boltzmann (1844–1906), who formulated the measures of entropy. Clausius was the first to introduce the term "entropy" to thermodynamics in 1876; Boltzmann gave a statistical interpretation to the term. Boltzmann considered that a macrostate of a given system is more entropic, as it is compatible with a greater number of microstates.

The classical example is that of a gas-filled box. The box has two compartments connected by a door. The system can be in a macrostate A, in which the temperature in one compartment is relevantly higher than in the other one, or in a macrostate B, with a uniform temperature. The macrostate A has a lower statistical probability, because it is compatible with fewer microstates than B.

Boltzmann proposes the following formula for measuring entropy:

$$S = K \ln W$$

where S is the entropy of a macrostate of a system, K is Boltzmann's constant, and W is the number of microstates compatible with the macrostate. As it can be easily observed, this equation is similar to the Shannonian formula for informational entropy.

Maxwell (1831–1879) took the next step with a thought experiment. If we place inside the box a demiurgic being who lets the rapidly moving particles pass to one compartment and the slower ones to the other, then the system evolves toward a less entropic state. Apparently, this situation is incompatible with the second law of thermodynamics.

Leo Szilard (1898–1964) found a sound answer to Maxwell's paradox. Maxwell's Demon—as the demiurge is now known—overcomes the universal tendency to entropy, thanks to the information he obtains about the speed of the particles. However, he had to measure the speed by means of whatever process secures the transaction of energy and the increase of entropy. It seems, therefore, that an (inverse) link exists between entropy and information.

Taking inspiration from this idea, Brouillin (1889–1969) developed the concept of *negentropy*, or negative entropy, as equivalent to information. More recently, Evans, Layzer, and Frautschi have attempted to equate information with a positive magnitude, the distance from thermodynamic equilibrium (see Brillouin 1962; Tribus, Shannon, and Evans 1966; Frautschi 1982; Brooks and Wiley 1986; Wicken 1987, 1988; Layzer 1990; Marcos 1991).

The last step prior to the emergence of the concept of information in biology was Schrödinger's (1944) classic *What Is Life?*, which claims that living things overcome the universal tendency to entropy by exporting entropy to their environment, as Maxwell's Demon does. Thus, a connection was made between thermodynamic order and biological complexity. Schrödinger contributes to the link between biological phenomena and physical entropy, and physical entropy had already been connected with information, so the stage was set for the encounter of information and biology. A slogan for this approach applied to biology could be: "A gain in (physical) entropy means a loss of (biological) information."

From the point of view proposed here, thermodynamic entropy conditions the information that the macrostate of a system can offer about its possible microstates to a receiver equipped with the right physical laws. If the particles of the system act together, the system as a whole is more dynamic. Correspondingly, the movement of the system offers a great deal of information about its elements. If entropy increases, the system is less dynamic and reflects less efficiently the positions and moments of its components. Thus, thermodynamic entropy is linked *specifically* with the information that a macrostate can give about a system's currently accessible microstates. So the basis for a general measure of information could not be entropy, negentropy, or distance from equilibrium (Marcos 1991).

Physical entropy is currently linked with (structural) order and (functional) organization, but order and organization are, respectively, relative to a structure and a function. Several types of order or organization may be identified even within the same system. Organization is also relative to a receiver connecting the message and the system of reference. A fragment of DNA is organized for the synthesis of a certain protein only if one knows how the cellular apparatus works. Physical entropy, therefore, should not be considered a general measure of organization; rather, it is a correct approach to *one* type of organization able to render work (Denbigh and Denbigh 1985; Nauta 1972). In biology, organization is always established with regard to a certain function. It is not just a question of structural regularity. This is why Schrödinger (1944) conjectured, before the discovery of the double helix, that genetic information must be contained in some kind of *aperiodic* crystal. Our concept and measure of information can bring us nearer to an estimation of biological organization than mere negative entropy. Furthermore, as we have seen, they facilitate a reasonable interpretation of the link between physical entropy and information.

Bioinformation and Shannon's Entropy

Chronologically, the next relevant domain is the transmission of signals. Here, the decisive contribution was Shannon's mathematical theory of communication, but Nyquist and Hartley may be mentioned as precedents. From this perspective, information is conceived of as surprise or uncertainty. There is a formal connection between physical entropy

(Boltzmann's entropy) and information entropy (Shannon's entropy), as von Neumann first noted. Although they both conform to the same formal schema, however, there is no reason why their behavior should follow the same laws (Marcos 1991).

The relation between Shannon's entropy and information as a triadic relation has been dealt with previously, including in mathematical terms, but some remarks are in order here. On the one hand, the structure of the system the message belongs to affects the information, but in the opposite way to that of the system of reference. When we try to pass information, we do not want the system to which the message belongs to impose any structural limitations on our communication, or at least we want them kept to a minimum. This is what Shannon calls *entropy* (freedom of choice within a source), and is recommended for a system acting as a symbolic one. This is why in some parts of biological systems—for example, in neuronal, genetic, immunological, and linguistic domains—unities can be combined in many different ways, for they must be flexible when representing other parts of the systems or external realities.

On the other hand, a higher level of structure, form, organization, or regularity in the system of reference brings about the possibility of transmitting more information about it, in line with common sense and philosophical tradition (Eco 1962; Moles 1972). Consequently, this matter is sometimes shrouded in confusion. It could be seen as a paradox that some authors correlate information with freedom of choice or low structural constraints, as Shannon does, and others, like Eco and Moles, correlate information with structural order, constraints, or regularity. But it is not paradoxical at all. One thing is the relative order of the system of reference, and another is that of the symbolic system. Shannon's entropy of the symbolic system correlates positively with information; in the object that the system informs about, the greater the order and organization, the more information that can be produced.

Finally, another factor limiting the amount of information is the correlation between the structure of the message and that of the referential system. If it is perfect, a maximum amount of information can be transmitted. No greater correlation exists than between a system and itself. In this regard, Shannon's measure is an absolute limit on the amount of information: no more information can be given about a system than is given by the system itself. Therefore, Shannon's measure is often referred to as a measure of possible information.

Bioinformation and Complexity

Another approach to information appeared recently, based on the work of Kolmogorov (1965) and Ray Solomonoff (2003): algorithmic or computational theory. Here, information is viewed as a special kind of complexity. Any sequence describing a text, image, musical composition, and so forth may be generated by means of a program and a suitable computer. If the sequence shows any regularity, symmetry, or redundancy, the program could be shorter than the sequence itself. If the sequence is more *complex*, or even chaotic, it will be less susceptible to compression, so the greater the complexity, the lesser the compressibility. Moreover, a general measure of complexity can be obtained by taking a universal Turing machine as the relevant computer, in which case a triadic relation could easily be detected: the algorithm or program is a message to a receiver based on the universal Turing machine, regarding a system of reference, namely, the compressed sequence.

It must be remembered that information, unlike complexity, is not a property of a single thing, but a relation among at least three entities, so some remarks may be made on the relationship between Kolmogorov's complexity and bioinformation. First, the relationship between information and the complexity of a sequence is not a direct one; that is, complexity cannot be simply equated with information. The need for a long program to generate a sequence does not translate directly into that sequence "having" a great deal of information. It would be counterintuitive, because chaotic sequences would be the most informational ones. Kolmogorov's measure of complexity can distinguish between a crystal and a protein, but a relevant concept of bioinformation must also distinguish between a functional protein and a random peptidic compound.

Second, Kolmogorov's notion of complexity has also been used to calculate the informational content of an individual object as a direct function of the length of the shortest program describing or producing it. Here, we must remember the difference between things and words. When complexity is assessed from the compressibility of a description encoded in a binary sequence, it could normally be referred to a universal Turing machine. The input into such a computer is a binary sequence, as is the output, so the computer cannot relate the description to the object itself. Therefore, a measure of the complexity of sequences is available, but this does not mean that we can calculate the complexity of the object described, because the information that a description gives about an object is always referred to a certain receiver in a concrete subject. For example, a DNA sequence is a good description of the three-dimensional structure of a protein *to certain cytoplasmic machinery*, but it would not make sense to say that it is generally, or for a Turing machine (see Rosen 1985).

Third, there are doubts as to whether natural selection can explain the increase in complexity throughout evolution (Marcos 1991a, 1992). After all, organisms exist that are very simple but seem perfectly adapted, a classic objection to Darwinism (von Bertalanffy 1968). The connection we have established between complexity and information may clarify the issue. Later variants in evolutionary succession may "take into account" those already existing, but not vice versa (Darwin 1859; Rosen 1985). Once an organism A is settled into its environment, any other organism B will adapt to this environment more effectively if it is equipped to relate informationally with A. This informational asymmetry means that both the environment and organisms become more and more complex, and thus maintain their adaptational dynamics throughout evolution. In this regard, complex systems could be indicative of a complex environment, for more information is required to adapt to a complex environment than to exist in a simple one. The existence of living beings that adapt to an environment in which others already exist may ensure the survival of the latter, rather than threaten it, because the environment to which the new system adapts is also that on which they depend. Humanity's acceptance of this idea is not unconnected with the increase in ecological awareness.

Bioinformation and Meaning

Information is intuitively linked with meaning, so some authors have considered some development of the theory of information necessary to understand this connection (Bar-Hillel 1964; Hintikka and Suppes 1970; Hintikka 1973; Barwise and Perry 1983; Hanson 1990; Villanueva 1990; Barwise and Seligman 1997; Seligman 2007). Carnap, Bar-Hillel, and Hintikka's basic idea is that the more a statement prohibits, the more it says (Carnap and Bar-Hillel 1952). The way of measuring this magnitude is as follows: we take a limited formal language and give equal probability to all the syntactically possible combinations of its elements; then we take a correct statement and the class of all its logical consequences. Some grammatically correct statements may be forbidden in this class, so the probability of the others increases. The amount of information (a number), is therefore a function of the number of prohibited statements, and the informational content (a class) depends on the statements still in play. It will be observed that contradictory statements would, somewhat counterintuitively, have the maximum content.

Even this approach could be reduced to a triadic relation scheme: the system of reference is the set of grammatically correct combinations of symbols. The message is the statement under consideration and we then need, in the words of Bar-Hillel, "a receiver with a perfect memory who 'knows' all of logic and mathematics, and together with any class of empirical sentences, all their logical consequences" (1964, 224). An ideal receiver is, of course, a receiver, and Bar-Hillel proposes treating his theory as an ideal pragmatics.

All that we can say is that the difference between linguistic and non-linguistic information lies in the nature of the link between the message and the system of reference. In one case, it is conventional and, in the other, natural. There is a natural link of cause and effect between fire and smoke (Devlin 1991), but the link between spirals of smoke and taking up the hatchet is conventional (Kampis and Csányi 1991; Kampis 1990). I would stress, however, that between a direct or strictly natural causality and a purely conventional nexus, there may be a zone of transition (see Millikan 1989), as we can see in biological contexts.

Bioinformation and Knowledge

Information is also related to knowledge. Dretske defines information as "a commodity that, given the right recipient, is capable of yielding knowledge" (1981, 47). So a triadic relation is also needed here: we have the message, the circumstances it informs about, and the "right recipient." Information is therefore related to knowledge in a dual way: it depends on the receiver's previous knowledge, and knowledge is an effect produced by information, at least a change in the receiver and at best a change in the receiver as regards the real state of affairs (see Dretske 1981). So knowledge itself can be viewed as the property of a subject (edification), or as a dyadic relation between subject and object (correspondence, correlation). It is easy to connect the first notion of knowledge with biology: bioinformation contributes to the construction of living beings themselves. It is more difficult to apply the demand for truth. Nevertheless, I think that even in biological contexts information somehow requires truth.

Bioinformation Location: Places and Levels

Where is bioinformation? In my opinion, a relative notion and measure of information could avoid the (pseudo)problem of finding the location of bioinformation. It could be (dis)solved by considering information, not as being already present somewhere (in the genes, cytoplasm, proteins, environment, ecosystem, brain, or wherever), but as being established by interactive relations.

The functioning of any living system (or part of a living system) depends on various factors. For example, the three-dimensional structure of a protein depends on DNA, but also on the very "machinery" of the cell. What the message is and what the receiver is are chosen conventionally but not arbitrarily. A message is usually defined as a small factor of great specificity in relation to a given function and displaying a high potential for variability. The DNA codifying a certain protein possesses these characteristics in relation to the function of synthesizing the protein in question, and the protein in relation to its biological function. In other words, the slightest alteration of the DNA could destroy the structure of the protein, and the slightest change in a protein could destroy its function.

Such an effect is unlikely to be the result of a similar change in an environmental factor. But this does not force us to identify the information with a property of the message. The information in a fragment of DNA about a protein obviously depends on its specificity, but only regarding a given receiver (Sattler 1986). Actually, the probability of any given protein arising in a prebiotic environment (Yockey 1977, 1981), even in the presence of a specific DNA, is minimal. Therefore, information is located neither before nor after the triadic relation. Kampis and Csányi state, "we have to give up the idea of a complete localization of information" (1991, 23).

On the other hand, any one fragment of DNA may, of course, produce information on more than one function, and not necessarily in the same quantity. For example, attempts could be made to calculate the amount of information in a fragment of DNA in relation to the transportation of oxygen, which is different from the function of producing a particular protein. The difference lies in the fact that the same function can be performed by different proteins or variants of a protein. This phenomenon of "synonymy" must not be ruled out from calculations on information.

Finally, let us deal very briefly with the location of information in living systems according to different hierarchical levels (Collier 2003). For all intents and purposes, in the absence of connecting principles, the amount of information obtained by an external observer—for example, a scientist—on a living system at different levels should be considered as amounts of information about different systems. Otherwise, more information would supposedly be derived about a living being from the knowledge of, for example, its atomic state than of its genetic makeup (see Atlan 1972). Information on the atomic state is not about the living being per se, unless we have theoretical principles connecting atomic states with some functional characteristics. Developing principles of connection between levels is like developing a receiver that allows us to obtain information about one level from another, acting as a message. This touches on the philosophical question of epistemological reductionism. We know that given certain principles of connection, one biological level can inform us about another, but we also know that a complete reduction is not viable, for any concrete informational relation is subject to imperfections.

Conclusion

The concept of information has become central in of our civilization, so much so that we call our society an "information society." The life sciences have been affected by this movement, and biology has developed what we might call an *informational paradigm*. Recently, a debate has arisen concerning the usefulness of this informational perspective in biology (Griffiths 2001; Queiroz, Emmeche, and El-Hani 2005).

As I have sought to establish, the understanding of many biological processes in terms of information is certainly useful and clarifying, provided that the very concept of information is submitted to scrutiny. It is necessary to clarify the notion of information, to derive from this notion a measure that is significant in biology, and to grasp the relationships between the concept of information and others akin to it commonly used in different biological theories. I tackled this task by presenting the outline of a theory of information.

First, I presented a concept of information as a triadic relation involving a message, a receiver, and a system of reference. The effect produced by this relation is a variation of the knowledge that the receiver has of the system of reference. Knowledge is understood in the wide sense, which does not require consciousness, and which is applicable to any living system and indeed to certain artifacts. Next, I derived from this concept of information a measure of information as a function of the magnitude of the change of knowledge produced by a given informational relation. I have shown that the results of this measure are coherent with our pretheoretical intuitions and that our measure of information may be formally linked with Shannon's.

Finally, I sought to clarify the relationship of the notion of information as a triadic relation with others that are also important in biology, like entropy (thermodynamic and informational), complexity, order, organization, meaning, and knowledge. The same concept of information as a relation has allowed us to tackle the recurring debate on the location of biological information. As it is a relation, it does not make sense to locate information before or after the informational relation in either a concrete place in living systems or their environment, or on a definite hierarchical level of biological organization.

References

Abramson, N. 1963. *Information theory and coding*. New York: McGraw-Hill.

Adriaans, Pieter. 2007. Philosophy of information, concepts and history (preliminary version). In *Handbook on the philosophy of information*, ed. Johan van Benthem and Pieter Adriaans. Available at http://www.illc.uva.nl/HPI/Draft_History_of_Ideas.pdf.

Alberts, Bruce, Alexander Johnson, Julian Lewis, Martin Raff, Keith Roberts, and Peter Walter, eds. 2002. *Molecular biology of the cell*. New York: Garland.

Albrecht-Buehler, Guenter. 1990. In defense of "nonmolecular" cell biology. *International Review of Cytology* 120: 191–241.

Allen, Colin. 1992. Mental content and evolutionary explanation. *Biology and Philosophy* 7 (1): 1–12.

Aristotle. 1995. Nicomachean ethics. In *The complete works of Aristotle*, 2 vols., ed. Jonathan Barnes, 1729–1867. Princeton: Princeton University Press.

Arp, Robert, Barry Smith, and Andrew D. Spear. 2011. *Building ontologies with basic formal ontology*. Cambridge, Mass.: MIT Press.

Atlan, Henri. 1972. *L'organisation biologique et la theorie de l'information*. Paris: Hermann.

Avery, John. 2003. *Information theory and evolution*. London: World Scientific Publishing.

Baddeley, Roland, Peter Hancock, and Peter Földiák, eds. 2000. *Information theory and the brain*. Cambridge: Cambridge University Press.

Bar-Hillel, Yehoshua. 1964. *Language and information*. Reading, Mass.: Addison-Wesley.

Barwise, Jon. 1986. Information and circumstance. *Notre Dame Journal of Formal Logic* 27 (3): 324–338.

Barwise, Jon, and John Perry. 1983. *Situations and attitudes*. Cambridge, Mass.: MIT Press.

Barwise, Jon, and Jerry Seligman. 1997. *Information flow: The logic of distributed systems*. Cambridge: Cambridge University Press.

Bond, Nigel W., and David Siddle, eds. 1989. *Psychobiology: Issues and applications*. Amsterdam: Elsevier.

Brandt, Christina. 2005. Genetic code, text, and scripture: Metaphors and narration in German molecular biology. *Science in Context* 18 (4): 629–648.

Brillouin, Léon. 1962. *Science and information theory*. London: Academic Press.

Brisken, Cathrin, Merav Socolovsky, Harvey F. Lodish, and Robert Weinberg. 2002. The signaling domain of the erythropoietin receptor rescues prolactin receptor-mutant mammary epithelium. *Proceedings of the National Academy of Sciences of the United States of America* 99 (22): 14241–14245.

Brooks, Daniel R., and E. O. Wiley. 1986. *Evolution as entropy: Toward a unified theory of biology*. Chicago: University of Chicago Press.

Burian, Richard M., and Marjorie Grene. 1992. Editorial introduction. *Synthese* 91 (1–2): 1–7.

Capurro, Rafael. 1978. *Information: Ein Beitrag zur etymologischen und ideengeschichtlichen Begründung des Informationsbegriffs*. Munich: Saur.

Capurro, Rafael, and Birger Hjørland. 2003. The concept of information. Available at http://www.capurro.de/infoconcept.html.

Carnap, Rudolph, and Yehoshua Bar-Hillel. 1952. An outline of a theory of semantic information. Technical Report No. 247, Research Laboratory of Electronics. Cambridge, Mass.: MIT. Reprinted in Yehoshua Bar-Hillel, *Language and information* (Reading, Mass.: Addison-Wesley, 1964).

Collier, John. 1988. The dynamics of biological order. In *Entropy, information and evolution*, ed. Bruce H. Weber, David J. Depew, and James D. Smith, 227–242. Cambridge, Mass.: MIT Press.

Collier, John. 2003. Hierarchical dynamical information systems with a focus on Biology. *Entropy* 5 (2): 100–124.

Collier, John. 2007. Information in biological systems (outline). In *Handbook on the philosophy of information*, ed. Johan van Benthem and Pieter Adriaans. Available at http://www.illc.uva.nl/HPI/Draft_ Information_in_Biological_Systems .pdf.

Darwin, Charles. 1859. *On the origin of species by means of natural selection*. London: Murray.

Denbigh, Kenneth George, and Jonathan Stafford Denbigh. 1985. *Entropy in relation to incomplete knowledge*. Cambridge: Cambridge University Press.

Dennett, Daniel. 1987. *The intentional stance*. Cambridge, Mass.: MIT Press.

Devlin, Keith. 1991. *Logic and information*. Cambridge: Cambridge University Press.

Dretske, Fred I. 1981. *Knowledge and the flow of information*. Oxford: Blackwell.

Dretske, Fred I. 2007. Epistemology and information. In *Handbook on the philosophy of information*, ed. Johan van Benthem and Pieter Adriaans, 29–47. Available at http://www.illc.uva.nl/HPI/Draft_Epistemology_and_Information.pdf.

Eco, Uumberto. 1962. *Opera aperta*. Milano: Bompiani.

Edelman, Gerald. 1987. *Neural Darwinism: The theory of neuronal group selection*. New York: Basic Books.

Elsasser, Walter M. 1975. *Chief abstractions in biology*. New York: American Elsevier.

Escarpit, Robert. 1976. *Theorie générale de l'information et de la communication*. Paris: Hachette.

Floridi, Luciano. 2007. Trends in the philosophy of information. In *Handbook on the philosophy of information*, ed. Johan van Benthem and Pieter Adriaans, 113–131. Available at http://www.illc.uva.nl/HPI/Modern_Trends_in_Philosophy_of_Information.pdf.

Fodor, Jerry A. 1986. Information and association. *Notre Dame Journal of Formal Logic* 27 (3): 307–323.

Forrest, Stephanie, and Steven A. Hofmeyr. 2000. Immunology as information processing. In *Design principles for immune systems and other distributed autonomous systems*, ed. Lee A. Segel and Irun R. Cohen, 361–387. New York: Oxford University Press.

Frautschi, Steven. 1982. Entropy in an expanding universe. *Science* 217 (4560): 593–599.

Gatlin, Lila L. 1972. *Information theory and the living systems*. New York: Columbia University Press.

Giere, Ronald N. 1988. *Explaining science*. Chicago: University of Chicago Press.

Gilbert, Scott F., and Sahotra Sarkar. 2000. Embracing complexity: Organicism for the 21st century. *Developmental Dynamics* 219 (1): 1–9.

Godfrey-Smith, Peter. 2000. Information, arbitrariness, and selection: Comments on Maynard Smith. *Philosophy of Science* 67 (2): 202–207.

Griffiths, Paul E. 2001. Genetic information: A metaphor in search of a theory. *Philosophy of Science* 68 (3): 394–412.

Grünwald, Peter D., and Paul M. B. Vitányi. 2003. Kolmogorov complexity and information theory. *Journal of Logic Language and Information* 12 (4): 497–529.

Guéroult, Martial. 1965. *La concept d'information dans la science contemporaine*. Paris: Minuit et Gauthier-Villars.

Hanson, Philip P., ed. 1990. *Information, language, and cognition*. Vancouver: University of British Columbia Press.

Hartley, Ralph. 1928. Transmission of information. *Bell System Technical Journal* 7 (July): 535–563.

Hintikka, Jaakko. 1973. *Logic, language games, and information: Kantian themes in the philosophy of logic*. Oxford: Clarendon Press.

Hintikka, Jaakko, and Patrick Suppes. 1970. *Information and inference*. Dordrecht: Reidel.

Hofstadter, Douglas. 1979. *Godel, Esher, Bach: An eternal golden braid*. New York: Basic Books.

Holzmüller, Werner. 1984. *Information in biological systems: The role of macromolecules*. Cambridge: Cambridge University Press.

Horan, Barbara L. 1992. What price optimality? *Biology and Philosophy* 7 (1): 89–109.

Jablonka, Eva. 2002. Information: Its interpretation, its inheritance, and its sharing. *Philosophy of Science* 69 (4): 578–605.

Janich, Peter. 1992. *Grenzen der Naturwissenschaft: Erkennen als Handeln*. Munich: Beck.

Kampis, George. 1990. *Self-modifying systems in biology and cognitive science*. Oxford: Pergamon Press.

Kampis, George, and Vilmos Csányi. 1991. Life, self-reproduction and information: Beyond the machine metaphor. *Journal of Theoretical Biology* 148 (1): 17–32.

Kitcher, Philip. 2001. Battling the undead: How (and how not) to resist genetic determinism. In *Thinking about evolution: Historical, philosophical and political perspectives (Festchrift for Richard Lewontin)*, ed. Rama S. Singh, Costas B. Krimbas, Diane B. Paul, and John Beatty, 396–414. Cambridge: Cambridge University Press.

Kjosavik, Frode. 2007. From symbolism to information? Decoding the gene code. *Biology and Philosophy* 22 (3): 333–349.

Kolmogorov, Andrey N. 1965. Three approaches to the quantitative definition of information. *Problems of Information Transmission* 1 (1): 1–7.

Kuhn, Thomas. 1996. *The structure of scientific revolutions*. Chicago: University of Chicago Press.

Kull, Kalevi. 1999. Biosemiotics in the twentieth century: A view from biology. *Semiotica* 127 (1/4): 385–414.

Küppers, Bernd-Olaf. 1990. *Information and the origin of life*. Cambridge, Mass.: MIT Press.

Layzer, David. 1990. *Cosmogenesis: The growth of order in the universe*. New York: Oxford University Press.

Li, Ming, and Paul Vitányi. 1997. *An introduction to Kolmogorov complexity and its applications*. New York: Springer-Verlag.

Lorenz, Konrad, and Franz M. Wuketits. 1983. *Die evolution des denkens.* Munich: Piper.

MacKay, Donald M. 1969. *Information, mechanism, and meaning.* Cambridge, Mass.: MIT Press.

MacLaurin, James. 1998. Reinventing molecular Weismannism: Information in evolution. *Biology and Philosophy* 13 (1): 37–59.

Marcos, Alfredo. 1991. Información y entropía. *Arbor* 549: 111–138.

Marcos, Alfredo. 1991a. Información y evolución. *Contextos* 17–18: 197–214.

Marcos, Alfredo. 1992. El Papel de la información en Biología. Ph.D. thesis, University of Barcelona.

Margalef, Ramon. 1980. *La Biosfera, entre la termodinámica y el juego.* Barcelona: Blume Ediciones.

Margalef, Ramon. 1968. *Perspectives in ecological theory.* Chicago: Chicago University Press.

Marijuán, Pedro C. 1989. La inteligencia natural: Introducción al estudio informacional de los sistemas biológicos. Ph.D. thesis, University of Barcelona.

Marijuán, Pedro C. 1991. Enzymes and theoretical biology: Sketch of an informational perspective of the cell. *BioSystems* 25 (4): 259–273.

Maynard Smith, John. 2000a. The concept of information in biology. *Philosophy of Science* 67 (2): 177–194.

Maynard Smith, John. 2000b. Reply to commentaries. *Philosophy of Science* 67 (2): 214–218.

Mayr, Ernst. 1982. *The growth of biological thought: Diversity, evolution, and inheritance.* Cambridge, Mass.: Harvard University Press.

Millikan, Ruth G. 1989. Biosemantics. *Journal of Philosophy* 86 (6): 281–297.

Moles, Abraham. 1972. *Théorie de l'information et la perception esthétique.* París: Éditions Denoël.

Moreno, Alavaro, and Kepa Ruiz-Mirazo. 2002. Key issues regarding the origin, nature, and evolution of complexity in nature: Information as a central concept to understand biological organization. *Emergence* 4 (1–2): 63–76.

Musgrave, Alan. 1993. *Common sense, science and skepticism: A historical introduction to the theory of knowledge.* Cambridge: Cambridge University Press.

Nauta, Doede, Jr. 1972. *The meaning of information.* The Hague: Mouton.

Nelson, David L., and Michael M. Cox. 2004. *Lehninger principles of biochemistry.* New York: Freeman.

Nishikawa, Ken. 2002. Information concept in biology. *Bioinformatics* 18 (5): 649–651.

Oyama, Susan. 2000. *The ontogeny of information: Developmental systems and evolution.* Durham: Duke University Press.

Paton, Ray C. 1992. Towards a metaphorical biology. *Biology and Philosophy* 7 (3): 279–294.

Peirce, Charles S. 1932–1935. *Collected papers of Charles Sanders Peirce*, 6 vols., ed. Charles Hartshorne and Paul Weiss. Cambridge, Mass.: Harvard University Press.

Pfeifer, Jessica. 2006. The use of information theory in biology: Lessons from social insects. *Biological Theory* 1 (3): 317–330.

Piaget, Jean. 1970. *L'épistemologie génétique*. Paris: PUF.

Popper, Karl. 1990. Towards an evolutionary theory of knowledge. In *A world of propensities*, ed. Karl Popper, 27–51. Bristol: Thoemmes.

Queiroz, João, Claus Emmeche, and Charbel Niño El-Hani. 2005. Information and semiosis in living systems: A semiotic approach. SEED 5 (1): 60–90. Available at http://www.library.utoronto.ca/see/SEED/Vol5-1/Queiroz_Emmeche_El-Hani_abstract.htm.

Roederer, Juan G. 2005. *Information and its role in nature*. Berlin: Springer-Verlag.

Rosen, Robert. 1985. *Anticipatory systems: Philosophical, mathematical and methodological foundations*. Oxford: Pergamon Press.

Salemi, Marco, and Anne-Mieke Vandamme, eds. 2003. *The phylogenetic handbook: A practical approach to DNA and protein phylogeny*. Cambridge: Cambridge University Press.

Sarkar, Sahotra. 1996. Biological information: A skeptical look at some central dogmas of molecular biology. In *The philosophy and history of molecular biology: New perspectives*, ed. Sahotra Sarkar, 187–232. Dordrecht: Kluwer.

Sarkar, Sahotra. 2000. Information in genetics and developmental biology: Comments on Maynard Smith. *Philosophy of Science* 67 (2): 208–213.

Sattler, Rolf. 1986. *Biophilosophy: Analytic and holistic perspectives*. Berlin: Springer-Verlag.

Schrödinger, Erwin. 1944. *What is life?* Cambridge: Cambridge University Press.

Seligman, Jeremy. 2007. Logic-semantic theories of information. In *Handbook on the philosophy of information*, ed. Johan van Benthem and Pieter Adriaans. Available at http://www.illc.uva.nl/HPI/Logico-Semantic_Theories_of_Information.pdf.

Shannon, Claude E., and Warren Weaver. 1949. The mathematical theory of communication. Urbana: University of Illinois Press.

Singh, Jagjit. 1966. *Great ideas in information theory, language, and cybernetics*. New York: Dover.

Solomonoff, Ray J. 2003. The universal distribution and machine learning. *Computer Journal* 46 (6): 598–601.

Stegmann, Ulrich E. 2005. John Maynard Smith's notion of animal signals. *Biology and Philosophy* 20 (5): 1011–1025.

Stuart, C. I. J. M. 1985. Bio-informational equivalence. *Journal of Theoretical Biology* 113 (4): 611–636.

Tipler, Frank J. 1995. *The physics of immortality: Modern cosmology, God and the resurrection of the dead.* New York: Anchor Books.

Tribus, Myron, Paul T. Shannon, and Robert B. Evans. 1966. Why thermodynamics is a logical consequence of information theory. *Journal of the American Institute of Chemical Engineering* 12 (2): 244–248.

van Benthem, Johan. 2007. Information is what it does. In *Handbook on the philosophy of information*, ed. Johan van Benthem and Pieter Adriaans. Available at http://www.illc.uva.nl/HPI/Edito.pdf.

Villanueva, Enrique, ed. 1990. *Information, semantics and epistemology.* Oxford: Blackwell.

von Bertalanffy, Ludwig. 1968. *General system theory: Foundations, development, applications.* New York: George Braziller.

Waddington, Conrad H., ed. 1968. *Towards a theoretical biology.* Edinburgh: Edinburgh University Press.

Wicken, Jeffrey S. 1987. *Evolution, thermodynamics and information: Extending the Darwinian program.* Oxford: Oxford University Press.

Wicken, Jeffrey S. 1988. Thermodynamics, evolution and emergence: Ingredients for a new synthesis. In *Entropy, information and, evolution: New perspectives on physical and biological evolution*, ed. Bruce H. Weber, David J. Depew, and James D. Smith, 139–188. Cambridge, Mass.: MIT Press.

Wiener, Norbert. 1961. *Cybernetics, or control and communication in animal and machine.* Cambridge, Mass.: MIT Press.

Yockey, Hubert P. 1977. A calculation of the probability of spontaneous biogenesis by information theory. *Journal of Theoretical Biology* 67 (3): 377–398.

Yockey, Hubert P. 1981. Self-organization origin of life scenarios and information theory. *Journal of Theoretical Biology* 91 (1): 13–31.

Yockey, Hubert P. 2005. *Information theory, evolution, and the origin of life.* Cambridge: Cambridge University Press.

4

The Biosemiotic Approach in Biology: Theoretical Bases and Applied Models

João Queiroz, Claus Emmeche, Kalevi Kull, and Charbel El-Hani

Biosemiotics is a growing field that investigates semiotic processes in the living realm in an attempt to combine the findings of the biological sciences and semiotics. Semiotic processes are more or less what biologists have typically referred to as "signals," "codes," and "information processing" in biosystems, but these processes are here understood under the more general notion of *semiosis*, that is, the production, action, and interpretation of signs. Thus, biosemiotics can be seen as biology interpreted as a study of living sign systems—which also means that semiosis or sign process can be seen as the very nature of life itself. In other words, biosemiotics is a field of research investigating semiotic processes (meaning, signification, communication, and habit formation in living systems) and the physicochemical preconditions for sign action and interpretation.

To treat biosemiotics as *biology interpreted as sign systems study* is to emphasize an important intertheoretical relation between biology as we know it (as a field of inquiry) and semiotics (the study of signs). Biosemiotics offers a way of understanding life in which it is considered not just from the perspectives of physics and chemistry, but also from a view of living systems that stresses the role of signs conveyed and interpreted by other signs in a variety of ways, including by means of molecules. In this sense, biosemiotics takes for granted and preserves the complexity of living processes as revealed by the existing fields of biology, from molecular biology to brain science and behavioral studies. However, biosemiotics attempts to bring together separate findings of the various disciplines of biology (including evolutionary biology) into a sign-theoretical perspective concerning the central phenomena of the living world, from the ribosome to the ecosystem and from the beginnings of life to its ultimate meanings. From this perspective, no positivist (i.e., theory-reductionist) form of unification is implied, but simply a broader approach to life processes in general, paying attention to the location of biology between the psychological (the humanities) and the physical (natural) sciences.

Furthermore, by incorporating new concepts, models, and theories from biology into the study of signs, biosemiotics attempts to shed new light on some of the unsolved questions within the general study of sign processes (semiotics), such as the question about the origins of signification in the universe (e.g., Hoffmeyer 1996), and the major thresholds in the levels and evolution of semiosis (Sebeok 1997; Deacon 1997; Kull 2000; Nöth 2000). Here, signification (and sign action) is understood in a broad sense, that is, not simply as the transfer of information, but also as the generation of the very content and meaning of that information in all living sign producers and sign receivers.

Sign processes are thus taken as real: they are governed by regularities (habits, or natural rules) that can be discovered and explained. They are intrinsic in living nature, but we can access them—not directly, but indirectly through other sign processes (e.g., scientific measurements and qualitative distinction methods)—even though the human representation and understanding of these processes in the construction of explanations is built up as a separate scientific sign system distinct from the organisms' own sign processes.

One of the central characteristics of living systems is the highly organized character of their physical and chemical processes, partly based upon informational and molecular properties of what has been described in the 1960s as the genetic code (or, more precisely, organic codes). Distinguished biologists, such as Ernst Mayr (1982), have seen these informational aspects as one of the emergent features of life, namely, as a set of processes that distinguishes life from everything else in the physical world, except perhaps human-made computers. However, while the informational teleology of computer programs are derived, qua being designed by humans to achieve specific goals, the teleology and informational characteristics of organisms are intrinsic, qua having evolved naturally, through adaptational and evolutionary processes. The reductionist and mechanistic tradition in biology (and philosophy of biology) has seen such processes as being purely physical and having to do with only efficient causation. Biosemiotics is an attempt to use the concepts of semiotics in the sense employed by Charles Sanders Peirce to answer questions about the biological emergence of meaning, intentionality, and a psychological world (CP 5:484).[1] Indeed, these are questions that are hard to answer within a purely mechanistic and reductionist framework.

The term "biosemiotic" was first used by F. S. Rothschild in 1962, but Thomas Sebeok has done much to popularize the term and the field.[2] Apart from Charles Peirce (1939–1914) and Charles Morris (1901–1979), early pioneers of biosemiotics were Jakob von Uexküll (1864–1944), Heini Hediger (1908–1992), and Giorgio Prodi (1928–1987), and the founding fathers were Thomas Sebeok (1920–2001) and Thure von Uexküll (1908–2004). After 2000, an institutionalization of biosemiotics can be noticed: since 2001, annual international meetings of biosemioticians have been taking place (initially organized by the Copenhagen and Tartu groups); in 2004, the International Society for Biosemiotic Studies was established (with Jesper Hoffmeyer as its first president; see Favareau 2005); the specialized publications *Journal of Biosemiotics* (Nova Science) and *Biosemiotics* (Springer) have appeared; several collections of papers have characterized the scope and recent projects in biosemiotics, such as a special issue of *Semiotica* 127 (1/4) (1999), *Sign Systems Studies* 30 (1) (2002), Sebeok and Umiker-Sebeok 1992, Witzany 2007, and Barbieri 2007.

Also, from the 1960s to the 1990s, the semiotic approach in biology was developed in various branches:

a. Zoosemiotics, the semiotics of animal behavior and communication
b. Cellular and molecular semiotics, the study of organic codes and protolinguistic features of cellular processes
c. Phytosemiotics, or sign processes in plant life
d. Endosemiotics, or sign processes in the organism's body
e. Semiotics in neurobiology
f. Origins of semiosis and semiotic thresholds

Biosemiotics sees the molecular evolution of life and the evolution of semiotic systems as two aspects of the same process. The scientific approach to the origin and evolution of life, partly due to the success of molecular biology, has given us highly valuable accounts of the outer aspects of the whole process, but has overlooked the inner qualitative aspects of sign action, and has led to a reductionist view of causality. Complex, self-organized, living systems are also governed by formal and final causality: formal causality in the sense of the downward causation from a whole structure (such as the organism) to its individual molecules, constraining their action, but also endowing them with functional meanings in relation to the whole metabolism, and final causality, in the sense of the tendency to take habits and to generate future interpretants[3] of the present sign actions.[4] Here, biosemiotics draws also upon the insights of fields like systems theory, theoretical biology, and the study of complex self-organized systems.

Particular scientific fields like molecular biology, cognitive ethology, cognitive science, robotics, and neurobiology deal with information processes at various levels, and thus spontaneously provide knowledge about biosemiosis (sign action in living systems). Biosemiotics, both as a research program and a general perspective on life, would attempt to integrate such findings, and to build a semiotic foundation for biology. By describing the continuity between body and mind, biosemiotics also helps us to understand the evolution of kinds of mind in living systems, thereby assisting in overcoming some forms of Cartesian dualism that are still haunting the philosophy of mind. In addition, it develops specific models of life processes (such as those proposed here for the genetic information system and signaling systems, discussed shortly), emphasizing their signifying nature, thus helping to enrich and complement the biological sciences as standardly understood.

In what follows, we will draw on Peirce's semiotics to construct two semiotic models: one of the cell's genetic sign system, the other of signal transduction in B cell activation. In this manner, we intend to shed light on the notion of information as employed in the biological sciences.

A Theoretical Basis for a Biosemiotic Approach to Living Systems

The concept of information and related notions in biology should be not only taken seriously, but also clarified by employing appropriate conceptual tools. The use of semiotic concepts and theories to interpret "information talk" can significantly contribute to a precise and coherent formulation of the whole set of notions related to the information concept in biology. A semiotic treatment of biological information can also help to clarify some misunderstandings about the role of genes in biological systems, avoiding much criticized notions such as genetic blueprints and programs (see, e.g., Oyama 2000; Nijhout 1990; Sarkar 1996; El-Hani 1997; Keller 2000), while preserving the concept of biological information, albeit radically reinterpreted. A semiotic approach also lends support to the now widely accepted idea that there is more to information in living systems than just genetic information.

Here, however, our primary aim is to demonstrate how some central notions of Peirce's semiotics can be used to model information processes in biological systems. We will specifically address works in applied semiotics in which we developed analyses of genetic information and signaling systems grounded in Peirce's theory of signs (Queiroz, Emmeche, and El-Hani 2005; El-Hani, Queiroz, and Emmeche 2006; El-Hani, Arnellos, and Queiroz 2007). In other papers, we presented semiotic analyses of communication processes in nonhuman animals (Queiroz and Ribeiro 2002; Ribeiro et al. 2007). The analyses offered here are not, however, the only way to apply semiotics to particular cases. It is important to notice that there are other ways of applying semiotics to biology that are not based on the semiotics of Peirce (e.g., Markoš 2002; Barbieri 2003). As a first step in our argument, we will discuss the status of information talk in biology.

Information Talk in Biology

During the 1950s and 1960s, genetics, cytology, and molecular biology were swamped by terms borrowed from information theory. This "information talk" or "quasi semiotics," still pervades these fields, including widely used terms such as "genetic code," "messenger RNA," "transcription," "translation," "transduction," "recognition," "genetic information," "chemical signals," "cell signaling," and so on. But as the concept of information and its plethora of associated notions were introduced in biology, so were several problems with which the tradition of biology was unprepared to cope. Instead of deepening the discussion about the problems involved in information talk, the trend in the biological sciences was one of treating information as merely sequence information in DNA or proteins.

Some researchers consider information talk as inadequate and merely metaphorical, thus expressing skepticism about the use of the term "information" and its derivatives in biology (Stuart 1985; Sarkar 1996). We disagree with this position, claiming instead that the notion of information and other related ideas grasp some fundamental features of biological systems and processes that might be otherwise neglected. The terms "code," "information," "signals," "message," "signaling," and so on can be seen as necessary to understand the organization of relations in living beings in such a way that makes it clear that what happens in such beings is much more than simple chemistry. Bray, for instance, argues that "organisms can be viewed as complex information-processing systems, where molecular analysis alone may not be sufficient" (cited by Williams 1997, 476–477). Ideker, Galitski, and Hood (2001), in a paper about systems biology, maintain that biology is an informational science. Indeed, since the early applications of cybernetic models in life sciences, biology has been increasingly conceptualized as a communication and information science (e.g., Keller 2005), even though in many cases it is not clear at all what is meant by "information" in biology (Emmeche 1994; Griffiths 2001; Jablonka 2002; Jablonka and Lamb 2005).

It is not surprising, then, that biologists felt the need to talk about information when delving into the molecular microstructure of living systems. Life scientists needed a way of conveying the idea that more than just physics and chemistry is going on there. Even though all cellular processes are physicochemical processes, they are *complexly organized* physicochemical processes interwoven in communication and information networks. In this context, it is quite difficult to see what would be the real advantage of stripping away information talk from biology, instead of making it more precise and exploring its consequences in more depth. Thus, the problem is not getting rid of information talk, but rather clarifying it by using an appropriate theoretical framework.

As Griffiths (2001) summarizes the problem, genetic information is a metaphor in search of a theory. We believe this applies, in general terms, to information talk in biology. One possibility for building a theory of information in biology is to rely on the mathematical theory of communication. This theory allows one to define the amount of information as the measure of the probability of selection of a particular message among the set of all possible messages. The probabilistic measure of information provided by this theory is nonsemantic, and hence indifferent to meaning (Shannon and Weaver 1949; Cover and Thomas 1999; Jablonka 2002). It is true that this meaning-free concept of information can be useful in biological research for several purposes (Adami 2004). Nevertheless, it has been argued that such a nonsemantic (and quantitative) understanding of information is not sufficient for a theory of biological information, and should be complemented by a semantic, pragmatic (and more qualitative) approach. Jablonka (2002), for instance, uses an example where a DNA sequence encoding a functional enzyme and a same-length sequence coding for a completely nonfunctional polypeptide (which can have only a single different nucleotide) would contain, according to the above-mentioned measure, the same amount of information. It is obvious, however, that these two messages do not mean the same to the cell. This indicates the necessity of a treatment of information in biology that includes a semantic and a pragmatic dimension. Or, to put it differently, a theory of biological information should also deal with the meaning of messages and the context in which they are interpreted. Here, we use semiotic concepts to build a semantic and pragmatic account of biological information. In particular, we propose a model of information as semiosis, grounded in Peirce's pragmatic theory of signs.

Peircean Semiotics: A Brief Introduction
Peirce is often considered the founder of modern semiotics (Weiss and Burks 1945, 386). Semiotics was defined by Peirce as "the doctrine of the essential and fundamental nature of all varieties of possible semioses" (CP 5:484). Semiotics describes and analyzes the structure of semiotic processes independently of their material bases, or of the conditions under which they can be observed: inside cells (cytosemiosis), among tissues and cell populations (vegetative semiosis), in animal communication (zoosemiosis), or in typically human activities (production of notations, metarepresentations, etc.). In other words, Peirce's concept of semiotics concerns a theory of signs in its most general sense. Peirce conceived general semiotics much like a formal science as mathematics is (CP 2:227). However, semiotics finds the objects of its investigation in the sign's concrete, natural environment and in "normal human experience" (CP 1:241).

Semiotics is subdivided into speculative grammar, critical logic, and speculative rhetoric (CP 2:229). The first division of this science is what interests us here. Its task is that of examining the "sign physiology of all kinds" (CP 2:83), that is, the concrete nature of signs as they emerge and develop, and the conditions that determine the sign's further development, nature, and interpretation. It is the branch that investigates (1) the conditions to which any and every kind of sign must be submitted, (2) the sign itself, and (3) its true nature (CP 1:444). As one of its tasks, speculative grammar elaborates on the classifications of signs or, in other words, the diversity of sign types and how they merge with one another to create complex semiotic processes. For Houser, the logician "who concentrates on speculative grammar investigates representation relations (signs), seeks to work out the necessary and sufficient conditions for representing, and classifies the different possible kinds of representation" (1997, 9). Between 1867 and 1911, Peirce developed a model of signs as processes, actions, and relations, and also elaborated divisions of signs in order to describe different kinds of semiotic processes.

Peirce's pragmatic model of meaning as the "action of signs" (semiosis) has had a deep impact (besides all branches of semiotics) on philosophy, psychology, theoretical biology, and cognitive sciences (see Freeman 1983; Fetzer 1997; Colapietro 1989; Tiercelin 1995; Hoffmeyer 1996; Deacon 1997; Freadman 2004; Hookway 2002). First and foremost, Peirce's semiotics is grounded in a list of categories—namely, Firstness, Secondness, and Thirdness—which corresponds to an exhaustive system of hierarchically organized classes of relations (Houser 1997). This system makes up the formal foundation of Peirce's philosophy (Parker 1998) and his model of semiotic action (Murphey 1993).

In brief, the categories can be defined as follows:

1. Firstness: what something is, without reference to anything else.
2. Secondness: what something is, in relation to something else, but without relation to any third entity.
3. Thirdness: what something is, insofar as it is capable of bringing a second entity into relation to a first one in the same way that it brings itself into relation to the first and the second entities.

Firstness is the category of vagueness and novelty: "firstness is the mode of being which consists in its subject's being positively such as it is regardless of anything else. That can only be a possibility" (CP 1:25). Secondness is the category of reaction, opposition, and differentiation: "generally speaking genuine secondness consists in one thing acting upon another, brute action. . . . I consider the idea of any dyadic relation not involving any third as an idea of secondness" (CP 8:330). Finally, Thirdness is the category of mediation, habit, generality, evolution, and conceptualization (CP 1:340).[5]

Semiosis and Information Processing
According to Peirce (CP 2:171, 2:274), any description of semiosis should necessarily treat it as a relation constituted by three irreducibly connected terms: sign-object-interpretant (S-O-I). Hereafter, we will refer to these terms of a triadic relation as S, O, and I, and to the triadic relation in itself, as "triad" (see figure 4.1). As the reader will note in figure 4.1, this triadic relationship communicates/conveys a form from the object to the interpretant through the sign (symbolized by the horizontal arrow). The other two arrows indicate that the form is conveyed from the object to the interpretant through a determination of the sign by the object, and a determination of the interpretant by the sign.

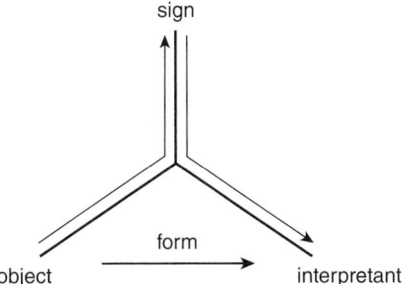

Figure 4.1
The semiotic relationship

For Peirce, a sign is something that stands for something other than itself. Peirce defined signs in several different ways (Marty and Lang 1997), but here we will highlight the definitions that will be useful in our work. He conceived a sign as a "First which stands in such a genuine triadic relation to a Second, called its Object, so as to be capable of determining a Third, called its Interpretant, to assume the same triadic relation to its Object in which it stands itself to the same Object" (CP 2:274; see also CP 2:303, 2:92, 1:541). The triadic relation between S, O, and I is regarded by Peirce as irreducible, in the sense that it is not decomposable into any simpler relation. Accordingly, the term "sign" was used by Peirce to designate the irreducible triadic process between S, O, and I, but he also used it to refer to the first term of this triadic relation. Some commentators have proposed that we should distinguish between the "sign in this strict sense" (*representamen*, or sign vehicle), when referring to the first term of the triad, and the "sign in a broad sense" (or sign process, sign as a whole) (e.g., Johansen 1993). Signs, conceived in the broad sense, are never alone.

In Peirce's definitions, we find several clues to understand how signs act. Any sign is something that stands for something else (its object) in such a way that it ends up producing a third relational entity (an interpretant), which is the effect a sign produces on an interpreter. In the context of biosemiotics, an interpreter is a biosystem such as a cell or an organism. In many biological informational processes, sign interpretation results in a new sign within the interpreter, which refers to the same object to which the former sign refers, or ultimately in an action, which can lead to the termination of an informational process. That the interpretant is often another sign, created by the action of a previous sign, is clear in the following statement by Peirce: a sign is "anything which determines something else (its interpretant) to refer to an object to which itself refers (its object) in the same way, the interpretant becoming in turn a sign, and so on, ad infinitum" (CP 2:303).

Accordingly, it is important to bear in mind always that the interpretant is not necessarily the product of a process that amounts to "interpretation" in the sense that we use this term to account for human cognitive processes. As explained previously, the fundamental character of the interpretant in many biological processes is that it is a new sign produced by the action of a previous sign in such a manner that both share the same referent, and indeed, refer to it in a similar way.

One of the most remarkable characteristics of Peirce's theory of signs is its commitment to a process philosophy. As a process thinker, it was quite natural that Peirce conceived semiosis as basically a process in which triads are systematically linked to one another so as to form a web (see Gomes et al. 2007). Peirce's theory of signs has a remarkable dynamical nature. According to Merrell, "Peirce's emphasis rests not on content, essence, or substance, but, more properly, on dynamics relations. Events, not things, are highlighted" (1995, 78). Thus, Hausman (1993) refers to the complex S-O-I as the focal factor of a dynamical process.

It is important not to lose sight of the distinction between the interpreter, which is the system that interprets the sign, and the interpretant. The interpreter is described by Peirce as a "Quasi-mind" (CP 4:536), a description that demands, for its proper interpretation, a clear recognition of Peirce's broad concept of mind (Ransdell 1977; Santaella-Braga 1994). It is not the case that only conscious beings can be interpreters in a Peircean framework. Rather, a translation machinery synthesizing proteins from a string of ribonucleic acid (RNA) or a membrane receptor recognizing a given hormone can be regarded as interpreters. A basic idea in a semiotic understanding of living systems is that these systems are interpreters of signs, that is, that they are constantly responding to selected signs in their surroundings. An interpreter is anything that carries on a sign process.

Thus, the interpreter does not have to be a conscious being, not even an organism, as it may be some part or subsystem within an organism, or a human-designed product. Nevertheless, because a sign process is itself an interpreter, the concept of interpreter appears to be secondary in Peirce's semiotics, even though it can play a heuristic role in building some models of semiotic processes.

We also need to consider here Peirce's distinctions regarding the nature of objects and interpretants, (For a review of these topics, see Savan 1988; Liszka 1990; Short 1996.) He distinguishes between the immediate and dynamical objects of a sign as follows:

> We must distinguish between the Immediate Object—i.e., the Object as represented in the sign—and . . . the Dynamical Object, which, from the nature of things, the Sign *cannot* express, which it can only *indicate* and leave the interpreter to find out by *collateral experience*. (CP 8:314; emphasis in the original)

and:

> We have to distinguish the Immediate Object, which is the Object as the Sign itself represents it, and whose Being is thus dependent upon the Representation of it in the Sign, from the Dynamical Object, which is the Reality which by some means contrives to determine the Sign to its Representation. (CP 4:536)

And we should also consider his distinction between three kinds of interpretants:

> The *Immediate Interpretant* is the immediate pertinent possible effect in its unanalyzed primitive entirety. . . . The *Dynamical Interpretant* is the actual effect produced upon a given interpreter on a given occasion in a given stage of his consideration of the Sign. (MS 339d:546–547; emphasis in the original)

and:

> The Final Interpretant is the one Interpretative result to which every Interpreter is destined to come if the Sign is sufficiently considered. . . . The Final Interpretant is that toward which the actual tends. (SS 110–111)

Let us first consider Peirce's distinction between the immediate and the dynamical objects of a sign. The dynamical object is something in reality that determines the sign, but can be represented by the sign only in some of its aspects. These aspects that the sign represents are the immediate object, that is, the dynamical object in its semiotically available form, that is, as immediately given to the sign. In another words, the immediate object is the dynamical object as the sign represents it (this is what we mean by "semiotic availability"). Because the sign represents the dynamical object in some of its features only, never in its totality, it can simply indicate that object, and it is left to an interpreter to establish what is the dynamical object through the interpreter's competence as a user of that sign, which, in turn, results from its previous experience and learning to become an interpreter.[6] This is why Peirce claims that the interpreter should find out what the dynamical object is by collateral experience. The system that is causally affected by the sign should establish which dynamical object the sign indicates through processes that have been selected for in the evolutionary history of that kind of system. In the ontogenetic timescale, the system will acquire its semiotic competence—that is, its competence as a sign interpreter—through development.

Peirce defines the dynamical interpretant as the actual effect of a sign, while the immediate interpretant is its "range of interpretability"—the range of possible effects that a sign is able to produce (see Johansen 1993, 166–167). The dynamical interpretant is thus the instantiation of one of the possible effects included in the immediate interpretant. The final interpretant in a semiotic process is, in turn, the final state of this process, understood as a tendency being realized when a given chain of triads is triggered, but not determined or bound to happen, because other final states can follow from the semiotic process, as in the case, for instance, of misinterpretation. In one way or another, the final interpretant can be seen as temporally solving the instability that is included into the sign process.

Peirce (CP 8:177) writes that a sign determines an interpretant in some "actual" or "potential" mind (in other passages, a "Quasi-mind"; see CP 4:536). It is indeed possible to differentiate between "potential" and "effective" semiosis. Potential semiosis is defined as a triadically structured process that is not actually taking place, but has a disposition to take place at a given moment; that is, it could occur under the appropriate conditions. Effective semiosis, in turn, concerns a sign that, by being actualized, has an actual effect on the interpreter. Semiosis necessarily entails the instantiation of chains of triadic relations, as a sign in a given triad will lead to the production of an interpretant, which is, in turn, a new sign. Therefore, an interpretant is both the third term of a previous triad and the first term (sign) of a subsequent triad (Savan 1988; see figure 4.2). Here, we have a first transition accounting for the dynamical nature of semiosis, namely, the interpretant-sign (I-S) transition. By this "transition," we simply mean that the same element that plays the role of the interpretant in a triad will play in a subsequent triad the role of the sign. After all, from a Peircean perspective, to perform sign processing and interpretation is to produce further (or, as Peirce says, more developed) signs.

Please also remember that the outline in this section is purely logical (or semiotic) and that within a particular physical, chemical, or biological system, the semiotic processes described here in general terms can be instantiated by different physical means, such as shifts in chemical con-

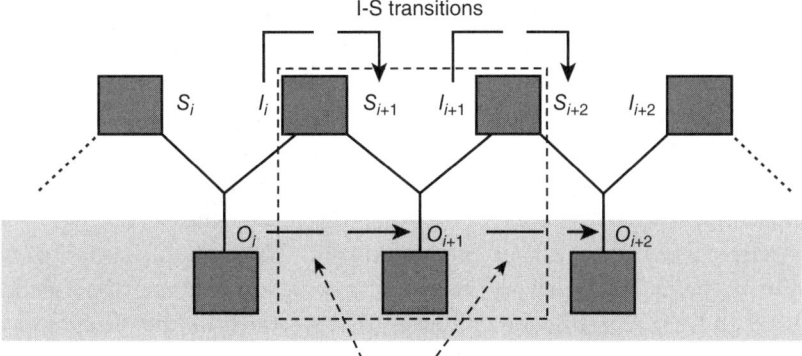

Figure 4.2
The triadic relation S-O-I forms a chain of triads. The grey area at the bottom of the figure shows that all signs in the chain of triads refer to the same dynamical object through a series of immediate objects. The arrows show the interpretant-sign (I-S) transition and the changes in the occupant of the functional role of the immediate object.

centrations or processes of molecular recognition. We will add this material aspect when we present our biosemiotic models.

When the I-S transition takes place, there is also a change in the occupant of the functional role of the immediate object (figure 4.2). When the interpretant becomes the sign of a new triad, the relation of reference to the same dynamical object depends on the fact that the new occupant of the role of immediate object stands for the same aspect of the dynamical object that the immediate object of a previous triad stood for. Thus, an object turns out to be a plural object via semiosis. We should stress, however, that instead of using the concept of *reference* in these models (as it is a highly debated and sometimes unclear concept that is not included in Peirce's theory of signs), it might be good to replace this concept in future works by a concept internal to Peirce's framework: the concept of *ground*.

As figure 4.2 shows, in a triad i a given sign S_i indicates a dynamical object by representing some aspect of it, the immediate object O_i. Through the triadic relation, an interpretant I_i is produced in the semiotic system. This interpretant becomes the sign in a subsequent triadic relation, S_{i+1}, which now indicates the same dynamical object. It should indicate this object through a new immediate object that corresponds to an aspect of the dynamical object represented in the sign. We now have a new occupant of the role of immediate object that stands for the same aspect of the dynamical object which was represented in the previous sign, S_i. It is in this sense that there is a change in the occupant of the functional role of the immediate object, from O_i in a previous triad to O_{i+1} in a subsequent triad. Through the triadic relation, a further interpretant, I_{i+1}, will be produced, which will then become the sign in a new triad, S_{i+2}, and thus successively, up to the end of that specific sign process.

Peirce also defines a sign as a medium for the communication of a form or habit embodied in the object to the interpretant (De Tienne 2003; Hulswit 2001; Bergman 2000), so as to constrain the interpretant as a sign or the interpreter's behavior (figure 4.1):

A Sign may be defined as a Medium for the communication of a Form. . . . As a medium, the Sign is essentially in a triadic relation, to its Object which determines it, and to its Interpretant which it determines. . . . That which is communicated from the Object through the Sign to the Interpretant is a Form; that is to say, it is nothing like an existent, but is a power, is the fact that something would happen under certain conditions. (MS 793:1–3) (See EP 2:544, n. 22, for a slightly different version.)

What is a form? There is a movement in Peirce's writings from "form as firstness" to "form as thirdness." Form is defined as having the "being of predicate" (EP 2:544), and it is also pragmatically formulated as a "conditional proposition" stating that certain things would happen under specific circumstances (EP 2:388). It is nothing like a thing (De Tienne 2003), but something that is embodied in the object (EP 2:544, n. 22) as a habit, a "rule of action" (CP 5:397), a "disposition" (CP 2:170), a "real potential" (EP 2:388) or, simply, a "permanence of some relation" (CP 1:415). Here, we would like to stress that the form communicated or conveyed from the object to the interpretant through the sign is not the particular shape of an object, or something alike, but a regularity, a habit that allows a given semiotic system to interpret that form as indicative of a particular class of entities, processes, or phenomena, and thus to answer to it in a similarly regular, lawful way. Otherwise, the semiotic system would not be really capable of interpretation.

The communication/conveyance of a form from the object to the interpretant constrains the behavior of an interpreter, in the sense that it brings about a constrained set of relations between the object and the interpretant through the mediation of the sign. We understand the "meaning" of a sign, thus, as an effect of the sign—conceived as a medium for the communication/conveyance of forms—on an interpreter by means of the triadic relation S-O-I. A meaning process can be thus defined as the action of a sign (semiosis).

In a Peircean approach, information can be strongly associated with the concepts of meaning and semiosis.[7] Peirce spoke of signs as "conveyers," as a "medium" (MS 793), as "embodying meaning." Accordingly, in his theory, the notions of meaning, information, and semiosis intersect and overlap in different ways (see Johansen 1993). Peirce defined "meaning" as the consequence of the triadic relation between sign, object, and interpretant (S-O-I) as a whole (EP 2:429), and also in terms of different correlates of a triad—e.g., object (MS 11, EP 2:274), interpretant (EP 2:496, EP 2:499; CP 4:536; see Fitzgerald 1966, 84; Bergman 2000). In turn, Peirce defined "information" at least ordinarily (CP 2.418) and metaphysically (CP 2.418) as a connection between form and matter, and logically (W 1.276) as the product of the extension and intension of a concept (Debrock 1996).

In the passage quoted earlier from MS 793, Peirce defines a sign both as "a Medium for the communication of a Form" and as "a triadic relation, to its Object which determines it, and to its Interpretant which it determines." If we consider both definitions of a sign, we can say, then, that semiosis is a triadic process of communication/conveyance of a form from the object to the interpretant by the sign mediation. And we can also stipulate that semiosis is, in a Peircean framework, information. For this reason, we systematically refer to *information* as the communication/conveyance of a *form* from O to I through S (Queiroz, Emmeche, and El-Hani. 2005; El-Hani, Queiroz, and Emmeche 2006; Queiroz and El-Hani 2006a, 2006b).

According to our interpretation of Peirce's ideas, information has the nature of a process: it is a process of communicating a form to the interpretant and operates as a constraining influence on possible patterns of interpretative behavior. When applying this general semiotic approach to biological systems, information will most often be an interpreter-dependent process. It cannot be dissociated from the notion of a situated (and actively distributed) communicational agent (potential or effective). It is interpreter-dependent in the sense that information triadically connects representation (sign), object, and an effect (interpretant) on the interpreter (which can be an organism or a part of an organism). In a biological system, information depends on both the interpreter and the object (in which the form communicated in information is embodied as a constraining factor of the interpretative process). Thus, a framework for thinking about information as a process can be constructed in Peircean terms by employing the following definitions:

• Information = semiosis: a triadic-dependent process through which a form embodied in the object in a regular way is communicated or conveyed to an interpretant through the mediation of a sign.
• Potential information = potential semiosis: a process of communicating or conveying a form from an object to an interpretant through the mediation of a sign that has a disposition to take place at a given moment, changing the state of the interpreter.
• Effective information = effective semiosis: the process by which a sign actually produces an effect (interpretant) on some system (an interpreter) by making the interpretant stand in a similar relation to the same object (the object of the sign) as that in which the sign itself stand. Thus, the sign mediates the relation between object and interpretant. The sign effectively communicates or conveys, in this way, a form from the object to the interpretant, changing the state of the interpreter.

Applied Biosemiotics: Modeling Two Semiotic Processes in Cells

Semiotic Analysis of Genes and Genetic Information
In the genetic information system, the synthesis of proteins and ribonucleic acids (RNAs) is related to deoxyribonucleic acid (DNA). Specific regions of this molecule act as templates for the transcription of RNAs by a multiprotein complex, including RNA polymerase. Messenger RNAs (mRNAs) act, in turn, as templates for the synthesis of proteins in the cytoplasm. DNA can act as a template for the synthesis of RNA, due to the specific base pairing of nucleotides, the monomers that constitute nucleic acids, and RNA in turn can act as a template for the synthesis of proteins due to specific relationships between sequences of three nucleotides (codons) and amino acids, the monomers of proteins. The set of these relationships amounts to the genetic code (which exists in several slightly different versions).

The effects of a protein-coding gene on a given cell or organism are regulated mainly by control of gene expression at the level of transcription initiation. This regulatory process results in the fact that only a subset of all genes present in any cell type in a multicellular organism is actually expressed.

The process by which the nucleotide sequence of mRNA serves as a template for the synthesis of proteins is called *translation* and is an essential part of protein synthesis. Messenger RNAs are often called the "vehicles" of the genetic information transcribed from DNA. The message at stake is "written" in the form of a series of codons, each specifying a particular amino acid. Another class of RNA molecules, transfer RNAs (tRNAs), play a fundamental role in the process of deciphering the codons in mRNA. Each type of amino acid has its own subset of tRNAs. They act as specific transporters, binding amino acids and carrying them to the growing end of a polypeptide chain in response to specific codons in the mRNA. The reason why the correct tRNA with its attached amino acid is selected at each step in protein synthesis is that each specific tRNA molecule contains a three-nucleotide sequence, called an *anticodon*, which base-pairs with its complementary codon in the mRNA. The specific relationship between tRNAs and amino acids, in turn, results from the attachment of the appropriate amino acid to a tRNA in a reaction catalyzed by a specific aminoacyl-tRNA synthetase. The specificity of the attachment between amino acids and tRNAs results from the capacity of each one of these enzymes to recognize one amino acid and all its compatible, or cognate, tRNAs. Therefore, the rules captured in the genetic code ultimately depend on the recognition activity of aminoacyl-tRNA synthetases.

If we now check the terms presented in the previous paragraphs, we will be able to see "information talk" in action. Our strategy was to use terms that are frequently employed in this same manner in biological papers and textbooks in order to highlight the importance of building a theory to give a precise meaning to this rather metaphorical language. As we mentioned previously, Griffiths (2001) wrote that genetic information is a metaphor in search of a theory. An analysis of molecular biology textbooks (Pitombo, Rocha de Almeida, and El-Hani 2008) shows that this is really so, as no idea related to information other than a reference to sequence information in DNA or proteins is offered in those textbooks as a ground for understanding the "information talk" that pervades them.

Against this background for understanding the genetic information system, we can move on to an analysis of genes and genetic information grounded in Peirce's theory of signs (the original sources for this analysis are Queiroz, Emmeche, and El-Hani. 2005; El-Hani, Queiroz, and Emmeche 2006). From the perspective of this theory, the action of a gene as a sign should be understood as a relationship between three elements (figure 4.3). By employing the definition of "information" put forward previously, genetic information can be thus described as a semiotic

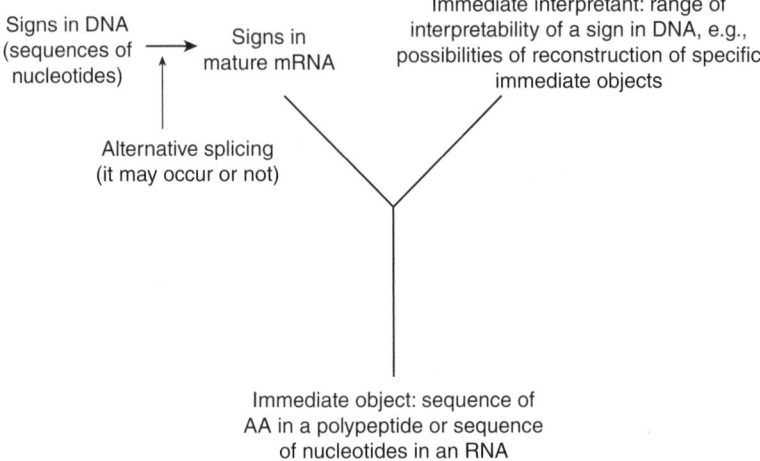

Figure 4.3
A general semiotic analysis of the gene as a sign. A sign is the mediating element in a semiotic process through which a form is communicated from an object to an interpretant. This is the reason why we consider the interpretant here as the reconstruction of a form (habit) embodied in an object. "Reconstruction" here amounts to a process by which the form of a protein in a cell generation is communicated through signs in DNA (in potency) to the form of a protein in the next cell generation. Thus, a regularity obtains in the three-dimensional structure and function of proteins over generations.

process. From this perspective, there is more to genetic information than just the sequence of nucleotides in a stretch of DNA. This is an important conclusion, as it goes against the treatment of genetic information as merely sequence information in DNA or proteins, and indicates a different path to conceptualize information, that is, in a theory of biological information grounded in Peircean semiotics.

In figure 4.3, a sequence of nucleotides in DNA is not treated as information in itself, but as the first correlate of information interpreted as semiosis—that is, a sign. Signs in DNA are transcribed into signs in mature mRNA, with or without the occurrence of alternative splicing, a process through which different patterns of RNA processing lead to a number of different mature RNA molecules, each coding for different, but related proteins (isoforms). If alternative splicing takes place, a sign in DNA will then be used to produce several different signs in mRNA. The immediate object of a gene as a sign in DNA is the sequence of amino acids or nucleotides represented in it. And as several different immediate objects can be represented in DNA (given processes such as alternative splicing), there is a range of interpretability of a sign in DNA, which amounts to the immediate interpretant.

A protein-coding gene, for instance, can become a sign in effective action in a cell only by standing—in a triadic-dependent relation—for a specific sequence of amino acids (immediate object) through a process of reconstruction of a specific form (interpretant). What is genetic information in this scheme? It can be understood as the whole process through which a gene acts as a sign in a given cell, mediating the reconstruction of a specific sequence of amino acids. Information is the triadic-dependent relation per se; it is a process, not something to be found in the first correlate of this process, a sign in DNA. In signs in DNA, we can find information only *in potency*. When this potential information becomes actual information, it is not something contained in isolated signs in DNA, but the very process through which those signs act.

The relationship between signs in DNA and sequences of amino acids in proteins is established by a complex mechanism of interpretation, involving transcription, RNA processing, and translation.[8] Thus, to interpret a string of DNA, more than one interpretative system is required, including, for instance, RNA polymerases, involved in the transcription of DNA into RNA, and ribosomes, involved in the translation of mRNA, into protein. These interpretative systems are parts or subsystems of a cell as a global interpreter, and their actions are subordinated to the latter. The idea that the cell can be seen as a global interpreter to which a series of interpretative subsystems in the genetic information system are subordinated is dramatically reinforced by recent analyses of the functional organization of proteomes. Consider, for instance, that the multicomponent cellular systems involved in transcription, RNA processing, and RNA transport do not form a simple linear assembly line, but a complex and extensively coupled network in which signals circulate in a nonlinear manner, involving several feedback loops (Maniatis and Reed 2002; Kornblihtt et al. 2004). It is this network structure that makes possible the coordination of the interpretative subsystems in the genetic information system by the cell.

It is clear, then, that we cannot easily move from claims at the cellular level to claims at the molecular level while pondering which system is interpreting genes as signs. It is becoming increasingly clear through recent advances in the understanding of cell systems that when a gene is interpreted, the interpretation process is indeed taking place at the cellular level, although multicomponent molecular subsystems are necessary to this endeavor. This idea that ultimately the whole cell participates in the network necessary for the interpretation that is demanded for the effect of a gene product to take place (Emmeche and Hoffmeyer 1991; Pardini and Guimarães 1992) is further supported by the role of an impressive array of signaling pathways regulating the interpretation of signs in DNA. As Fogle observes, "DNA action and function become meaningful in the context of a cellular system. Coding information in the DNA is necessary but insufficient for the operation of living systems" (2000, 19).

Accordingly, a Peircean approach to genes and genetic information entails that genetic structures should not be seen in isolation from the larger system by which they are interpreted. From this perspective, the meaning of a gene to its interpreter, the cell—or, to put it differently, the biological meaningfulness of a gene—is found not only in DNA sequences in a chromosome. That there is more to genetic information than just a sequence of nucleotides in DNA means that we will have to include in our models of information the effect of the gene-as-a-sign on the cell or organism, and, in fact, the very role of cellular subsystems as interpreters of strings of DNA, in such a way that they relate signs to specific dynamical objects, proteins that play a function inside the cellular system and have an effect on it or on the organism of which the cell is a part.

In a Peircean framework, we move from an identification of genetic information with sequential information in DNA to its understanding as a triadic-dependent, semiotic process. As a way of stressing the difference between an account of information as a process and more usual explanations about what is information, consider, for instance, Maynard Smith and Szathmáry's (1999, 9–10) argument that information is "that something" that is conserved throughout a series of changes in the material medium underlying a communication process. According to the model developed previously, "that conserved something" is not information, but rather the permanence of a certain relation in the reconstructed form. Information is rather the process by which a form is conveyed through several different media (signs) in such a way that a relational structure is conserved throughout the process, even though the significant aspects of the object's form are continually reconstructed. Applying this idea in the context of the analysis offered in this section, it is not genetic information that is conserved throughout the different tokens of DNA molecules (different material media) in different organisms and generations, but rather a relational structure in the reconstructed form, that is, a habit or a tendency to build tokens of the same kind of protein (in the case of a protein-coding gene) based on the signs available in DNA. These signs themselves can only harbor potential information. Genetic information, in turn, is taken to be the process by which the permanence of a relational structure is conveyed to a new token of a protein, that is, the whole process through which genes as signs in DNA (entailing the potentiality of genetic information as a process) are irreducibly related to objects and interpretants.

Even if one concedes that this argument shows an advantage of our account in relation to the one given by Maynard Smith and Szathmáry, there is a further issue to be considered. Both accounts must meet the important requirement of providing an explanation of the representational character of signs. Prior to its application to biological processes, one might get the impression that the Peircean model of semiosis as a triadic relationship is simply a formal, uninterpreted schema. How, then, when applied to biological phenomena, does the representative character of a sign take on a specific meaning? On Maynard Smith's (2000) view, the representational nature of signs is explained by bringing in the idea of formal and/or final causation. We are also sympathetic with these causal notions. Therefore, despite the difference pointed out earlier, there are also similarities between our account and Maynard Smith's with regard to the explanation of the representational character of a sign in the biological sphere. However, a crucial advantage of the Peircean model is that what superficially looks like simply a formal definition of the logic structure of the sign is in fact also a pragmatic definition of the structure of meaning, because the meaning of a sign can be accessed only through that sign's effect (interpretant) upon some interpreting system, such as a cell or an organism. Thus, if we offer an answer, for specific biological cases, to the problem of how a sign takes on a specific meaning, this does not mean that we should be committed to look for any mysterious additional emergence of the representative character of the sign. The origin of semantic-pragmatic meaning is embedded within the same triadic formal structure of the sign interpretation process, that is, as the process by which new interpretants are generated.[9]

Transcription, RNA processing, and protein synthesis can be understood, in semiotic terms, as processes of actualization of potential signs in protein-coding genes. When put into action, a protein-coding gene becomes part of effective semiosis, a triadic-dependent process by means of which the gene as a sign indicates a given functional product, synthesized after splicing, mRNA edition, or any other complexity involved in the path from a DNA stretch to a protein. This functional product has, in turn, an effect on the organism in which it is expressed (its final interpretant), participating in its adaptive interactions with its surroundings, and thus contributing to the presence of those potential signs in the next generation in a high frequency. Notice that we are not postulating any inversion of the central dogma (as if sequences of amino acids in proteins might determine sequences of nucleotides in DNA). We are referring, rather, to the effect of functional proteins on the likelihood that certain genes—certain signs mediating the process of the synthesis of those proteins—will be present in future generations.

The actualization of a gene depends on boundary conditions established by a higher-level semiotic network, a network of signaling processes that regulate gene expression, ultimately determining the likelihood of transcription of a given gene, or splicing of a given pre-mRNA according to a particular pattern, or chemical modification of a given protein in a manner that modulates its function in a particular way (e.g., by phosphorylation), and so on. A variety of regulatory mechanisms studied in cellular and molecular biology can be thus understood as composing a macrosemiotic environment, establishing boundary conditions that will downwardly determine which potential genes in a string of DNA will be actualized, entering into effective action in a cell.

This shows how several complexities involved in gene expression can be introduced in our analysis: boundary conditions established by this macrosemiotic environment will determine, for instance, which stretch of DNA will be read (e.g., allowing for an analysis of transcription of overlapped or nested genes), which pattern of RNA splicing or RNA editing will be instantiated in order to produce a particular mature mRNA (allowing for the subtleties of alternative RNA splicing or RNA editing to be taken into account), which functional protein will be effectively constructed by the cell (allowing for chemical and/or structural modifications suffered by the primary amino acid sequence of a protein to be considered), and so on.

The regulatory influence of the macrosemiotic level—that is, of the network of signaling processes on interpretative subsystems, and, thus, on transcription, splicing, translation—shows that we have to ultimately consider the whole cell as participating in the network necessary for the actualization of potential genes in DNA. The cellular network of semiotic processes is, in turn, highly responsive to environmental factors, given the semi-open nature of living systems. Accordingly, genes, as potential signs in DNA, are actualized in response to regulatory dispositions arising from a network of signaling pathways that elicit cellular specific responses to other signs arising from a hierarchy of contexts, environments, or, in our own terms, semiotic levels that can direct gene expression (i.e., establish boundary conditions for the selection of potential genes in DNA), ranging from systems of gene-gene interactions to organisms, and passing through nucleus, cytoplasm, cell, cell surface, extracellular matrix, morphogenetic fields, collective condensations of cells (blastemas), organs, and so on (see, for example, Hall, 2001). Thus, the cell, as an interpreter, answers to an environmental cue or sign by means of a specific alteration of its internal states, triggered by a whole network of signal transduction culminating in a change at some level of gene regulation. (A semiotic analysis of signal transduction systems follows. See also Bruni 2003, Queiroz and El-Hani 2006b, and El-Hani, Arnellos, and Queiroz 2007.) These relations cannot be understood only in terms of molecular interactions taking place in networks of signal transduction, because this latter process crucially involves semiotic events, as the widespread usage of information talk in modeling and explaining signaling pathways clearly suggests.

This semiotic analysis also allows us to offer an interesting account of the "transmission" of information. It is not effective information that is being communicated when one observes, for instance, "vertical transmission" from parent to offspring. From the perspective of the model explained earlier, what is being communicated is only potential information, that is, the potentiality of a process called "information." It is only this potentiality that can be said, as explained prior, to be carried by stretches of DNA. Signs in DNA will become elements in effective information only when interpreted by the cell. Effective information itself cannot be carried from one system to another; only potential information can be carried by the first correlates of triads, signs (which, in biological systems, are typically physicochemical entities).

This biosemiotic analysis of the genetic information system leads to the following conclusions:

1. Genes should be treated as signs in DNA, which can have an effect on a cell only through a triadic dependent process (semiosis).
2. This process *is* genetic information and involves more than just genes as signs in DNA but also objects and interpretants.
3. Genetic information is the process by means of which a form in a dynamical object (a functional protein) is communicated to an interpretant (the reconstruction of a specific sequence of amino acids in a cell) through signs in DNA.

Let us now turn to the semiotic modeling of signaling systems to which we referred earlier when discussing the macrosemiotic environment within which genes, as potential signs in DNA, are downwardly selected to be actualized.

A Semiotic Model of Signal Transduction in B Cell Activation

The B cell antigen receptor (BCR) is a multiprotein complex consisting of a membrane-bound immunoglobulin molecule (mIg), the ligand-binding part, and an Ig-α/Ig-β heterodimer associated with mIg, which acts as a signaling subunit and couples the receptor to intracellular signal transducer elements (Reth and Wienands 1997). BCR has two functions in B cell activation (Pierce 2002): it initiates signaling pathways that result in a series of intracellular processes in B cells, including changes in gene expression patterns, which lead, in turn, to the activated B cell phenotype, and it plays a role in the uptake and processing of antigens to be presented to T helper cells, which will assist B cells in achieving full activation.

Reth and Wienands (1997) proposed a model of molecular interactions in signaling pathways based on functional definitions, intended to express the roles played by several elements in such pathways, acknowledging (as it is proper of functional definitions) that different elements can fulfill those roles, or, to put it differently, be the occupants of the functional roles described in the model in different signaling processes. Such a functional model has the important characteristic of being general, in contrast to molecular, mechanistic models of particular signaling pathways. Reth and Wienands characterize eight functional categories of signaling elements (figure 4.4).

Through signal transduction, living systems are capable of internalizing a cue to a certain aspect of the environment, by producing intracellular signs in response to an extracellular sign. *Receptors* play a central role in the processes through which a cell shows the capacity of answering to its surroundings. A receptor is in most cases a transmembrane protein that undergoes, when bound by an extracellular ligand, a conformational or topological (e.g., receptor aggregation) change that is, according to Reth and Wienands, "transmitted into the cell" (1997, 456). But how is the molecular change suffered by the receptor communicated to the intracellular milieu? Here, *transducers* enter into action. (But the issue of how the reference to the same cue or signal is maintained in the several changes in the material basis of the message remains open, and is indeed the matter to be dealt with in semiotic models.) Receptors

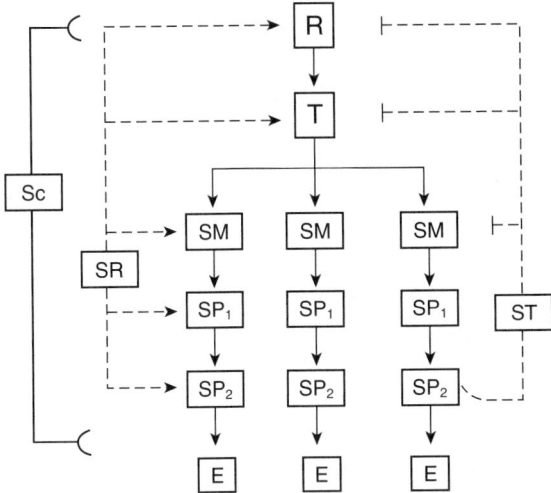

Figure 4.4
Reth and Wienands's (1997) functional model of signaling pathways. Arrows represent different types of functional connections between signaling elements. Dashed arrows represent regulatory relationships: R, receptor; T, transducer; SM, signal manager; SP, signal processor; SR, signal regulator; ST, signal terminator; Sc, scaffold protein; E, effector.

usually do not have an intracellular catalytic domain and are thus dependent on transducer elements to carry out their signaling function. In most cases, transducers are enzymes physically associated with the intracellular part of the receptor. In its resting state, the receptor often represses signaling activity of the associated transducer, but when it is activated by ligand binding, it suffers a conformational or topological change that leads to the activation of the transducer.

Each signaling pathway is switched on by the activity of the transducer and controlled by a *signal manager*, the third category in Reth and Wienands' (1997) model, located at the start of a particular signaling route. There can be several signaling pathways arising from the same receptor. There are cases in which a signal manager interacts directly with an effector, which instantiates an action under the regulation of the signaling pathway. When this is not the case, the signal manager activates a signal cascade consisting of one or several *signal processors*. *Signal regulators*, in turn, modify the efficiency and duration of signals traveling down a signaling pathway, by amplifying or decreasing the signal. Such changes in the intensity of a signal can have major biological effects. As we can see in figure 4.4, signal regulators can act at the level of receptors, transducers, signal managers or signal processors, that is, at all functional levels of the signaling system.

Signal transduction occurs in an organized microenvironment, in which different elements of a signaling pathway are connected both functionally and spatially. This architecture of signaling elements can be established before or after the activation of a receptor. In the former case, *scaffold* or *adaptor proteins* play an important role in organizing the spatial and functional architecture of signaling elements, by bringing them together in a preformed protein complex.

Even if the stimulus is persistent, signal transduction through many receptors is terminated after some time, due to the activity of *signal terminators*, which can be phosphatases as well as kinases or GTPases. They establish a feedback loop that changes the activity of the receptor, transducer, and/or a particular signal manager. At the endpoint of a signaling pathway, one finds one or several *effectors*, which can be enzymes, transcription factors, or cytoskeletal elements. They are the elements whose behavior is modulated by the signaling pathway.

We stated previously that signal transduction is a process through which living systems can answer in a regular and (usually but not always) adaptive manner to the environment, by producing intracellular signs in response to extracellular signs. The mechanistic interactions involved in this process are aptly modeled by Reth and Wienands in functional (and, thus, properly general) terms, but, if the series of mechanistic interactions that take place in a signaling pathway amounts to a process of *signal transduction*, a description in terms of molecular interactions or even functional definitions will not be enough. It is not that some additional element, besides the molecules themselves, should be added to the mechanistic and material aspect of the signaling pathway; rather, what should be added to the picture is the kind of relation explained earlier, a semiotic relation by means of which a molecule such as an antigen can be a sign that stands for something else—say, a pathogen—and, in turn, lead to the production, within the living system, of other (signaling) molecules that stand in the same relation to that object in which the antigen itself stood. Only in this manner we will be able to explain not only the molecular interactions and functional roles in a pathway, but also the maintenance of the reference to the same object, namely the pathogen, while several different signaling molecules are engaged in the pathway. This is clearly a fundamental property to account for, if we want to explain why this is a signal transduction process.

In more detail, to model in Peircean terms the maintenance of the reference to an extracellular sign throughout the several changes in intracellular signs that characterize a signaling pathway, one should consider how the processes described by Reth and Wienands instantiate a triadic relation in which a receptor that acts as an interpreting system recognizes a sign (the extracellular signal, an antigen), which refers to an object in the world (a dynamical object, such as a pathogen) through a feature semiotically available in its representation (the molecular form of the antigen, as an immediate object that indicates the pathogen, as a dynamical object). Receptors act as interpreting systems by activating transducers in response to ligands (signs). That is, the receptor communicates the sign process to the interior by coupling to transducers, catalytic molecules that trigger the production of another sign inside the cell in response to the extracellular sign. This subsequent sign is the interpretant of a first triadic relation, and it takes the role of a sign for a subsequent triadic relation, allowing signaling to proceed. This happens through a series of intracellular signs that can diverge, if several signaling pathways are triggered by the transduction of a single extracellular sign, and are amplified by signal regulators along the pathways. Each pathway ends in an effector, which produces the final interpretant in the process, an action through which sign interpretation has an effect on the cell phenotype.

Let us take now a closer look at initiation events at the BCR signaling system. Figure 4.5 presents a model of the main events at stake. In resting B cells, BCR is excluded from membrane domains (lipid rafts) that concentrate the transducer *Lyn*. In the absence of antigen, the BCR monomer has a weak affinity for lipid rafts, but antigen binding makes BCR molecules associate with each other, increasing affinity for those domains. Stable residency in lipid rafts results in association with *Lyn*, which phosphorylates BCR, initiating several signaling pathways. In figure 4.5, another kinase is shown, named *Syk*, which initiates one of the signaling pathways resulting from BCR activation.

As we saw earlier, when interpreted from a Peircean perspective, an antigen is a sign that stands for something else, such as a pathogen, and a receptor such as BCR acts as an interpreting system in the cell membrane, triggering processes by means of which new signs—that is, interpretants—are produced inside the B cell. The first interpretant in this case is the phosphorylated state of BCR, which is a sign that stands for the pathogen as the antigen itself stood for it. This generates a new triad, linked to the previous one by the double role played by the

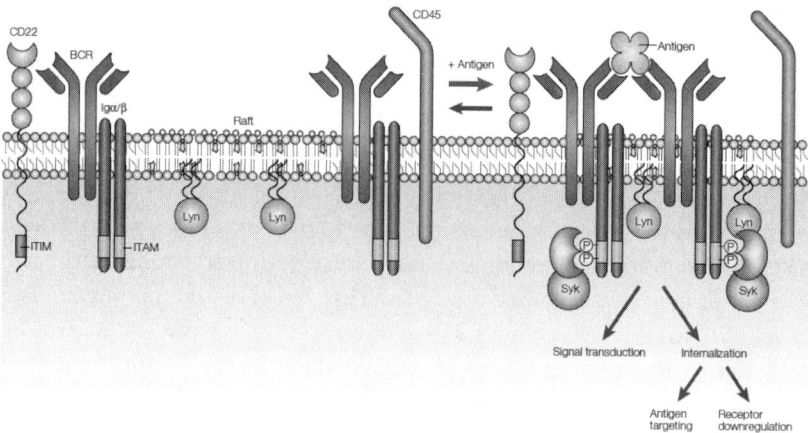

Figure 4.5
Model of the initiating events in the signal-transduction pathways leading to B cell activation (from Pierce 2002)

phosphorylated state of BCR, which is both the interpretant of a first triad, and the sign of a second triad (figure 4.6). We are dealing, thus, with the I-S transition, a basic process underlying the generation of chains of triads. When the I-S transition takes place, the aspect of the pathogen which was represented in the antigen (O_i) is now represented in the phosphorylated state of BCR (O_{i+1}). To put it differently, following the I-S transition, there is a change in the occupant of the functional role of O (figure 4.6).

It is this latter change that makes it possible for the same entity or process to be kept as a stable referent throughout the signaling process, despite the several changes in the material bases of signaling, that is, in the signs involved. The maintenance of the reference to the pathogen in a signaling pathway can be modeled as such changes of occupants because all the immediate objects in a chain of triads stand for the same dynamical object—the pathogen. The fact that the reference to the same dynamical object is maintained can be explained on the basis that the latter is, in a Peircean framework, the primary constraining factor in semiosis, because its form—understood as a regularity or habit—is communicated through several semiotic, triadic relations. Such a communication of the form of the dynamical object, as semiotically available in a series of immediate objects, is *information* in a signaling pathway. After all, information is conceived, in the Peircean framework developed here, as a triadic-dependent process through which a form embodied in the

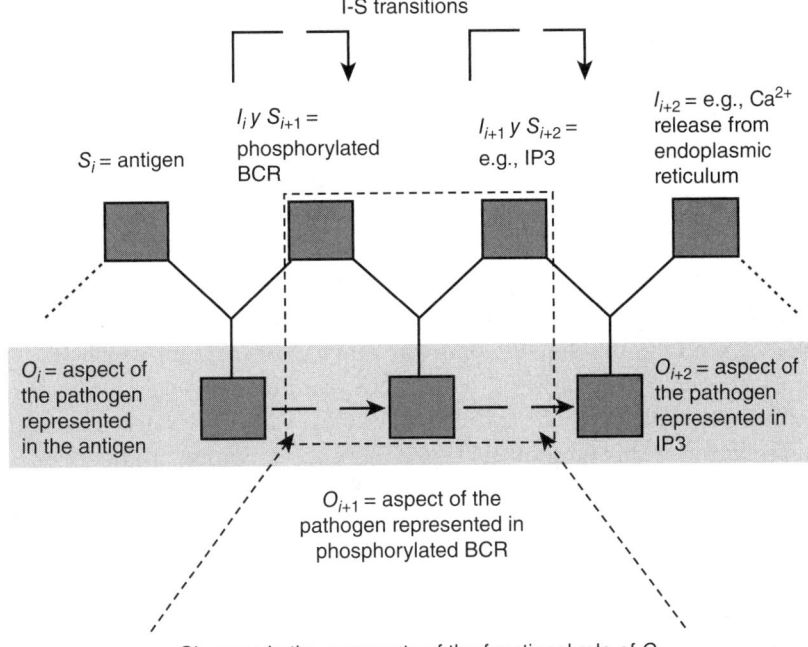

Figure 4.6
A model of one of the signaling pathways triggered by activated BCR as a chain of triads. Notice the I-S transition and the changes in the occupants of the functional role of O. The maintenance of the reference to the pathogen in a signaling pathway is modeled in terms of these changes of occupants, because all the immediate objects stand for the same dynamical object, the pathogen, throughout semiotic, triadic relations that communicates the form of the object and are conceived, according to the theoretical framework developed here, as *information* in a signaling pathway.

object in a regular way is communicated to an interpretant through the mediation of a sign.

Biochemical and genetic evidence has shown that *Syk* has a key role in a well-defined pathway of B cell activation, which results in the release of Ca^{2+} from the endoplasmic reticulum (Reth and Wienands 1997). In this case, the binding of *Syk* to the phosphorylated BCR makes a specific interpretative process proceed. When *Syk* is activated, it leads to the activation of another enzyme, phospholipase Cγ (*PLC-γ*), which is an effector, converting the membrane component phosphatidylinositol 4,5-biphosphate into the two second messengers diacylglycerol (DAG) and inositol 1,4,5-triphosphate (IP3). This illustrates a case of divergence of intracellular signals, modeled in semiotic terms by means of the production of more than one interpretant from a single sign, namely, the phosphorylated state of BCR.

DAG remains attached to the inner side of the plasma membrane and recruits and activates the cytosolic protein kinase C (PKC). IP3 binds to receptors on the endoplasmic reticulum, causing the release of Ca^{2+} ions. The release of Ca^{2+} ions is a new interpretant in the signaling pathway managed by *Syk*. The number of different PKC substrates (for example, CD20, c-Raf, IκB) and the multifunctional role of Ca^{2+} ions in cell metabolism, and also in signaling, make it clear how an original sign-response can be broadly diversified by the signaling systems of a cell. As we can see in figure 4.7, the pathway managed by *Syk* in which IP3 is involved does not end in Ca^{2+} ions, but continues through further I-S transitions, which we will not model here for reasons of space. The final interpretant of this (and other) signaling process amounts to the regulation of gene expression so as to lead to B cell activation.

DAG and IP3 stand for the pathogen in the same way as the antigen and the phosphorylated state of BCR stood, maintaining the reference

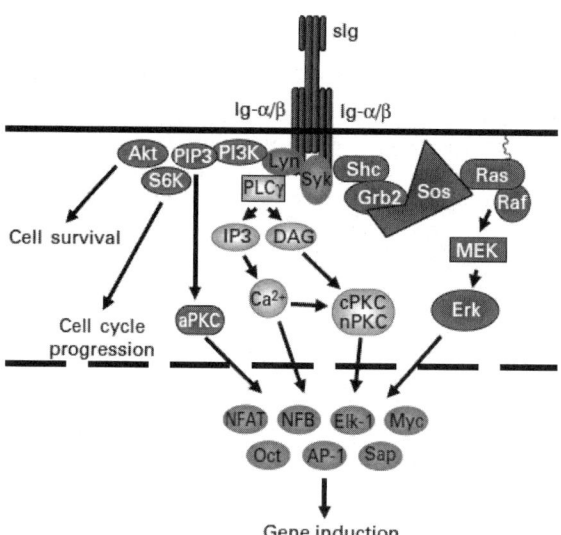

Figure 4.7
Several intracellular signaling pathways are initiated by the cross-linking of B cell receptors by antigen (Goodridge and Harnett 2005). In the center of the figure, one can see the signaling pathways modeled above, involving *Syk*, *PLCγ*, IP3, and Ca^{2+} release. Notice the integration between this signaling pathway and the one involving DAG, which leads to the activation of *cPKC* and *nPKC*. Notice, also, that the pathway involving IP3 and Ca^{2+} regulates in the end patterns of gene expression in B cells.

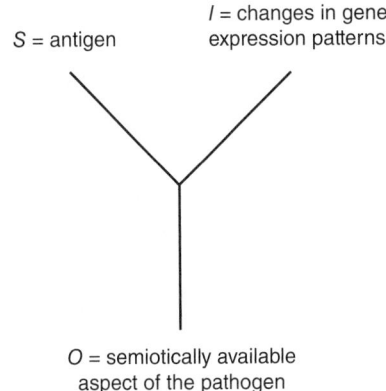

Figure 4.8
A global semiotic analysis of a semiotic process triggered by antigen-binding to BCR

of the signaling process through changes of the occupants of the functional role of the immediate object. IP3, for instance, acts as a sign to a subsequent triad, triggering the production of Ca^{2+}, which in turn will occupy the role of sign in a further triad, up to the final interpretant of this particular semiotic process.

From a global perspective, the overall result of the semiotic process modeled previously can be grasped in terms of a triad containing the antigen as a sign, the pathogen as represented, say, in the three-dimensional form of the antigen as an immediate object, and changes in the pattern of gene expression in B cells, as an interpretant (figure 4.8).

To stress the necessity of semiotic modeling of signaling processes, we can ask why molecules such as DAG and IP3 can be called "second messengers." What is the "message" and how is it preserved in them? The message refers to the presence of a non-self entity—for instance, a pathogen—within the organism. But how is the reference to such an entity preserved in the messengers? In order to successfully model the maintenance of reference throughout the process, we should go beyond the pairwise or dyadic interactions between molecules and their substrates, and build a semiotic, triadic model capable of showing how the reference to a non-self entity external to the cell can be maintained during the processing of signs within the cell.

A semiotic analysis allows us to go beyond a metaphorical usage of the expression "second messenger": DAG and IP3 are second messengers precisely because they are interpretants produced as a result of the processing of an extracellular sign (a "first messenger"), in this case, an antigen. In turn, the changes of the occupants of the functional role of O in chains of triads corresponding to the signaling pathways managed by *Syk* show how the reference to the pathogen is maintained, while the material bases of the message—namely, the signs—keep changing throughout the process.

To put it differently, we argue that to understand signaling processes, we need at least three properly connected, but different models:

1. Molecular, mechanistic models of particular signaling pathways, in which the molecular interactions that take place in them are properly represented and explained
2. General, functional models, such as the one proposed by Reth and Wienands (1997), which represent and explain in general terms how different occupants can play the several functional roles in a signaling pathway
3. Semiotic models, such as the one proposed by El-Hani, Arnellos, and Queiroz (2007), and reviewed and extended here, which represent and explain in semiotic terms how different occupants can play the same semiotic roles in a signaling pathway

Finally, consider the role of signaling processes, as a higher-level semiotic network, in the actualization of genes as potential signs, by affecting the likelihood of their transcription, or the patterns of splicing of their premRNA, or posttranslational changes of their functional products. Accordingly, the next step in our research will be to employ the basic framework developed herein to model signal transduction in connection with gene actualization, combining in a single model the accounts we developed in separate papers.

Concluding Remarks

The framework for building a theory of biological information presented here is consistent with the general picture of genetic information and signaling processes in genetics and molecular biology, with the fundamental differences that first, a concept of information is explicitly formulated within a heuristically powerful theoretical framework, and second, information is on these grounds conceptualized as a process. Consequently, to make this semiotic framework and the current structure of genetics and molecular biology compatible, it is necessary to conceive the latter in more process-oriented terms. This is a fruitful avenue to be pursued in order to build a framework for biology that is more compatible with the increasingly complex and dynamic nature of biological systems revealed by recent advances in the biological sciences (for some suggestions to the same effect, see, e.g., Neumann-Held 2001; Keller 2005). We consider the compatibility of our semiotic analyses with the framework of genetics and molecular biology as a strong feature. Of course, the pros and cons of complementing the current models in genetics and molecular biology with a semiotic concept of information is a matter for further investigation.

We have argued that semiotic modeling is a necessary counterpart to functional and mechanistic models of genetic and signaling systems. The conceptual tools offered by Peircean semiotics, along with the biosemiotic models that they enable us to construct, can deepen our understanding of biological phenomena that are described by a communicational and informational vocabulary. This is particularly important in a time in which biology is increasingly seen as a science of information. It is always useful to remind ourselves that at present we do not have an established general notion of biological information (despite the roles that the meaning-free concept of information offered by the mathematical theory of communication can play in biological research), and it is a basic contention of this work that biosemiotics can help in building a semantic/pragmatic concept of biological information.

Notes

1. In the body of the chapter, Peirce's works will be referred to as CP (followed by volume and paragraph number) for quotations from the *Collected Papers of Charles Sanders Peirce, 1866–1913* (Peirce 1932–1935); EP (followed by volume and page number) for quotations from *The essential Peirce: Selected philosophical writings, 1893–1913* (Peirce 1998); MS (followed by the number of the manuscript) for quotations from *The Annotated Catalogue of the Papers of Charles S. Peirce* (Peirce 1967); and SS (followed by page number) for quotations from *Semiotic and Significs: The Correspondence between Charles S. Peirce and Victoria Lady Welby* (Peirce 1977).

2. A review of the history of biosemiotics can be found in Favareau 2007 and Kull 1999. See also Emmeche, Kull, and Stjernfelt 2002.

3. In the Peircean framework, the interpretant is the effect produced by a sign. The concept will be explained in more detail later in this chapter in "Semiosis and Information Processing."

4. If one prefers not to use the category of causality to explain these aspects of living systems, one may speak of formal and final *determination*, in a framework

that acknowledges the existence of other kinds of determination in natural systems, besides causal determination (see El-Hani and Queiroz 2005).

5. For further discussion of the categories, see Hookway 1985, Murphey 1993, and Potter 1967/1997.

6. By "learning" here, we mean the result of both ontogenetic and phylogenetic processes that can lead a system to improve its sign-interpreter abilities.

7. Also, with the concept of experience (CP 1:537).

8. A detailed semiotic model of transcription and translation is put forward by El-Hani, Queiroz, and Emmeche (2006).

9. Much more could be said about Maynard Smith's views about biological information and the concept of representation, generally speaking, but this would require more space than we can devote to this issue in the framework of the current chapter.

References

Adami, Christoph. 2004. Information theory in molecular biology. *Physics of Life Reviews* 1 (1): 3–22.

Barbieri, Marcello. 2003. *The organic codes: An introduction to semantic biology*. Cambridge: Cambridge University Press.

Barbieri, Marcello, ed. 2007. *Introduction to biosemiotics: The new biological synthesis*. Berlin: Springer.

Bergman, Mats. 2000. Reflections on the role of the communicative sign in semeiotic. *Transactions of the Charles S. Peirce Society: A Quarterly Journal in American Philosophy* 36 (2): 225–254.

Bruni, Luis E. 2003. A sign-theoretic approach to biotechnology. Unpublished Ph.D. dissertation, Institute of Molecular Biology, University of Copenhagen.

Colapietro, Vincent Michael. 1989. *Peirce's approach to the self: A semiotic perspective on human subjectivity*. Binghamton: SUNY Press.

Cover, Thomas M., and Joy A. Thomas. 1999. Information theory. In *The MIT encyclopedia of the cognitive sciences*, ed. Robert A. Wilson and Frank C. Keil, 404–406. Cambridge, Mass.: MIT Press.

Deacon, Terrence W. 1997. *The symbolic species: The co-evolution of language and the brain*. New York: W. W. Norton.

Debrock, Guy. 1996. Information and the metaphysical status of the sign. In *Peirce's doctrine of signs: Theory, applications, and connections*, ed. Vincent Michael Colapietro and Thomas M. Olshewsky, 80–89. Berlin: Mouton de Gruyter.

De Tienne, André. 2003. Learning *qua* semiosis. *SEED* 3 (3): 37–53.

El-Hani, Charbel Niño. 1997. Explicações causais do desenvolvimento: são os genes suficientes? *Cadernos de História e Filosofia da Ciência* 7 (1): 121–167.

El-Hani, Charbel Niño, and João Queiroz. 2005. Downward determination. *Abstracta* 1 (2): 162–192.

El-Hani, Charbel Niño, João Queiroz, and Claus Emmeche. 2006. A semiotic analysis of the genetic information system. *Semiotica* 160 (1/4): 1–68.

El-Hani, Charbel Niño, Argyris Arnellos, and João Queiroz. 2007. Modeling a semiotic process in the immune system: Signal transduction in B-cells activation. *TripleC: Cognition, Communication, Co-operation* 5 (2): 24–36.

Emmeche, Claus 1994. The computational notion of life. *Theoria—Segunda Época* 9 (21): 1–30.

Emmeche, Claus, and Jesper Hoffmeyer. 1991. From language to nature: The semiotic metaphor in biology. *Semiotica* 84 (1/2): 1–42.

Emmeche, Claus, Kalevi Kull, and Frederik Stjernfelt. 2002. *Reading Hoffmeyer, rethinking biology*. Tartu: Tartu University Press.

Favareau, Donald. 2005. Founding a world biosemiotics institution: The International Society for Biosemiotic Studies. *Sign Systems Studies* 33 (2): 481–485.

Favareau, Donald. 2007. The evolutionary history of biosemiotics. In *Introduction to biosemiotics: The new biological synthesis*, ed. Marcello Barbieri, 1–67. Berlin: Springer.

Fetzer, James H. 1997. Thinking and computing: Computers as special kinds of signs. *Minds and Machines* 7 (3): 345–364.

Fitzgerald, John J. 1966. *Peirce's theory of signs as foundation for pragmatism*. Notre Dame, Ind.: Mouton & Co.

Fogle, Thomas. 2000. The dissolution of protein coding genes in molecular biology. In *The concept of the gene in development and evolution: Historical and epistemological perspectives*, ed. Peter J. Beurton, Raphael Falk, and Hans-Jörg Rheinberger, 3–25. Cambridge: Cambridge University Press.

Freadman, Anne. 2004. *The machinery of talk: Charles Peirce and the sign hypothesis*. Stanford: Stanford University Press.

Freeman, Eugene. 1983. *The relevance of Charles Peirce*. La Salle, Ill.: Monist Library of Philosophy.

Gomes, Antônio, Charbel Niño El-Hani, Ricardo Gudwin, and João Queiroz. 2007. Towards the emergence of meaning processes in computers from Peircean semiotics. *Mind & Society—Cognitive Studies in Economics and Social Sciences* 6 (2): 173–187.

Goodridge, Helen S., and Margaret M. Harnett. 2005. Introduction to immune cell signaling. *Parasitology* 130 (S1): S3–S9.

Griffiths, Paul E. 2001. Genetic information: A metaphor in search of a theory. *Philosophy of Science* 68 (3): 394–403.

Hall, Brian K. 2001. The gene is not dead, merely orphaned and seeking a home. *Evolution & Development* 3 (4): 225–228.

Hausman, Carl R. 1993. *Charles S. Peirce's evolutionary philosophy*. Cambridge: Cambridge University Press.

Hoffmeyer, Jesper. 1996. *Signs of meaning in the universe.* Bloomington: Indiana University Press.

Hookway, Christopher. 1985. *Peirce.* London: Routledge & Kegan Paul.

Hookway, Christopher. 2002. *Truth, rationality, and pragmatism: Themes from Peirce.* Oxford: Oxford University Press.

Houser, Nathan. 1997. Introduction: Peirce as logician. In *Studies in the logic of Charles Sanders Peirce*, ed. Nathan Houser, Don D. Roberts, and James Van Evra, 1–22. Bloomington: Indiana University Press.

Hulswit, Menno. 2001. Semeiotic and the cement of the universe: A Peircean process approach to causation. *Transactions of the Charles S. Peirce Society: A Quarterly Journal in American Philosophy* 37 (3): 339–363.

Ideker, Trey, Timothy Galitski, and Leroy Hood. 2001. A new approach to decoding life: Systems biology. *Annual Review of Genomics and Human Genetics* 2 (September): 343–372.

Jablonka, Eva. 2002. Information: Its interpretation, its inheritance, and its sharing. *Philosophy of Science* 69 (4): 578–605.

Jablonka, Eva, and Marion J. Lamb. 2005. *Evolution in four dimensions, genetic, epigenetic, behavioral, and symbolic: Variation in the history of life.* Cambridge, Mass.: MIT Press.

Johansen, Jørgen Dines. 1993. *Dialogic semiosis: An essay on signs and meanings.* Bloomington: Indiana University Press.

Keller, Evelyn Fox. 2000. *The century of the gene.* Cambridge, Mass.: Harvard University Press.

Keller, Evelyn Fox. 2005. The century beyond the gene. *Journal of Biosciences* 30 (1): 3–10.

Kornblihtt, Alberto R., Manuel de la Mata, Juan Pablo Fededa, Manuel J. Muñoz, and Guadalupe Nogués. 2004. Multiple links between transcription and splicing. *RNA* 10 (10): 1489–1498.

Kull, Kalevi. 1999. Biosemiotics in the twentieth century: A view from biology. *Semiotica* 127 (1/4): 385–414.

Kull, Kalevi. 2000. An introduction to phytosemiotics: Semiotic botany and vegetative sign systems. *Sign Systems Studies* 28: 326–350.

Liszka, James Jakob. 1990. Peirce's interpretant. *Transactions of the Charles S. Peirce Society: A Quarterly Journal in American Philosophy* 26 (1): 17–62.

Maniatis, Tom, and Robin Reed. 2002. An extensive network of coupling among gene expression machines. *Nature* 416 (6880): 499–506.

Markoš, Anton. 2002. *Readers of the book of life: Contextualizing developmental evolutionary biology.* Oxford: Oxford University Press.

Marty, Robert, and Alfred Lang. 1997. 76 definitions of the sign by C. S. Peirce, with an appendix of 12 further definitions or equivalents proposed by Alfred

Lang. Available at: http://www.cspeirce.com/menu/library/rsources/76defs/76defs.htm.

Maynard Smith, John. 2000. The concept of information in biology. *Philosophy of Science* 67 (2): 177–194.

Maynard Smith, John, and Eörs Szathmáry. 1999. *The origins of life: From the birth of life to the origins of language.* Oxford: Oxford University Press.

Mayr, Ernst. 1982. *The growth of biological thought: Diversity, evolution, and inheritance.* Cambridge, Mass.: Harvard University Press.

Merrell, Floyd. 1995. *Peirce's semiotics now.* Toronto: Canadian Scholar's Press.

Murphey, Murray G. 1993. *The development of Peirce's philosophy.* Cambridge, Mass.: Harvard University Press.

Neumann-Held, Eva M. 2001. Let's talk about genes: The process molecular gene concept and its context. In *Cycles of contingency: Developmental systems and evolution,* ed. Susan Oyama, Paul E. Griffiths, and Russell D. Gray, 69–84. Cambridge, Mass.: MIT Press.

Nijhout, Herman F. 1990. Problems and paradigms: Metaphors and the role of genes in development. *BioEssays* 12 (9): 441–446.

Nöth, Winfried. 2000. Umberto Eco's semiotic threshold. *Sign Systems Studies* 28:49–61.

Oyama, Susan. 2000. *The ontogeny of information: Developmental systems and evolution.* Durham: Duke University Press.

Pardini, Maria Inês de Moura Campos, and Romeu Cardoso Guimarães. 1992. A systemic concept of the gene. *Genetics and Molecular Biology* 15 (3): 713–721.

Parker, Kelly A. 1998. *The continuity of Peirce's thought.* Nashville: Vanderbilt University Press.

Peirce, Charles S. 1932–1935. *Collected papers of Charles Sanders Peirce,* 1866–1913, 6 vols., ed. Charles Hartshorne and Paul Weiss. Cambridge, Mass.: Harvard University Press.

Peirce, Charles S. 1967. *Annotated catalogue of the papers of Charles Sanders Peirce,* ed. Richard S. Robin. Amherst: University of Massachusetts Press.

Peirce, Charles S. 1977. *Semiotic and significs: The correspondence between Charles S. Peirce and Victoria Lady Welby,* ed. Charles S. Hardwick. Bloomington: Indiana University Press.

Peirce, Charles S. 1998. *The essential Peirce: Selected philosophical writings, 1893–1913,* ed. Peirce Edition Project. Bloomington: Indiana University Press.

Pierce, Susan K. 2002. Lipid rafts and B-cell activation. *Nature Reviews: Immunology* 2 (2): 96–105.

Pitombo, Maiana Albuquerque, Ana Maria Rocha de Almeida, and Charbel Niño El-Hani. 2008. Gene concepts in higher education cell and molecular biology textbooks. *Science Education International* 19 (2): 219–234.

Potter, Vincent G. 1967/1997. *Charles S. Peirce: On norms and ideals.* New York: Fordham University Press.

Queiroz, João, and Charbel Niño El-Hani. 2006a. Semiosis as an emergent process. *Transactions of the Charles S. Peirce Society: A Quarterly Journal in American Philosophy* 42 (1): 78–116.

Queiroz, João, and Charbel Niño El-Hani. 2006b. Towards a multi-level approach to the emergence of meaning processes in living systems. *Acta Biotheoretica* 54 (3): 174–206.

Queiroz, João, and Sidarta Ribeiro. 2002. The biological substrate of icons, indexes, and symbols in animal communication: A neurosemiotic analysis of Vervet monkey alarm-calls. In *The Peirce seminar papers: Essays in semiotic analysis*, vol. 5, ed. Michael Shapiro, 69–78. New York: Berghahn Books.

Queiroz, João, Claus Emmeche, and Charbel Niño El-Hani. 2005. Information and semiosis in living systems: A semiotic approach. *SEED* 5 (1): 60–90.

Ransdell, Joseph. 1977. Some leadings ideas of Peirce's semiotic. *Semiotica* 19 (3/4): 157–178.

Reth, Michael, and Jürgen Wienands. 1997. Initiation and processing of signals from the B cell antigen receptor. *Annual Review of Immunology* 15: 453–479.

Ribeiro, Sidarta, Angelo Loula, Ivan Araújo, Ricardo Gudwin, and João Queiroz. 2007. Symbols are not uniquely human. *BioSystems* 90 (1): 263–272.

Santaella-Braga, Lucia. 1994. Peirce's broad concept of mind. *European Journal for Semiotic Studies* 6 (3–4): 399–411.

Sarkar, Sahotra. 1996. Biological information: A skeptical look at some central dogmas of molecular biology. In *The philosophy and history of molecular biology: New perspectives*, ed. Sahotra Sarkar, 187–231. Dordrecht: Kluwer.

Savan, David. 1988. *An introduction to C. S. Peirce's full system of semiotic.* Toronto: Toronto Semiotic Circle.

Sebeok, Thomas A. 1997. The evolution of semiosis. In *Semiotics: A handbook on the sign-theoretic foundations of nature and culture*, vol. 1, ed. Roland Posner, Klaus Robering, and Thomas A. Sebeok, 436–446. Berlin: Walter de Gruyter.

Sebeok, Thomas A., and Jean Umiker-Sebeok, eds. 1992. *Biosemiotics: The semiotic web 1991.* Berlin: Mouton de Gruyter.

Shannon, Claude E., and Warren Weaver. 1949. *The mathematical theory of communication.* Urbana: University of Illinois Press.

Short, Thomas L. 1996. Interpreting Peirce's interpretant: A response to Lalor, Liszka, and Meyers. *Transactions of the Charles S. Peirce Society: A Quarterly Journal in American Philosophy* 32 (4): 488–541.

Stuart, C. I. J. M. 1985. Bio-informational equivalence. *Journal of Theoretical Biology* 113 (4): 611–636.

Tiercelin, Claudine. 1995. The relevance of Peirce's semiotic for contemporary issues in cognitive science. In *Mind and cognition: Philosophical perspectives on cognitive science and artificial intelligence*, ed. Leila Haaparanta and Sara Heinämaa, 37–74. Cambridge, Mass.: MIT Press.

Weiss, Paul, and Arthur Burks. 1945. Peirce's sixty-six signs. *Journal of Philosophy* 42 (14): 383–388.

Williams, Nigel. 1997. Biologists cut reductionist approach down to size. *Science* 277 (5325): 476–477.

Witzany, Günther, ed. 2007. *Biosemiotics in transdisciplinary contexts: Proceedings of the gathering in biosemiotics 6, Salzburg 2006.* Helsinki: Umweb.

5

Problem Solving in the Life Cycles of Multicellular Organisms: Immunology and Cancer

Niall Shanks and Rebecca A. Pyles

The aim of the present inquiry is to incorporate our current understanding of information processing and problem solving in biological systems into an evolutionary perspective. Because our discussion centers on biological phenomena in multicellular organisms, and those especially relevant to human medicine, we also present perspectives most commonly assigned to the field of Darwinian medicine. We present two detailed case studies concerning: (1) adaptive immunity, whereby the immune system has to process information to solve problems in the face of host-parasite coevolution, and (2) the dynamics of cancer, whereby lineages of cancer cells evolve adaptations in a complex and hostile biotic environment, ultimately finding ways to solve problems in the face of immunological assault. We conclude with a discussion of why the cases discussed here require a critical reevaluation of the typical accounts of the relationship between (functional) evolutionary biology on the one hand, and mechanistic biology on the other.

A significant portion of this essay is dedicated to discussion of important aspects of the mechanisms and function of information processing in the vertebrate immune system. The immune system has been described as a complex, autonomous, distributed information processing system (Orosz 2001). That the immune system is vastly complex needs no further comment here. That the immune system is an autonomous and distributed system means that there is no "central immunological processor" by analogy with the CPU in a standard computer. This, in turn, means that it is a system composed of many parts (e.g., cells and molecules) whose behavior produces and reflects dynamical patterns of interaction among the parts. The study of the complex dynamical relationships among the parts of the immune system has been termed *immunoecology*. The study of the mechanisms by which the immune system recognizes, processes, and reacts to stimuli has been termed *immunoinformatics*

(Orosz 2001, 126). These matters are of interest from both the standpoint of the history of biology and contemporary analyses of immunological function.

The great nineteenth-century biologist Claude Bernard (1865/1957) observed that organisms are complex, internally organized systems that maintain orderly exchanges with the environments in which they are embedded. Consequently, they are constantly impinged upon by external forces capable of disrupting the internal biological order essential for health and life. For Bernard, the issue was one of homeostasis: the achievement and maintenance of a nonequilibrium steady state internal to the organism in the face of external perturbation (Bernard 1865/1957). The achievement and maintenance of homeostasis requires mechanisms, especially regulatory mechanisms, able to generate rapid physiological responses to external stimuli. In short, homeostasis requires physiological problem solving.

Consider, for example, biochemistry. The early history of biochemistry was concerned mostly with the characterization of basic chemical types (lipids, carbohydrates, proteins, and so on), but as the science developed, attention turned to investigations of the mechanisms of chemical dynamics and the characterization of reaction pathways. After the Second World War, dynamical investigations came to focus on how pathways are integrated into circuits, and how circuits, in turn, are integrated into complex networks involving cycles and regulatory mechanisms (positive and negative feedback). Thus, biochemical systems can be characterized as cybernetic systems with multiple inputs, complex processing, and complex distribution of outputs. The description of systems-level biochemical complexity is the concern of *metabolic control analysis* with a focus on the origins and maintenance of nonequilibrium dynamical steady states (Fell 1992). The issue has become biochemical problem solving.

As noted by Wilkins (2002), a similar progression has happened in the study of the genetic basis for developmental pathways in developmental biology. The early focus on genes and their products shifted to an analysis of genetic pathways and their role in the processing of genetic and molecular signals. Pathways can be assembled into circuits, and these, in turn, can be integrated into complex networks (e.g., see Carroll, Grenier, and Weatherbee 2001). The modern understanding of organismal development reflects, not the disconnected behavior of individual genes, but genes organized into networks whose behavior is modulated by complex patterns of inter- and intracellular signaling.

In the study of immunology, we find the familiar pattern in which the early interest in the characterization of immunological players is now supplemented with the characterization of the dynamical interactions among the players. In the immune system, interactions occur by means of recognizing, processing, and responded to signals, ultimately producing a highly networked system capable of generating and maintaining immunological homeostasis (Segel 2001). Invasion by pathogens from without generates threats to homeostasis, as do perturbations from within. Examples of the latter include perturbations arising from *self*-reactive cells in the context of autoimmunity, as well as perturbations from cells that violate the internal physiological *social contract*, such as cancer cells. The immune system can be depicted as an informational problem-solving system evolved to cope with perturbations—those arising from without, as well as from within, the multicellular individual.

Theorists have drawn analogies between the immune system and systems composed of insects, for example, collectives of ants or wasps. The interest here has been focused primarily on analogies with respect to organization, adaptive problem solving, signaling, and communication in the context of functional collective behavior (e.g., Mitchell 2001; Gordon 2001; Bonabeau 2001; Segel 2001). Orosz (2001) has explicitly compared the ability of colonies of ants to perform complicated tasks to that of the immune system's "leukocyte swarms." Indeed, as with the functionally specialized castes in social insect colonies, so too are there functionally specialized cell types in the leukocyte swarm. How might these analogies be useful? There are at least three useful points of comparison:

1. *Distributed autonomous problem solving* This phenomenon refers to the ability of collectives of insects to solve problems without centralized control or directives originating either external or internal to the collective. For example, wasps construct nest structures of a complexity and relative scale roughly equivalent to that of the human pyramids. What is it that governs the construction operations? What creates blueprints, regulates the behavior of the agents engaged in construction, and sees ahead with an eye to the completion of the program? Where are the analogs of the human architects, engineers, and foremen? Nest construction in wasps is not guided by any "intelligence" internal or external to the insect collective; there is no single "smart wasp" or even a collective of such with a "group mind." The individuals that are constitutive of

the collective are, in essence, dumb agents operating and interacting in accord with simple rules. The appearance of intelligent guidance, or even teleology, is illusory. Instead, individual wasps engaged in nest construction respond to chemical stimuli provided by the emerging nest structure itself—work most recently completed, rather than direct communication among the builders, directs future construction and construction behaviors, namely, "work in progress" (Shanks and Karsai 2004). In the study of social insects, this phenomenon is known as *stigmergy*. In our discussion of adaptive immunity, we will present examples of *immunological stigmergy* in which contingent environmental cues direct complex behaviors of otherwise dumb leukocyte swarms.

2. *Adaptive problem solving* Analogies based on ant foraging behavior have also been proposed to illuminate features of problem solving that are manifested during the process of adaptive immunity (Gerhart and Kirschner 1997). Some ant species employ foraging mechanisms based on behavioral variation and selection, along with chemical signaling, to improve the efficiency of foraging behaviors. Ants in search of food explore the environment in a random manner, leaving behind volatile pheromone trails. If an ant comes across a food source, it follows its own trail back to the colony, reinforcing the pheromone trail as it goes. As other ants leave the colony, they respond to the strength of signal of the successful trail, repeating the process and continually reinforcing the signal of this trajectory (Gerhart and Kirschner 1997). This analogy will be useful in our description of specialized leukocyte swarms that exploit variation-and-selective reinforcement mechanisms which also involve chemical (and molecular) signal recognition. The details are, of course, very different.

3. *Behavioral regulation* In some social insect societies, conflicts of interest arise from a tendency of some individuals to reproduce at the expense of the colony. Uncontrolled worker reproduction can upend the balance in the division of labor between workers and the queen. To prevent such exploitation, some workers kill eggs laid by other workers —a phenomenon termed "worker policing" by Ratnieks and Wenseleers (2005, 54). By analogy, it has been suggested that specialized "castes" of cells in the immune system perform a similar "police role" (Wickelgren 2004, 596). Although talk of *policing* strikes us as unhelpfully anthropomorphic in this context, a growing body of recent research suggests that there are indeed specialized cells in the immune system—namely, regulator T cells—whose dynamics *regulate* the behavior of the immune system, especially in the context of phenomena related to auto-

immunity and cancer (Sakaguchi 2004; Sakaguchi and Powrie 2007; Knutson, Disis, and Salazar 2007).

The focus of our discussion is not on analogies with insect systems; at some point, all analogies break down. Rather, in ways derived from these analogies, we are concerned with features of the immune system and of the phenomenon of cancer that are relevant to the recognition and processing of biological information and to adaptive problem solving. However, the most general context of our discussions of these diverse biological dynamic systems must be, by definition, evolutionary.

Darwin Reexamined

Charles Darwin's great achievement was to explain how populations of organisms over successive generations could, in effect, solve problems by confronting them in their environments. The resulting solutions are known as *adaptations*. The classical Darwinian description of problem solving rests on three basic ideas: (1) populations of organisms show variation with respect to certain inherited characteristics of their members; (2) individuals in such populations differ in rates of survival and reproduction by virtue of their characteristics, thereby manifesting differential reproductive success; and (3) the heritable characteristics that contribute to differential reproductive success will often be inherited by the progeny of successful individuals. Over successive generations, all other things being equal, those characteristics will manifest themselves as *adaptations*. The identification of an adaptation is stringent: a characteristic qualifies only if a case can made that it conferred a fitness advantage historically, and thus was selected for (Sober 2000). Because the inheritance of characteristics is involved in the selection of adaptations, problem-solving adaptive evolution requires the generation, preservation, and transmission of specific kinds of molecular information.

This application of the Darwinian explanation emphasizes the mechanism of biological problem solving in populations of organisms over time and many generations, an application we call *phylogenetic Darwinism*. In this discussion, we consider how classical Darwinian principles also may be extended to explain biological problem solving as resulting from the generation, preservation, and transmission of specific kinds of molecular information that occur within populations of somatic cells during the life-cycles of individual multicellular organisms—an application termed *ontogenetic Darwinism* (Shanks 2004).

Consider the *biological hierarchy of organization* as the thesis that describes the constitution and organization of the living world. The top of the hierarchy is the *biosphere* itself, considered to be composed of modules in descending orders of magnitude, such as biomes, communities, populations, individuals, organs, tissues, cells, intracellular structures, and macromolecules. *Phylogenetic Darwinism* plays a role in biological explanations at all of these levels, but most commonly has been excluded from consideration of the internal somatic dynamics of individual organisms.

By contrast, ontogenetic Darwinism provides explanations of biological problem solving that can occur in populations *of cells* within the life cycle of an individual multicellular organism. That is, it gives an account of *somatic adaptation within an individual* (i.e., adaptations that are inherited by cells or subunits within an individual, rather than inherited by the offspring of the individual organism in question). We apply these explanations to phenomena that depend on changes capable of producing *adaptations* in somatic cell lineages within an individual (multicellular) organism, and give rise to what often are considered medically significant phenomena. The emphasis here on somatic cells, then, is somewhat different from the *selfish gene hypothesis* (Dawkins 1989), in which germ-line genes (representing a portion of the molecular constitution of germ-line cells) were considered to be units of selection in an account of phylogenetic Darwinism.

We begin with a discussion of evolved somatic adaptations in the immune system, for as Parham has observed:

At some point [in] this century the experimental biologists, in an echo of Henry Ford, divorced themselves intellectually from the evolutionary biologists. This artificial and regrettable separation remains with us today. For the immunologists it was always a sham, for the very foundations of their subject are built upon stimulation, selection and adaptive change. Now we see clearly the immune system for what it is, a vast laboratory of high speed evolution. By recombination, mutation, insertion and deletion, gene fragments are packaged by lymphocytes, forming populations of receptors that compete to grab hold of antigen. Those that succeed get to reproduce and their progeny, if antibodies, submit to further rounds of selection. . . . There is no going back and the destiny of each and every immune system is to become unique, the product of its encounters with antigen and the order in which they happen. This all happens in somatic tissues, in a time frame of weeks and is perhaps too vulgar, too fast, for traditional tastes to be even called evolution. (1994, 373)

Thus, each and every unique immune system conveys information about how antigenic (molecular) information, for example, was recognized, processed, and responded to. The immune system does indeed operate on evolutionary principles, but, as usual, the devil is in the details, and we must examine some mechanistic issues to make the point clearly.

The Basic Fabric of Immunity

Our understanding of the immune system is a complex fabric of ideas and concepts. Here, we propose to unravel a few strands of that fabric so that the nature of adaptive immunity may be revealed. The reader also is directed to valuable discussions found in Parham 2000, Hofmeyr 2001, Frank 2002, and Yates and Lyczak 2004.

The immune system consists of a mosaic of molecules, cells, tissues, and organs that function codependently to provide protection from infectious diseases. Although organized tissues and organs, such as the skin, thymus, and bone marrow, are centrally involved in the immune system, the principal players are the cells that carry out responses to infection. It is these cells, resident in blood, spleen, and lymphatic tissues, that are ready to disperse and respond to infection anywhere in the body. The organization required to direct the dispersal and response of these cells depends on an intricate network of communication—information in the form of chemical signals passed internally as well as externally between and among cells. Indeed, these signals pass not only laterally between existing cells, but also vertically across generations of particular cell populations.

The broad evolutionary context in which the immune system operates, then, is that of host-parasite/pathogen coevolution. The immune system provides multicellular organisms with a means to respond to invasion by parasites and pathogens, and, in turn, these invaders evolve traits that enable them to evade the immunological surveillance of the host.

Pathogens include bacteria, viruses, fungi, and internal parasites (the latter includes unicellular protozoa and multicellular invertebrates, such as many types of worms). With short generation times (e.g., some bacteria reproduce every twenty minutes), some pathogenic microorganisms have enormous potential for rapid evolution. High mutation rates observed in some pathogens (e.g., influenza and HIV viruses) and in some parasites (e.g., malarial parasites whose genes coding for binding proteins mutate at the rate of 2 percent per generation [Nesse and Williams 1996]) contribute to the possibility of rapid evolution when coupled with the effects of natural selection. Numerous examples of evolved pathogen and parasite strategies to escape immunological surveillance can be found in Nesse and Williams 1996. The immune system represents an evolutionary response on the part of hosts to these rapidly changing pathological and parasitic targets.

Multicellular organisms, such as humans, have two basic types of immune response: the *innate immune response* and the *adaptive immune response*. The innate immune response consists of a range of nonspecific mechanisms that can either prevent host invasion in the first place, or act to reduce or eliminate it once host invasion has occurred.

General mechanisms of innate immunity include anatomical barriers (skin, mucous membrane linings, and mucus) and physiological variables such as acidity and temperature (there is evidence that fever, e.g., is an adaptive response to infection [Nesse and Williams 1996]). Additionally, innate immunity also includes two reactive response systems, that is, a scavenger response (*phagocytic*) and a chemical response (*complement*). Phagocytic responses involve roaming "scavenger" cells, such as *macrophages*, that detect and engulf extracellular molecules and materials, clearing the system of both debris and pathogens. Complement responses include the production of enzymes inducing bacterial lysis, of chemical defenses such as superoxide and nitric oxide, and of antimicrobial peptides such as defensins (Yates and Lyczak 2004). Both the phagocyctic and complement system are activated by signals from foreign pathogens or parasites that "label" them as foreign. In turn, the actions of the innate immune system serve as signals that unleash the power of the entire immune response. However, the innate immune system is determined at birth and does not change during the life cycle of the multicellular organism. Although pivotal, the innate immune response system does not adapt to specific pathogens or parasites, unlike the adaptive immune system.

In the following descriptions, we will be concerned primarily with the adaptive immune response. Accounts of this form of immune response were pioneered in the 1950s by Niels Jerne and Sir Frank Macfarlane Burnet, but our understanding has expanded enormously since that time. Not only have many of the molecular mechanisms of adaptive immunity been elucidated, but rigorous mathematical models now exist to describe the dynamics of the immune system (e.g., Nowak and May 2000). Most multicellular animals have types of cells that respond to signals of invaders. To date, however, true *adaptive* immunity appears to be found only in jawed vertebrates. There is some tantalizing evidence that certain genes (e.g., *Rag1* and *Rag2*) that play a role in antibody composition and recombination within jawed vertebrates are also found in some invertebrates. However, the function of these genes within invertebrates currently is uncertain (Travis 2006).

In the course of the adaptive immune response within an individual, portions and products of the immune system adapt to target specific invaders. The cells primarily responsible for the adaptive immune response are the many specialized types of leukocytes (white blood cells) and their associated tissues. Invaders are identified as such because their constitutive proteins and/or chemicals they release are foreign (non-self). The invaders' chemistry is, in effect, treated as information in the form of signals that are processed and identified by the immune system. Once identified as foreign, these signals are termed *antigens*. Responses to antigens are initiated by both the innate and the adaptive immune systems, but here we are most concerned with the players in the adaptive response. Of primary importance are the information processing and resultant actions of populations of specialized leukocytes, the T and B lymphocytes (hereafter referred to simply as T cells and B cells). These cell types are each represented by subspecialized populations with diverse roles. In general, types of T cells drive the initial and ongoing adaptive immune response, and B cells produce specific molecules called *antibodies*. Antibodies are immune information signals that bind to antigen and either mark the invader for elimination or obstruct the invader by blocking access to healthy cells.

A good account of adaptive immunity, in the Darwinian sense, must include mechanisms that generate heritable variation and that provide for selection via differential rates of survival and/or reproduction. Therefore, to consider a Darwinian description of adaptive immune, the following five characteristics must be explained: (1) increasing rates of antibody production in early stages of infection; (2) an immunity "memory" that enables fast response to subsequent infections by a given antigen; (3) the persistence of antibodies in a given infectious agent long after that specific antigen has been eliminated; (4) the fact that antibodies produced in the later stages of a given infection are more efficient at binding than those initially produced; and (5) that the immune system is generally good at recognizing the distinction between *self* and *nonself*. It is appropriate here to note that errors by the immune system with respect to this distinction may result in autoimmune disorders (see Cziko 1995). But before attempting a more detailed Darwinian presentation of adaptive immunity, we must first describe the players in the system, their actions, and their changes over time.

The Fabric of Adaptive Immunity

For this discussion, our focus rests with our major cellular players: the B cells, which are derived from precursor cells residing in bone marrow, and the T cells, which also are derived from bone marrow cells but undergo changes through maturation in the thymus gland. Both of these lymphocyte types are represented by subspecialized populations that perform diverse tasks and functions in the immune response. For example, B cells can respond directly to particular antigens, can produce a specific antibody, or can serve as the organism's "memory" of previous invasions. The activation of these diverse cell types and their successful function depends on an extensive information network that provides for the processing of invader signals, as well as the production of new allied information that is passed among the cellular players. For example, antibody production from B cells cannot occur without *activation* by a specialized population of T cells, called *T helper cells*. T helper cells themselves are identified by receiver signals in the form of particular proteins (CD4 proteins) that exist on their cell membranes and serve as receptors of incoming data from other cells.

To understand the diverse functions of T cells, it is first necessary to consider the information provided by the *major histocompatibility complex* (MHC), which is present in two different types. *Class I MHC* molecules are produced by most cells in the body and represent an internal label that enables an organism to initially identify *self*. *Class II MHC* molecules are used as label for nonself, that is, whole or parts of foreign invading cells, viruses, particles, or even abnormally changed native cells. The presence of *Class II MHC* molecules, when bound to such antigens, act as a signal to other immune cells. Antigen-processing cells include the macrophages and specific B cells that may recognize the specific antigen of the invader. Macrophage cells engulf and "digest" invading cells and particles. The digested parts—those constitutive proteins from the invader (now antigen fragments)—are then bound to molecules of the *Class II MHC* group.

This MHCII+antigen complex is then transported to the surface of the processing cell and is thus "presented" to the specialized members of the leukocyte swarm as information. T cells recognize these combinations of MHCII+antigen by binding to them, and the act of binding stimulates each T cell to reproduce, producing more cells with copies of the antigen-specific signal. These signals can activate any B cells that possess a signal receiver (receptor) that reacts to the specific antigen signal presented by the T cell. The activation of B cells also results in the production of an internal signal that induces reproduction, thus producing more of the antibody that reacted to the particular antigen. In this way, the later stages of an immune response reflect "work in progress" that manifests a form of *immunological stigmergy*. Interestingly, HIV is a particularly insidious virus, because it kills helper T cells, short-circuiting the signal network and gradually leaving the victim ever more open to opportunistic infection.

The antibody molecules that recognize antigens directly are known as *immunoglobulins*, and these are carried on the surface of B cells or are secreted by them to circulate freely in the body. Antibodies can bind to antigens that are simply the exposed parts of a whole invader. T cells have recognition molecules carried on their surfaces (*receptors*) that enable a T cell to bind with chemical signals from its environment, including the MHCII+antigen complex. Note, then, that T cells are limited to recognizing nonself only as antigen fragments (i.e., "processed") in the company of MHCII.

For T cells, a variety of specific receptor types are present and are indicative of the specialized function of that particular T cell—such as CD4 receptors present on T helper cells. These CD4 or helper T cells respond to peptide antigens presented by MHCII+antigen complexes by secreting protein signals that activate other types of immune system cells. Cytotoxic ("killer") T cells express CD8 glycoprotein and, once activated by binding, act to produce cytotoxins that kill invading cells. Another population of CD8 T cells, the effector T cells, kill any self cells that are infected with viruses or other intracellular pathogens. In order to identify self, these T cells recognize cells that present MHCI molecules. This action of effector T cells prevents the replication of the pathogen that may occur within self cells and reduces the spread of infection. Thus, the receptors on T cells act as signal receivers, responding to complex signals derived from multiple sources. Once received, the information is processed by corresponding internal signals that activate the cell to perform specialized functions, again, primarily in the form of the production of new signals. This complex of processing and reaction to shared information—producing actions and interactions of networks of immune cells to fight invaders and infections—is often called *cellular* or *cell-mediated immunity* to distinguish it from the actions of antibodies (Parham 2000).

Ontogenetic Darwinism and Adaptive Immunity

Adaptive immunity thus refers to the property of the vertebrate immune system to tailor a specific immune response to a specific antigen. The process involved can best be understood in terms of Darwin's principles of adaptive evolution as applied to the populations of lymphocytes. The adaptations in question are abilities of those lymphocytes to bind to antigen, whether in its natural state (B cells) or to complexes of antigen fragments bound to MHC molecules (T cells). Unlike explanations in phylogenetic Darwinism, these adaptations generated by selective forces occur within the body of an individual multicellular organism and are tailored on a time frame of hours, days, or weeks.

What particular adaptations occur in the adaptive immune response? One major problem confronting the vertebrate immune system is the enormous diversity of potential foreign (antigen) molecules. How can such a system respond effectively in the face of such diversity? The answer lies in the production of populations of B cells that produce antibodies that match the diversity of possible molecular shapes of antigen. The human immune system, as a whole, is capable of recognizing well in excess of one hundred million different antigen shapes. Because genes are required to make specific antibody molecules and to make T cell receptor molecules, the expression of such enormous diversity requires a corresponding foundation of genetic diversity. In point of fact, there is not enough information in the inherited genome (containing some 25,000 genes) to code directly for this expressed diversity. How, then, does it arise?

Let us focus on antibody molecules or immunoglobulins. These molecules display a Y-shaped structure. The upper tips form binding sites for antigen and vary tremendously in composition from one antibody to the next, and the tail serves as a link to interact with other cells and tissues in the multicellular organism. The chemical structure of the tail is the same for all antibodies in a particular immunoglobulin class (e.g., IgG, IgA, IgM, IgD, and IgE in humans). The actual Y-shape of each antibody molecule is formed by four polypeptide molecules—two identical light chains and two identical heavy chains. Both light and heavy chains exhibit variable and constant regions. The variable regions of these chains constitute the antigen-binding sites and the constant regions relate to the immunoglobulin class.

Antibodies are produced by the immunoglobulin genes of the B cells, which code for the light and heavy polypeptide chains. The polypeptide chains are actually coded by information from different genes that is then spliced together—a mix-and-match approach. Light chains are produced by combinations of variable (V) and junction (J) regions, which are then combined with a constant (C) region. Heavy chains are likewise spliced, with information from V, J, and D (diversity) regions combined and then spliced to a C region. The fact that there are many different codes contained in the V, J, and D regions multiplies (by magnitudes) the final combinations possible in the antibody produced.

In humans, there are approximately one hundred genes that code for the V region, twelve D-coding genes, and four J-coding genes (Frank 2002); the mixing results in 4,800 different V-D-J combinations. But a separate event produces the V-J combination for the light chain, with 400 different combinations possible (100 V × 4 J). Thus, 1,920,000 (4,800 × 400) different antibody constructions are created by the independent formation of heavy and light chains. This mix-and-match process is called *somatic recombination*. Additional variety also is produced by the following: (1) randomly chosen DNA bases are added between segments brought together during somatic recombination, and (2) somatic recombination events are subject to mutation or other changes by insertions and deletions of gene fragments. So, although the genes that code for antibodies are inherited in typical Mendelian fashion during mitotic cell reproduction, the antibody finally configured varies tremendously within the population of B cells and among B cells *over time*. The typical human has approximately 10^{12} active lymphocytes. This population size, together with the recombinant variation possible, provides the potential for an enormously diverse population of antibodies. *As required in a Darwinian explanation of adaptive immunity, we have a (highly) variable population of cells.*

With a highly diverse population of B cells, there is a good chance that at least one will bear an antibody molecule that can bind to antigen presented by a specific infectious agent. How quickly or tightly the antibody binds to antigen is a reflection of its specificity or "fit." When an antibody molecule binds to antigen and is recognized by a helper T cell, the T cell sends a signal that activates the particular B cell, stimulating it to reproduce. Reproduction is by simple mitotic division so that the stimulated "parent" B cell essentially makes clones of itself. By this selective process, cells that possess that specific antibody are greatly increased in frequency within the total population of B cells in the individual organism. B cells with antibodies effective for a given antigen present in the body reproduce at greater rates compared to those without bound antibody. This selection process also results in increased rates of specific antibody production in early stages of infection. Thus, the descendents of successful B cells inherit the properties (molecular adaptations) that were possessed by the successful ancestor (i.e., the parent B cell). *As required for a Darwinian explanation of adaptive immunity, differential reproductive success of heritable variation provides an explanation for increased antibody production early in a particular infection.*

Moreover, some of these same descendant B cells will persist in the organism long after the current invader has been defeated. These cells bear information concerning the earlier encounter with antigen. Thus, these persistent B cells function as the memory of the immune system, priming it for a fast response to the similar antigens presented upon reinfection with the same pathogenic or parasitic agent. The longevity of a particular memory B cell is not known, but the persistence of the antibody in the organism suggests that a "parental" memory B cells gives rise to its own lineage of daughter cells, which ensure the presence of each successful antibody over time. *As required for a Darwinian explanation, differential reproductive success of heritable variation provides an explanation for the persistence of specific antibodies and a "memory" of immunity that enables fast response to subsequent infections by a given antigen.*

Thus far, we have described three of the five features of the adaptive immune response identified in the previous section as necessary for a Darwinian explanation. We now see how antibody production rises in the early stages of infection, how the immune system has a memory, and why appropriate antibodies persist long after a given encounter with antigen is over.

What of the fourth characteristic, that antibodies produced later in the course of a given infection are better at binding to antigen than those initially produced? This, too, relates to the huge diversity of antibodies, but in a slightly different way. Although tremendously variable, most antibodies present in the body won't be a perfect fit for a particular antigen. In all probability, an antibody's "affinity" to a new antigen will be low; although binding will occur, the fit won't be perfect and binding won't occur quickly. At the beginning of an infection, antibodies probably have low "affinity thresholds" to a particular antigen. Affinity maturation is a time-dependent process that results from differential (very high) mutation rates in the B cells that have been selected to undergo reproduction. This mutation process, called *somatic hypermutation*, represents the high rate of nucleotide substitutions in the immunoglobulin heavy- and light-chain genes within the activated and dividing B cells (Parham 2000).

Somatic hypermutation results in daughter cells with slight variations in immunoglobulin genes. Antibodies produced by these variant daughter cells may bind to pathogen more tightly and more quickly than earlier antibodies. Over time periods of hours and days, B cells with high affinity will effectively compete for antigen present. B cells that produce better fit antibodies are then preferentially signaled to reproduce more quickly, increasing the changes that the best-fit antibodies will be produced in greater quantity. *In this Darwinian explanation, somatic hypermutation provides heritable variation at an especially rapid pace, and differential reproduction results in highly "fit" antibodies and competition for antigen, which in turn also reduces pathogen load in the individual multicellular organism.*

As referenced earlier, some of the B cells with a high affinity for a given antigen will be targeted as "memory" B cells. Although maintained at low population levels, the lineages of memory B cells with highly specific antibodies ensure that, if reinfection by the same pathogen occurs, a very specific and rapid secondary response is "armed" and ready to go. In fact, this secondary response in humans may often occur without their having any knowledge that they have been reinfected.

And, finally, what of the fifth feature that requires an evolutionary explanation of the adaptive immune response? How do evolutionary mechanisms relate to the fact that the immune system is generally good at recognizing the distinction between self and nonself? We left this discussion for last in this section, as the mechanisms involved are rather different from differential reproduction. Here we find that differential survivorship is the key.

The common theme in our descriptions of adaptive immune response has focused on a diversity of mechanisms that produce variation, whether in antibodies or in receptors. In generating this diversity, some cells inevitably will be produced with receptors that are self-reactive, that is, they have an affinity for proteins of native cells within the individual's own body. Without mechanisms to eliminate or inactivate self-reactive lymphocytes, the immune system attacks self, resulting in autoimmune disorders. No system is perfect—as hay-fever and asthma sufferers will attest. Some autoimmune disorders can even be triggered by infection; for example, infection by *Streptococcus pyogenes* ("strep throat" bacteria) can induce the production of antibodies that cross-react with self, resulting in sometimes catastrophic cases of rheumatic fever.

When discussing how the immune system creates tolerance of self, immunologists distinguish between mechanisms of central tolerance and peripheral tolerance. Central tolerance mechanisms are confined to central maturation locations: bone marrow for B cells and the thymus gland for T cells. Maturation of T cells in the thymus gland exposes these cells to almost all self proteins of the individual. If a maturing T cell should bind with a self protein, it is marked for elimination. Signals produced by self-reactive cells generate internal signals that induce programmed cell death (*apoptosis*) of the cell. A similar exposure process occurs in bone marrow for maturing B cells. This process of programmed cell death of self-reactive cells, typically called *clonal deletion*, is well-documented but may not be the only mechanism of central tolerance. Maturing T cells that bind appropriately to MHC complex may produce signals that protect cells from signals that would induce apoptosis (Miller 2001). The mechanism of apoptosis is relatively common in vertebrate organisms, and it plays an important role in many developmental phenomena by selectively reducing an initial overabundance of cells (e.g., the sculpting of fingers and toes). The mechanism of programmed cell death, whether in immune response or development, is a form of Darwinian cull at work on cellular populations.

However, central tolerance is not enough (Pier et al. 2004). Inevitably, some self-reactive lymphocytes will escape. In addition, other processes—such as somatic hypermutation in B cells—inevitably produce variants that may react to self. Peripheral tolerance represents mechanisms that occur throughout the body to suppress or eliminate self-reactive cells. T helper cells play an important role in peripheral tolerance via the mechanism known as *costimulation*. As described previously, the complete activation of a B cell to reproduce requires two signals: one derived from the bond with a particular antigen and the second provided by a T helper cell. A B cell bound to a self-protein will not be marked by the appropriate MHC, will not receive costimulation from a T cell, and will die (Hofmeyr 2001). *Thus, as required for our Darwinian explanation, differential survivorship of heritable variation—in this instance, reduced survival of intolerant variation—provides a mechanism for the ability of the adaptive immune response to distinguish self from nonself.*

But the story does not end here. Tolerance of *self* in the adaptive immune response also involves dynamics (e.g., Pier et al. 2004) that extend the application of ontogenetic Darwinian into the arena of ecology, writ small. As noted by Gallagher (1997) over a decade ago, principles derived from population biology may aid our understanding of the mechanisms that provide for homeostatic control of lymphocyte numbers, life span, and population dynamics. If we consider the interactions of lymphocytes with their environment and among their various populations, are we not indeed seeking an ecological perspective (albeit one internal to an individual)? The application of evolutionary mechanisms that contribute to explanations of cooperation and cooperative dynamics of organisms (e.g., predator-prey interactions) may likely add explanatory power to understanding the interactive dynamics among populations of cells that constitute those same organisms. The use of ecological explanations, appropriately recast using the language and concepts of immunology, should further understanding of the cellular and molecular levels of organization that occur within the individual multicellular organism.

Some of the details of such an account have gradually become clearer, especially with respect to the generation and maintenance of immunological homeostasis. Part of the solution to the problem of maintaining immunological homeostasis lies in the actions of a special class or "caste" of T cells in the leukocyte swarm. CD4+ Regulatory T cells (T_{reg} cells) play an active regulatory role by inhibiting actions of *self*-reactive cells. Genetic errors or environmental insults that upset the balance between self-reactive T cells and T_{reg} cells in favor of the former can cause a variety of autoimmune diseases (Sakaguchi 2004). Indeed, T_{reg} cells appear to be key players in the maintenance of immunological homeostasis. Accidental by-products of the normal antigen that attack the T cell system are self-reactive T cells, which, when left to their own actions, will significantly disrupt immunological homeostasis. The actions of T_{reg} cells to down-regulate self-reactive T cells thus serve as an important regulatory mechanism. But T_{reg} cells themselves have the potential to disrupt homeostatic balance in the other direction by overregulation of T cell populations, thereby inhibiting effective immune responses to invaders from without. Thus, the study of T_{reg} cells has become essential not only to those interested in infectious disease, but also to immunologists interested in transplantation and to oncologists concerned with rogue cancer cells from within (Sakaguchi and Powrie 2007).

With this in mind, we will proceed with another case study that serves our main argument: that evolutionary mechanisms contribute to an understanding of the ontogenetic progressions of an individual multicellular organism. Our previous descriptions of the basic biological properties and informational behavior of the immune system will serve as a foundation to investigate the dynamics of cancer, whereby lineages of cancer cells either evolve Darwinian adaptations or are eliminated by natural selection of their environment.

Cancer: Diseases with Darwinian Dynamics

Many of the themes of ontogenetic Darwinian we have discussed in relation to the immune system recur in the context of cancer. As British oncologist Mel Greaves has pointed out:

> The analogy between the development of cancer clones and the evolution of new species has been made many times in the past. However, the new molecular discoveries in cancer research have greatly reinforced the validity and value of the comparison at the level of cells and gene, to the extent that we can now see that cancer doesn't just parody evolution, it is a form of evolution played by the same Darwinian ground rules as apply to evolution in general and particularly for asexually propagating species. The essential game plan is progressive genetic diversification by mutation within a clone, coupled with selection of individual cells on the basis of reproductive and survival fitness, endorsed by their particular mutant gene set. It's evolution on the fast track. (2000, 39)

In this section, we explore these ideas. Before proceeding further, however, it will be helpful to say something about the nature of cancer, viewed as a class of diseases with a common evolutionary dynamic.

Cancers are diseases that involve disorderly cell propagation often accompanied by subsequent destruction of healthy tissue. Cancers are diseases that violate the multicellular "social" contract, as observed by Alberts et al.:

> The body of an animal operates as a society or ecosystem whose individual members are cells, reproducing by cell division and organized into collaborative assemblies or tissues. In our earlier discussion of the maintenance of tissues, our interests were similar to those of the ecologist: cell births, deaths, habitats, territorial limitations, and the maintenance of population sizes. The one ecological topic conspicuously absent was that of natural selection: we said nothing of competition or mutation among somatic cells. The reason is that a healthy body is in this respect a very peculiar society, where self-sacrifice—as opposed to survival of the fittest—is the rule. (2002, 1313)

The authors go on to point out that somatic cells are doomed to die with the organism they collectively constitute. Almost all cells, both somatic and germ cells (egg and sperm), carry copies of the same genes. If the multicellular individual reproduces successfully, germ cell genes identical to genes found in somatic cells will travel down the generations. Somatic cells, in effect, cooperate in order to propagate germ-line cells.

Cooperation may be disrupted when somatic cells acquire (during their lifetime) a suite of mutations that permit selfish propagation at the expense of other cells constitutive of the multicellular collective. A human body consists of more than 10^{14} cells. Billions of cells undergo mutations every day. Cancer, however, is not inevitable. As noted by Nesse and Williams (1996), DNA editing and repair mechanisms correct genetic flaws and tumor-suppressor genes inhibit inappropriate cell division. For example, activation of the gene *p53* in a particular cell inhibits cell division and can even initiate apoptosis (programmed cell death). Mutations in gene *p53* have been observed with respect to fifty-one types of human cancers, including 70 percent of colorectal cancer cases, 50 percent of lung cancer cases, and 40 percent of breast cancer cases (Nesse and Williams 1996).

In addition to intracellular mechanisms that correct for mutations, there also are intercellular mechanisms that act to guard against the selfish propagation of cells. Cells in the neighborhood of a deviant cell can emit chemical signals that inhibit cell division. The immune system, too, guards the neighborhood. When cancer cells accumulate sufficient changes to appear as nonself, the immune system can mount an attack on the growing problem. There are multiple layers of protection to control the selfish behavior of deviant somatic cells.

So, for cancer to develop, a somatic cell must accumulate a highly improbable suite of molecular changes that enable it to evade all of these mechanisms—it must acquire, to use Greaves's gambling metaphor, a "full house" of genetic changes. For example, the initiation of colorectal cancer requires mutation events that inactivate a particular tumor suppressor gene pathway (*adenomatous polyposis coli* or *APC* pathway; Nowak 2006). The folded region of the colon wall in which the *APC* mutant cell first appears becomes dysplastic and, as the mutant cell reproduces, the accumulation of abnormal cells creates a polyp. However, malignancy of these polyp cells requires subsequent mutations that activate oncogenes (e.g., RAS or BRAF) and inactivate tumor-suppressor genes (e.g., *p53*). Of polyps that appear, approximately 10 to 20 percent progress to full malignant colorectal cancer (Nowak 2006). With the "full house" of changes, the result is invasive carcinoma.

Once a cell acquires the "full house" of mutations, it can divide at the expense of its neighbors. As with any reproductive cycle, DNA replication is not entirely faithful, and as a consequence, reproduction of the deviant cell gives rise to a population of descendent cells that display genetic variation. Some of these variants will be able to divide more successfully than others, and their descendents will inherit the reproductive advantage. Over time, the outcome of repeated rounds of mutation, competition, and natural selection cause matters to go from bad to worse (Alberts et al. 2002). Ultimately, cellular adaptations emerge that enable cancer cells to propagate to other tissues and organs of the multicellular organism. In the end, the success of the diseased mutant clones ultimately destroys the entire organism. Cancer, as a disease with a Darwinian dynamic, operates with no eye to the future.

Though basically accurate, this account ignores informational aspects of the ontogeny of the disease. Cancer cells proliferate and prosper only by using chemical signals to subvert normal physiological processes. In effect, these signals turn crucial features of the body's internal ecology against itself. But again, principles of ecology come into play. Once a colony of cancer cells becomes a clump about 1 mm in size, it confronts a resource acquisition problem. To expand the colony further, a system is needed to provide nutrients and oxygen and to remove wastes. At this point, then, colony size is limited by anoxia and metabolic poisoning. As observed by Weinberg (1999), unless something changes, the tumor cells cannot break out of a futile cycle of mitosis followed by starvation and asphyxiation.

To expand, the colony needs a system of blood flow, a capillary system similar to that enjoyed by adjacent normal cells. Successful colonies will be those in which some of the (variable) cancer cells acquire the ability to emit chemical signals that induce angiogenesis. Angiogenesis, the process leading to development of capillaries, is a normal function essential for organismal development and wound healing, for example. The normal signaling network, involving VEGF (vascular endothelial growth factor) and bFGF (basic fibroblast growth factor), is recruited to serve the colony of cancer cells (Weinberg 1999). To the extent that the colony is successful in this campaign of chemical misinformation, it can expand and invade surrounding tissue.

This, however, is not the only place where cancer colonies exploit chemical misinformation. As noted previously, another line of defense against cancer is the immune system, and here, again, cancer colonies exploit normal chemical signaling networks. We now reconsider the role of T_{reg} cells, introduced at the end of the last section, as both players in and potential disruptors of the maintenance of immunological homeostasis.

So how do tumors evade a successful immune response? One way in which the immune system prevents tumor growth is through immunosurveillance by special types of self-reactive T cells called *tumor antigen-specific T cells*. As a tumor grows, it subverts successful immunosurveillance by emitting chemical signals that have the effect of increasing numbers of T_{reg} cells in the lymphatic and peripheral circulation. Because T_{reg} cells down-regulate numbers of self-reactive T cells, the tumor-induced population increase in T_{reg} cells will likely reduce numbers of tumor antigen-specific T cells (Knutson, Disis, and Salazar 2007), thereby reducing effective immunosurveillance. To make matters worse, tumors can exploit normal chemical signaling pathways to recruit or induce T_{reg} cells into the microenvironment of the tumor itself. Once there, T_{reg} cells may act to block actions of T cells that penetrate the tumor, effectively reducing normal immune response pathways. Details of the dynamical actions of and reactions to T_{reg} cells are currently the subject of much research on both humans and nonhuman animals (Knutson, Disis, and Salazar 2007). Nevertheless, evidence clearly supports the view that this is another example of how chemical misinformation can be employed by pathological cells to serve the invasion (and often destruction) of the healthy cellular ecology.

In closing this section, we cannot neglect to note the fact that cancer operates with a Darwinian dynamic that has unfortunate implications for chemotherapeutic modalities. The problem parallels the case of the evolution of drug resistance in bacteria, such as *Staphylococcus aureus*. Here the population of bacteria displays genetic variation with implications for susceptibility to antibiotics. At clinical or subclinical doses of antibiotics, susceptible bacteria do not survive or reproduce. The bacteria that survive and reproduce are those with genotypes that permit tolerance of the administered doses of antibiotics. Further genetic changes produce variants on these successful bacterial "themes," and the result is that successive rounds of mutation and selection produce populations of bacteria that display resistance to antibiotics, with devastating consequences for those infected by such strains.

Similarly, chemotherapy for cancer involves the administration of "differential poisons," that is, substances that are more toxic to rapidly dividing cancers cells (and, unfortunately, rapidly dividing normal cells) than to the other cells constitutive of the organism under treatment. In advanced metastatic cancer, there may be as many as 10^{12} cancer cells present, representing a genetically diverse population in which some cells are less sensitive to a particular chemotherapy than others. If "insensitive" cells survive a therapeutic assault, they become the parents of a descendant population of cells that are selectively resistant. In essence, the descendents inherit the molecular adaptations that enabled their ancestors to survive a particular "differential poison" (Greaves 2000). The genetic basis for resistance to one type of chemotherapy also means that these cells have potential to be resistant to others as well. This is at least part of the reason why cancer is difficult to cure. It also returns us to a point made earlier in this chapter. Just as the immune system is a reflection of its historic encounters with antigen, and the order of those encounters, so too an extant community of cancer cells after chemotherapy is a historical community that bears information about previous encounters with chemotherapeutic agents. In the worst cases, as in the best, Darwinian problem-solving results in the preservation of highly specific kinds of chemical or molecular information.

Ontogenetic Darwinism, then, not only plays a crucial role in the operations of the vertebrate immune system, but also lies at the heart of our current best understanding of the dynamics of cancer. The study of ontogenetic Darwinism is not just of medical interest, however, for it has important implications for our understanding of evolution itself.

Lessons

As we have seen, the formation of explanations of adaptive phenomena in immunology and oncology involves the modification of evolutionary explanations originally formulated to cope with adaptive phenomena in populations of organisms. We have further observed how explanations of these ontogenetic adaptations involve reference to information processes in the form of chemical communication and signaling. In other words, there is an important informational dimension to these explanations. To appreciate more fully this dimension, we consider its implications as regards the issue of reductionism in biology.

Ernst Mayr (2004), one of the architects of modern evolutionary biology, observed that biology actually consists of two rather different fields: *functional biology* and *historical biology*. Functional biology deals with activities of the living organism, such as its genome, cellular processes, and integrated systems. Explanations in functional biology lie, ultimately, with the mechanics of physics and chemistry. On the other hand, historical biology is concerned with the continuity of function over time. Rather than attempting to explain *how* particular functions work in life, historical biology is concerned with *why* particular functions have (or have not) persisted over time. These questions lie in the realm of evolutionary biology.

In a similar vein, medical applications tend to be concerned with two different approaches that describe the causality of disease (Nesse and Williams 1996). Genetic predisposition to atherosclerosis and eating fatty foods would be considered proximate or "near" causes of a heart attack. But why are the genes that promote fat craving and cholesterol deposition present in humans? Historical or evolutionary approaches ask why we are designed the way we are. Questions of functional biology in medicine align with "how" questions, the search for proximate causes of disease within a particular individual. In contrast, evolutionary medicine focuses on the search for why humans are susceptible to some diseases and not to others. It is here, then, where we wish to expand our notion of the value of evolutionary medicine.

It is true that looking at medical phenomena from the standpoint of phylogenetic Darwinism typically means taking a long-term perspective, and it certainly plays a legitimate role in enriching our understanding of biomedical phenomena. It is also true that phylogenetic Darwinists recognize that rapid evolution is possible for organisms with short generation times, such as viruses, bacteria, and insects. Here, the relevant history may concern events occurring over the course of a few months. But we now see that phylogenetic Darwinism is only a part of the medical story, albeit a very important part. Consideration also needs to be given to the role of ontogenetic Darwinism. Such consideration means not looking at the history of a species, but focusing instead on events in the life cycle of an individual. Through our examination of adaptive immunity and cancer, we have found that it is not easy to draw a sharp distinction between mechanistic explanations, on the one hand, and evolutionary explanations, on the other. For adaptive immunity and cancer, important aspects of the mechanistic explanation are actually evolutionary in nature. This means that mechanistic (proximate) explanations in immunology and oncology cannot be drawn solely from (nonbiological) physics and chemistry. Of course, chemistry plays a crucial role here, for the communicative relationships involved in ontogenetic Darwinism involve information processes in the form of chemical signaling. But as we have seen, chemistry is not enough, because the explanation concerns how chemical signaling processes adapt as a consequence of their having evolved in an inter- or intracellular ecology on a rapid time scale. And if so, then this sort of mechanistic explanation is *non*reductionistic, as it also depends on evolutionary biological ideas.

As we saw in the introduction, the biological world exhibits a hierarchy of organization. From the standpoint of descriptive epistemology, each level in the biological hierarchy is characterized by types or kinds of explanatory generalizations. As Dawkins has observed, "At every level the units interact with each other following laws appropriate to that level, laws which are not conveniently reducible to laws at lower levels" (1982, 113). Although Dawkins is correct that some concepts are not reducible to lower levels, it does not follow that this is true for all biological explanations. On the other hand, our discussion suggests that such reductionism is false in the case of ontogenetic Darwinism.

Evolution on the fast track, be it in immunology or oncology, *is* the explanation of *how* crucial events occur. This strongly supports the need to expand the scope of Darwinian medicine to include not just phylogenetic Darwinism, but also ontogenetic Darwinism. Proponents of Darwinian medicine—noting with Dobzhansky (1973) that nothing in biology makes sense except in the light of evolution—have frequently bemoaned the lack of bridges to medicine (e.g., Ewald 1994). But although it is true that medical researchers and educators have not given phylogenetic Darwinism its due, they have developed the data on which ontogenetic Darwinism is founded. Evolutionary biologists and educators could usefully incorporate these medical insights into their own work, just as medical researchers and educators could usefully benefit from biomedical perspectives derived from phylogenetic Darwinism.

Acknowledgments

Although any errors in this text are solely the products of the authors, we wish to acknowledge the value of many discussions with our colleagues, Dr. Michael T. Gallagher and Dr. Dan M. Johnson, and the inspiration created from the many questions of our students.

References

Alberts, Bruce, Alexander Johnson, Julian Lewis, Martin Raff, Keith Roberts, and Peter Walter. 2002. *Molecular biology of the cell*. New York: Garland Science.

Bernard, Claude. 1865/1957. *An introduction to the study of experimental medicine*, trans. Henry Copley Greene. New York: Dover.

Bonabeau, Eric. 2001. Control mechanisms for distributed autonomous systems: Insights from the social insects. In *Design principles for the immune system and other distributed autonomous systems*, ed. Lee A. Segel and Irun R. Cohen, 281–292. New York: Oxford University Press.

Carroll, Sean B., Jennifer K. Grenier, and Scott D. Weatherbee. 2001. *From DNA to diversity: Molecular genetics and the evolution of life's diversity*. Oxford: Blackwell.

Cziko, Gary. 1995. *Without miracles: Universal selection theory and the second Darwinian revolution*. Cambridge, Mass.: MIT Press.

Dawkins, Richard. 1982. *The extended phenotype: The long reach of the gene*. Oxford: Oxford University Press.

Dawkins, Richard. 1989. *The selfish gene*. Oxford: Oxford University Press.

Dobzhansky, Theodosius. 1973. Nothing in biology makes sense except in the light of evolution. *American Biology Teacher* 35 (3): 125–129.

Ewald, Paul W. 1994. *Evolution of infectious disease*. Oxford: Oxford University Press.

Fell, David A. 1992. Metabolic Control Analysis: a survey of its theoretical and experimental development. *Biochemical Journal* 286 (September): 313–330.

Frank, Steven A. 2002. *Immunology and evolution of infectious disease*. Oxford: Oxford University Press.

Gallagher, Richard B. 1997. Population biology of lymphocytes. *Science* 276 (5320): 1817.

Gerhart, John, and Marc Kirschner. 1997. *Cells, embryos, and evolution: Toward a cellular and developmental understanding of phenotypic variation and evolutionary adaptability*. Boston: Blackwell.

Gordon, Deborah M. 2001. Task allocation in ant colonies. In *Design principles for the immune system*, ed. Lee A. Segel and Irun R. Cohen, 293–301. New York: Oxford University Press.

Greaves, Mel. 2000. *Cancer: The evolutionary legacy*. Oxford: Oxford University Press.

Hofmeyr, Steven A. 2001. An interpretative introduction to the immune system. In *Design principles for the immune system*, ed. Lee A. Segel and Irun R. Cohen, 3–26. New York: Oxford University Press.

Knutson, Keith L., Mary L. Disis, and Lupe G. Salazar. 2007. CD4 regulatory T cells in human pathogenesis. *Cancer Immunology, Immunotherapy* 56 (3): 271–285.

Mayr, Ernst. 2004. *What makes biology unique: Considerations on the autonomy of a scientific discipline*. Cambridge: Cambridge University Press.

Miller, Jacques F. A. P. 2001. Immune self-tolerance mechanisms. *Transplantation* 72 (8): S5–S9.

Mitchell, Melanie. 2001. Analogy making as a complex adaptive system. In *Design principles for the immune system*, ed. Lee A. Segel and Irun R. Cohen, 335–360. New York: Oxford University Press.

Nesse, Randolph M., and George C. Williams. 1996. *Why we get sick: The new science of Darwinian medicine*. New York: Vintage Books.

Nowak, Martin A., and Robert M. May. 2000. *Virus dynamics: Mathematical principles of immunology and virology*. Oxford: Oxford University Press.

Nowak, Martin A. 2006. *Evolutionary dynamics: Exploring the equations of life*. Cambridge, Mass.: Harvard University Press.

Orosz, Charles G. 2001. An introduction to immuno-ecology and immunoinformatics. In *Design principles for the immune system*, ed. Lee A. Segel and Irun R. Cohen, 125–149. New York: Oxford University Press.

Parham, Peter. 1994. The rise and fall of great class I genes. *Seminars in Immunology* 6 (6): 373–382.

Parham, Peter. 2000. *The immune system*. New York: Garland Science.

Pier, Gerald B., Howard Ceri, Chris Mody, and Michael Preston. 2004. T cell maturation and activation. In *Immunology, infection and immunity*, ed. Gerald B. Pier, Jeffrey B. Lyczak, and Lee M. Wetzler, 125–146. Washington, D.C.: ASM Press.

Ratnieks, F., and T. Wenseleers. 2005. Policing insect societies. *Science* 307 (5706): 54–56.

Sakaguchi, Shimon. 2004. Naturally arising CD4+ regulatory T cells for immunologic self-tolerance and negative control of immune responses. *Annual Review of Immunology* 22 (April): 531–562.

Sakaguchi, Shimon, and Fiona Powrie. 2007. Emerging challenges in regulatory T cell function and biology. *Science* 317 (5838): 627–629.

Segel, Lee A. 2001. Diffuse feedback from a diffuse informational network: In the immune system and other distributed autonomous systems. In *Design principles for the immune system*, ed. Lee A. Segel and Irun R. Cohen, 203–226. New York: Oxford University Press.

Shanks, Niall. 2004. *God, the Devil and Darwin: A critique of intelligent design theory*. Oxford: Oxford University Press.

Shanks, Niall, and Istvan Karsai. 2004. Self-organization and the origin of complexity. In *Why intelligent design fails: A scientific critique of the new Creationism*, ed. Matt Young and Taner Edis, 85–106. New Brunswick, N.J.: Rutgers University Press.

Sober, Elliott. 2000. *Philosophy of biology*. Boulder, Colo.: Westview Press.

Weinberg, Robert A. 1999. *One renegade cell: How cancer begins*. New York: Basic Books.

Wickelgren, Ingrid. 2004. Policing the immune system. *Science* 306 (5696): 596–599.

Wilkins, Adam S. 2002. *The evolution of developmental pathways*. Sunderland, Mass.: Sinauer Associates.

Yates, Karen E., and Jeffrey B. Lyczak. 2004. Overview of immunity. In *Immunology, infection and immunity*, ed. Gerald B. Pier, Jeffrey B. Lyczak, and Lee M. Wetzler, 1–23. Washington, D.C.: ASM Press.

6

The Informational Nature of Biological Causality

Alvaro Moreno and Kepa Ruiz-Mirazo

Biological entities constitute highly complex systems. Since Darwin, this complexity has been thought to result from a long-term historical and collective process that goes far beyond the ontogenetic trajectories of individual living beings. In that framework, one in which biological activity takes place in a physically limited environment but conveys a spatial and temporal dimension larger than that of a single organism, natural selection would explain how the diversity and complexity of living phenomena may increase over time. Furthermore, this Darwinian evolutionary process, which involves a whole population of systems undergoing many subsequent reproductive cycles, seems to not have a predetermined upper bound. In contrast to what happens in both nonliving and artificial systems, living systems create and maintain new complexity in an indefinite, open-ended way.[1]

It is important to realize that this capacity requires a certain form of *organization*. Modern biology has stressed the explanatory power of Darwinian theory to account for the phenomenology of life. Indeed, few present-day biologists would disagree with Dobzhansky's famous claim that nothing in biology makes sense except in light of evolution (1973). It is quite reasonable to ask how this capacity could arise in the first place, that is, how it could develop in systems that emerge from an inert physicochemical domain (see Ruiz-Mirazo, Umerez, and Moreno 2008).

Despite certain widespread views (e.g., Maynard Smith 1986), Darwinian evolution requires more than the properties of multiplication (reproduction), variability, and heredity. In fact, evolution, as we know it, depends on the competitive coexistence of self-reproducing entities whose organization shows a very distinctive genotype-phenotype separation. This separation requires a wide enough functional domain where selective forces may act: namely, a minimal, but sufficiently rich, concept of *organism*. Therefore, we cannot claim to explain the appearance of the first organisms by Darwinian evolution. On the contrary, it seems more plausible to suppose that the establishment of evolution by natural selection must have been preceded by simpler forms of evolution, in which survival and fitness were replaced by persistence, capacity for self-maintenance, or bare stability/robustness (Keller 2007), and that these simpler mechanisms of evolution developed together with the different systems and forms of organization that could appear and be maintained in the conditions of primitive Earth.

In this chapter, we will first discuss the kind of organization that is required to support Darwinian evolution, with particular interest in the type of material mechanisms (or *constraints*) and causal relationships underlying that organization, as well as the possible precursor forms of organization that led to it. Then, we will argue that the causal structure of cellular organization, as displayed by present living beings, comprises a very specific type of causal action, which is based on a strong dynamic decoupling and should be regarded as informational. Finally, we shall explain why this form of causation is precisely what allows the establishment of a link between two different spheres or levels in the biological domain (the level of individual, autonomous systems or organisms, and the level of historical, collective processes), which is crucial to understanding proper Darwinian evolution, that is, an evolutionary process with true open-ended potential.

Organizational Requirements for Darwinian Evolution

In natural conditions, only certain types of chemical systems provide grounds for a potentially indefinite increase in molecular and organizational complexity. In contrast with purely physical systems, a chemical reaction network allows for a rich variety and plasticity of relations among its components, and the possibility that these will change and reshape over time.[2] Trying to go beyond the spontaneous phenomena of self-organization observed experimentally in systems with different chemically reacting species (B-Z reactions, reaction-diffusion patterns, etc.), several theoretical models in the literature aim to capture that basic property of some chemical networks, supposedly on the way toward protometabolic systems (Rosen 1973; Varela, Maturana, and Uribe 1974; Kauffman 1986; Fontana 1992; Segré, Ben-Eli, and Lancet 2000). All these models, despite the differences in their conceptual premises and tools, share a core idea: the causal closure (which, depending on the model, may be catalytic, operational, or efficient closure) that should be reached among the components and reaction processes constituting the system. Even if it is quite clear that any real system will be intrinsically dissipative—and thus, by necessity, materially and thermodynamically open—the self-maintaining dynamics of these special chemical systems would be based on a cyclic network of (reenforcing) productive and catalytic relations.[3]

In these conditions, chemical systems may start retaining part of the complexity they generate: the new components and regulation mechanisms would be kept in the network insofar as they contribute to its global stability. However, given that the self-maintenance of this kind of system depends on the distributed/holistic structure of its organization, when the degree of complexity increases, the system becomes increasingly fragile. Although redundancies could partially compensate or buffer perturbations, because the preservation and regeneration of complex components depend on many statistically interactive processes, they are bound to disappear, sooner or later. So the robust self-maintenance of increasingly complex chemical networks cannot be based solely on global correlations (like in the most typical, self-organization phenomena [Nicolis and Prigogine 1977]), or homeostatic feedbacks between different sets of reactions, but require also some form of regulatory (self-constraining) mechanisms based on local, molecularly specific, and reliable mechanisms of interaction among the components of the system.[4] The development of these mechanisms is crucial to the building of new types of reaction couplings/decouplings and catalytic control loops in a way that the self-maintenance of the system becomes *functionally regulated* or, in Bickhard's terms, it becomes a sort of "recursive self-maintenance" (2004).

As Bickhard himself states,

Recursive self-maintenance requires some way of differentiating environmental conditions, two or more differing ways of contributing to conditions for stability, and appropriate switching relationships between the differentiations and the alternative potential kinds of interaction. Such properties require that the system have infra-structure: process (sub)organizations that are on a sufficiently slower time scale that they can be taken as fixed, as structural, during the differentiations, switching, and interactions. (2004, 82)

In other words, the appearance of functional regulatory mechanisms implies that the organization of the system generates within itself a set of processes partially decoupled from the dynamics of the remaining processes, thus creating different temporal scales in the operational dynamics of the system (modularity for regulatory control). This involves not only the regulation between different internal parts of the system (so that each can be identified by means of its distinctive contribution to the stability of the whole), but also of the boundary conditions that define the viability range and relationship between system and environment (Ruiz-Mirazo and Moreno 2000, 2004).

The problem, then, is how these mechanisms might appear and become fixed or recorded in the organization. This issue overlaps with an argument raised by many authors who are critical of *metabolism-first* approaches and who claim that functionally interesting chemical systems cannot be based merely on self-organizing phenomena, but that polymers with a specific sequence and template properties (like RNA molecules) would be required (Pross 2004). According to Di Paolo (2005), for instance, the presence of templates would make a clear difference between minimal forms of autopoiesis (Maturana and Varela 1973) and other self-producing systems with proper "adaptive control" mechanisms.[5] And, from a wider evolutionary perspective, we should also recall Maynard Smith and Szathmáry's (1995) theoretical scheme of transitions, according to which open-ended evolution would require "unlimited hereditary replicators." But how do these replicators actually come about?

In contrast with the standard view (Eigen and Winkler-Oswatitsch 1992; Maynard Smith and Szathmáry 1995), we maintain that natural selection (NS) alone cannot explain the appearance of those template structures and their interactions from more simple molecular population dynamics. In fact, for NS to occur, driving systems toward different, potentially higher, forms of complexity are required so that they can be selected over lower forms of complexity (Wicken 1987). Such a scenario probably already involves a combination of self-organization, self-assembly, and self-reproduction processes (Ruiz-Mirazo, Umerez, and Moreno 2008). Otherwise, evolution would lead only to local optima with minimal or very low complexity levels (recall Spiegelman's [Kacian et al. 1972] monster).

Thus, we propose an alternative scenario in which, from self-reproducing protocellular forms of recursively self-maintaining organization, the incipient action of NS may help to find and preserve higher levels of organization, opening the way to new forms of protometabolism. One can see this as a process in which NS leads to new and occasionally more complex organizations that in turn provide a wider functional variety to be selected for, enlarging the range of action and consequences of NS in a kind of mutual enhancing effect. So, from very early stages, we should think in terms of an intertwined collection of systems whose individual organizations generate and are being generated by a larger populational and transgenerational web of relations, even if (lacking reliable heredity and clear geno-phenotype distinctions) such a global supranetwork is still very far from what we understand, nowadays, as an evolving ecosystem/biosphere.

Accordingly, the appearance and maintenance of autonomously regulated chemical networks—namely, metabolic organizations—goes together with the unfolding of an evolutionary process. Indeed, in these conditions, it is much more likely that metabolic systems based on a single type of polymer (a modular template, like RNA) develop, establishing quite elaborate hereditary and catalytic mechanisms. However, if we compare these hypothetical systems with the simplest forms of present-day life, their limitations will become quite apparent. We can advance several reasons for this: for instance, in a single-polymer scenario, as the length of the polymer increased, it is clear that *folding* would become more and more important, and the three-dimensional (3D) shape of the molecule could be used for catalytic and regulatory tasks. But, at the same time, the probability of making errors in the copying of the exact sequence of monomers that define the polymer (or its complementary chain) would also increase.

In addition, and more significantly, these systems would face a deep structural bottleneck: as the capacity for expressing sequential variety into functional variety increases, the capacity for storing and copying sequential order decreases, and vice versa. The reason is that the material basis of a truly efficient modular template must be different from that of the catalytic function (Moreno and Fernández 1990; Benner 1999). Whereas catalytic versatility depends on shape versatility (which, in turn, depends on the potential to map 1D sequential variety into 3D variety), the reliable storage and replication of different forms of sequential order depend on the ability to maintain the same 3D configuration, *despite* changes in the former 1D configuration.[6]

Therefore, even if one-polymer (let us call them) "hereditary metabolisms" should allow the transmission and maintenance of highly complex molecular patterns (e.g., the ordered sequences of RNA-like molecules), the evolutionary potential of these systems would still not be fully open-ended. We agree that it would be suitable, even at this phase, to speak of the unfolding of a new domain, beyond individual metabolisms, where selection mechanisms actually operate; or to interpret particular sequences of the components found in particular systems as the result of a collective and historical process; or, at the individual system level, to describe the organization in terms of minimal forms of *dynamic decoupling*, as it is possible to establish metabolic control subsystems, which involve different characteristic/operational time scales within the same system. However, for open-ended evolution to begin, a stronger type of dynamic decoupling has to take place in the system, articulated on two distinct, though complementary, types of macromolecular components (and corresponding modes of operation): on the one hand, *rate-dependent catalysts* (proteins), and on the other, *rate-independent records* (DNA) (Pattee 1977). In that way, as we will explain more extensively in the next section, highly specific catalytic units (enzymes) specialize in the fine-tuned regulation of metabolic reactions, whereas pure templates (now "freed" from catalytic tasks) specialize in the reliable storage and transmission of particular sequences through successive generations.

The Causal Structure of the Living Cellular Organization

Present-day cells appear as self-enclosed, highly complex webs of chemical reactions made up of many classes of components. However, the extreme precision and metabolic efficiency of this organization lies in the fact that practically all the biological reactions are controlled by enzymatic proteins.[7] These proteins are made of strings of twenty different building blocks (amino acids) whose specific ordering determines the identity of each protein. So a particular protein is synthesized when amino acids are correctly linked to form a specific string, which folds to become operational. But this process of protein synthesis requires, in turn, the participation of other proteins; it is driven by peptidyltransferases and tRNAs, both involved in the formation of peptide strings, plus other enzymes responsible for their polymerization. Accordingly, and using broadly the Aristotelian (1995) distinction between *material* and *efficient* causes, we could say that the amino acids constituting the proteins are their material cause, whereas the specific proteins that catalyze the polymerization of amino acids would be the efficient causes in the synthesis of each protein.

However, this is not the whole story, as the specific sequences of amino acids in each protein could be neither generated, nor maintained, by the dynamical structure of the metabolic network, but depend on the specific sequences of the nucleotides of DNA molecules. The reason is that, as we explained in the previous section, highly efficient catalytic function requires a certain type of ordered polymer, not suited for template function; thus, the system should also have a way to store those specific sequences that are metabolically functional for it. DNA molecules do this job. They can record and propagate complex sequential order because of two related features: first, they are constituted by energetically quasi-degenerated strings of building blocks (i.e., nearly thermodynamically equivalent sequences), and, second, the chemical composition and 3D shape of a DNA molecule allows it to act as a template, making copies of its own sequential order (namely, its primary, 1D structure).

Because this capacity for storing and copying order is at odds with enzymatic capacity, there must be at least two different types of polymer responsible for (a) substrate-specific catalytic function and (b) the function of storage of specific sequences. Thus, it follows that an *indirect* mechanism is necessary to link the sequence of nucleotides (of the DNA/RNA molecules) to the sequence of amino acids. This *translation* process requires a complex number of steps in which (some of) the proteins themselves are involved. Actually, extremely low reactive DNA can play such an important role in the cell—that of guiding the synthesis of proteins—only with the aid of proteins themselves, the very result of that process. This means that DNA can act causally only in the context of an organizationally closed network.

However, this is a highly elaborate form of closure, based on the complementary relationship (Pattee 1977, 1982) between two dynamically decoupled levels, which takes place through the aforementioned indirect mechanism. These two levels are complementary, in the sense that each of them depends on the other one: the higher level, constituted by a set of discrete, meta-stable states (the sequences of DNA), drives the process of selective construction of dynamic functional structures (proteins), which in turn harness the dynamic relations of the lower-level components, and maintain the whole organization as a cyclic, recursive network. Therefore, biochemical mechanisms depend critically on each other: for example, the construction of proteins depends on the translation mechanism, which in turn depends on (certain) proteins.

Overall, then, there is a double form of regulatory control in the organization of the cell. On the one hand, enzymes act as regulatory controls on an underlying web of chemical reactions; on the other hand, enzymes are fabricated under a purely formal genetic control. And all these processes are mutually connected in a recursive, operationally closed network. Thus, what we see when we try to understand the causal structure of the cellular organization is that each component is caused by other functional components within the system.

Interestingly, this causally circular organization is the only way to ensure that templates efficiently play the role of storing the sequential order necessary to the construction of highly specific catalytic components.[8] DNA is, in this way, kept away from the muddle of metabolic reactions. It is certainly involved in processes of change and recombination, but at its own *rate-independent* pace (Pattee 1977). In the actual synthesis of proteins, it exists as an almost inert component, providing a template—a specific sequence of units—which is a pattern to be recognized and (after several steps) mapped on to another sequence made of different building blocks (amino acids), because, in the synthesis of proteins, not even the specific materiality of DNA's building blocks (the nucleotides) matters—only their sequential order. Thus, in Aristotelian (1995) terms, DNA molecules are the *formal cause* of proteins,[9] because the specific sequence of nucleotides in DNA conveys the specific identity of the latter (what makes a protein different from others is not so much its material composition, but rather the specific order of their building blocks).[10]

By *formal cause/causation*, we can understand several things. Generically speaking, a cause can be said to be formal if it produces or contributes to a process by which sets of aggregates acquire a given structure that they did not previously have (Moreno 2000). There are many processes that lead to some kind of restructuring of matter: the formation of dissipative structures, hetero- and autocatalysis, self-assembling processes, template replication, and so on. In this sense, the whole prebiotic chemical evolution can be considered a result of multiple kinds of formal causation. During this process, many kinds of organization may emerge, maintaining their cohesion through open cycles of interaction, both internally and with the environment (Collier and Hooker 1999). In other words, as Van Gulick (1993) pointed out, most patterns in the world exist because they are self-maintaining. These patterns are the result of circular processes: emerging organized structures exerting *selective action* on the lower-level dynamics of their constitutive parts, which recursively contributes to the maintenance and preservation of those very organized systems.

In this generic sense, many types of what we have already called efficient causes can also be considered as formal ones. However, here we are dealing with a kind of causation in which this re-structuring of matter is not an implicit process, but the consequence of an explicit, thermodynamically quasi-conservative pattern (see below). Hence, we shall call "formal causation" only the specific case where something produces or contributes to a process of matter restructuring in virtue of an already existing pattern (Moreno and Umerez 2000). For instance, the replication of a modular template satisfies this concept of formal causation, as the actual sequence of the original template determines the sequence of the offspring (resulting template strings). But probably the simplest possibility of formal causation *acting in an organizational context* is the role played by modular (RNA-like) templates in what we previously called "hereditary metabolisms."

Let us analyze this scenario in terms of causal relations. During the process of reproduction of those single-polymer metabolic organizations, a pattern would be transmitted and, therefore, maintained: the ordered sequence of the RNA-like molecules. This is what would allow the creation of a new populational level, where selection mechanisms operate, guiding evolution in an already immense "sequence space."[11] As a consequence, each selected sequence in an individual organization can be seen as the result of a collective and historical process. The causal connection between the sequential domain and the individual organization is therefore a transfer of *form*, not in the sense of a 3D shape, but in terms of 1D sequences. Although still a limited case (mainly because of the existence, at this one-polymer stage, of direct links between sequential 1D patterns and 3D molecular shapes), this would be the first time that a high enough level of complexity is achieved, so that the emergence of this kind of intrinsic *formal* causation is made possible.

However, this early formal causation within possible primitive metabolic organizations is not only quantitatively but *qualitatively* different from what we observe in the role played by DNA in present-day life. First, the action of DNA is largely independent of rates and thermodynamic/kinetic considerations: it ultimately depends only on its linear sequence, which is an energetically degenerate order that can be rewritten again and again (as happens in transcription and editing processes). Instead of what would happen with the sequential order of RNAs in "hereditary metabolisms," DNA's 1D structure is wholly decoupled from processes involving intrinsic, physicochemical time, charge, and energy barriers. As a consequence of this, a second important difference between the causal role of DNA and RNA appears. Because the causal role of DNA in the cell depends on its linear 1D form, which is dynamically inert, it turns out that changes in this domain become largely independent of their dynamical effects. In other words, these changes give rise to a new, *compositional* domain: a new universe where changes producing new sequences take place independently of the dynamics of the level on which these sequences causally act.

All this leads us to the question of the original cause of this sequential domain. If we ask ourselves about the ultimate cause of the specific form that DNA molecules carry, the answer will necessarily send us to a more encompassing framework than that of the individual cell. Although, materially speaking, DNA is made up of building blocks in much the same way as any other component of the system, and, in terms of efficient cause, DNA is also copied and repaired (like other components) by the action of certain enzymes, the specific order of the nucleotide sequence of a given DNA—namely, its form—is really a consequence of the evolutionary process, which brings us beyond the level of the cell. DNA represents a material connection between the evolutionary and the organismic levels, between the collective-historical dimension and the individual organization. This is why the formal cause of the DNA *escapes*, in a certain sense, the causal closure that characterizes the organization of living cells.

DNA's Causal Action as Information

In the type of RNA (or RNA-like) world described earlier, embedded in a protometabolic cellular organization, a sequential domain that transcends the individual organism level starts unfolding. As we mentioned, the causal link between this collective sequential domain and the functioning of individual organisms can already be characterized, at that stage, as a *transfer of form*. Nevertheless, the creation of a new type of cellular organization, based on two (or, rather, three!)[12] different types of polymer related through a code implies a significant transformation of the sequential domain. Because the changes of the sequences are no longer restricted by the ever-present possibility of affecting the template capacity, the sequential domain becomes fully compositional; that is, sequences of nucleotides can now be freely recombined.

This change is of paramount importance. For if the sequential domain of the templates were not fully decoupled from the metabolic dynamics, self-reproducing populations of organisms would not be able to explore without limits the space of sequential patterns that are potentially functional. Therefore, the appearance of this free compositional domain of linear sequences, whose rate of change is decoupled from the dynamics of the organization of the individual systems, introduces a radically new type of formal causation. Instead of the sequential form in the one-polymer world scenario, which very often acts in a rate-dependent, dynamic way, the sequential form of DNA is dynamically inert. DNA, as has been stressed, acts as a formal cause only indirectly, through a code-like mechanism. In other words, the formal content of DNA, when considered within the organization of the cell, is equivalent to an *informational* text, which can be indefinitely rewritten out of the dynamical domain in which it causally operates.

Thus, in order for information to come about in the system, there must be some sort of material structure whose different configurations can be recursively reshaped, triggering some sort of "self-processing events." In other words, we identify something as informational only when it is possible to recognize parts—components in the system that are made of discrete units—subject to being recombined to produce new components in a way that both the reordering and the causal consequences of it are independent of the time and energy spent in the immediate processing steps (compositionality). As already explained, virtually the only relevant feature for those steps is the *form* of the component/structure. In physical terms, informational structures can be easily stored, reshaped, and transmitted, because they constitute a quasi-conservative domain. The process of construction of new informational structures comes out as a "rewriting" process, independent of physical-dynamical laws (Pattee 1977, 1982). In fact, it is that high degree of independence that allows those structures to operate by virtue of their form. At the same time, they cannot do without the underlying physical-dynamic level, which ultimately sustains them and which is where their informational content is actually expressed. In sum, the concept of information involves a *syntactic* dimension (a rate-independent domain of processing operations and sequence reconfigurations) and a *semantic* or *referential*[13] dimension (i.e., a causal connection with particular physical objects or events that is, precisely, independent of physical/material causality).

Accordingly, the difference of information from other possible types of formal causation lies in the fact that information implies a domain strongly decoupled from the dynamical domain in which it causally operates (Ruiz-Mirazo and Moreno 2006). Thus, the transition from (proto) metabolic systems based on a single type of polymer, carrying out both catalytic and template tasks, toward "genetically instructed metabolisms," where these two tasks are performed by two separate polymers (DNA and proteins), allows the formation of an unlimited domain in which rate-independent recursive operations can take place. Furthermore, the fact that DNA is decoupled from the metabolic dynamics opens as well the possibility of processing and editing the sequential domain itself within the cell. Thus, within a given organism, the recursive interpretation of genetic information is, in principle, possible through the construction of components devoted to *(re)processing the very sequential content of DNA*. Of course, this task might be performed by proteins, but—as has been pointed out (Mattick 2004)—also, *and mainly*, by RNAs. Therefore, by conserving RNA components and using their double-function capacity, these systems are bound to develop (as eukaryotic cells have clearly done) a much more elaborate sequential domain, articulated in new forms of editing and managing regulatory controls.[14] But what we want to stress here is that all these new functions, which are crucial for understanding the appearance of higher levels of complexity during evolution, are necessarily grounded in the DNA/protein functional separation.

Concluding Remarks

As we have explained, informational causation both requires and supports two interrelated but different phenomenological domains. One is the world of physiological processes taking place in the cell (or, even in more complex multicellular organisms); the other one is a metanetwork of processes that interconnects populations of organisms in both synchronic and diachronic ways. These two domains depend on each other, because informationally organized metabolisms cannot be maintained in the long run without the more encompassing, ecological network of Darwinian evolution processes, and the latter, as we previously argued, cannot take place without the former.

This entangled, cross-level relation shows a novel kind of openness. Unlike the limited evolutionary universe of a single-polymer (RNA-like) world, the appearance of informational, genetically instructed organizations generates a new form of self-preservation. This is a process in which a set of individual organizations reproduce their basic functional dynamics, bringing about an unlimited variety of equivalent systems, of ways of expressing those dynamics that are not subject to any predetermined upper bound of organizational complexity, even if they are subject to the energetic-material restrictions imposed by a finite environment and by universal physicochemical laws (Ruiz-Mirazo 2000; Ruiz-Mirazo and Moreno 2004; Moreno 2007). Thus, the key to this new kind of *global* self-preservation is the following distinction. On the one hand, self-reproducing systems can adopt very diverse forms of functional organization (metabolic and developmental diversity), provided that they do not lose their capacity to be reliably transmitted through generations. On the other hand, all of them share a core common organizational structure based on a strong form of dynamic decoupling: the code-mediated genotype-phenotype decoupling whose long-term preservation depends precisely on that capacity of continuous, unlimited variation and production of diversity.

Thus, the specific way in which the individual organizations achieve self-maintenance and self-reproduction is deeply linked with the historical-collective interactions among them. At the population level, a huge sequence domain appears where NS operates, which changes much more slowly than individual metabolic organizations. This is why genes can explore—free from the constraints of the metabolic dynamics—many possible patterns, until a functional outcome is found and fixed. Thus, informational causation is a dynamically decoupled link between the organismic and evolutionary levels. Although it is naturally embedded in the organizational closure of each organism, when considered genealogically, information sends us beyond the level of individual organisms (because the generation of new information requires NS acting at the evolutionary level).

Information thus acts by transferring the slow processes of change that take place at the evolutionary scale to the sequential (genetic) domain of individual metabolisms. For many authors, these collective-evolutionary processes cannot be understood as part of a wider form of organization. (See, e.g., Bickhard's [2004] criticism of the etiological view on functions, as not formulated in terms of "current organization.") However, the traditional, sharp opposition between the historical, evolutionary explanations and the causally proximate, physiological ones (Mayr 1961) may be overcome by a new way to interpret evolution in global *organizational* terms. From this new perspective, in some sense there is no difference between *why* and *how*. Changes occurring in the genetic pool of a population that shape the 1D structure of the genetic material in the course of many generations do have a real causal power in terms of current organization: they contribute to the slower processes of self-persistence of that collective metanetwork, in which living organisms achieve self-maintenance and self-reproduction. This metanetwork is organized in a globally self-sustaining way, so that the different species change according to the transformations of the boundary conditions for the self-maintenance and self-reproduction of their units (individual organisms), which in turn are the result of multiple interactions at the populational and ecological level.

However, there is an important asymmetry in this individual-collective organizational scheme, first, in the degree of functional integration: only the individual self-maintaining/self-producing systems are autonomous-metabolic agents with a self-produced, active physical border; second, in the degree of openness: although organisms already involve an open, spatiotemporal domain in which their individual lifelines are drawn, it is the self-preservation dynamic of the global organization that unfolds in a truly open-ended historical domain that we call *evolution*.

In conclusion, the evolutionary growth of complexity is possible thanks to the emergence of a new type of causal action—information—which connects two dynamically decoupled levels, creating a multifarious, collective, and historical super-organization: *life*.

Acknowledgments

Funding for this work was provided by grants FFI2008-06348-C02-01/FISO (KRM and AM) and FFU2009-12895-C02-02 (AM), from the Spanish MICINN, together with IT 505-10 (KRM and AM), from the Basque Government. Kepa Ruiz-Mirazo holds a *Ramon y Cajal* research fellowship.

Notes

1. For a more extensive and detailed treatment of the idea of "open-ended evolution," see our recent article (Ruiz-Mirazo, Umerez, and Moreno 2008).

2. Physical systems do not possess the same capacity as chemical ones for converting structural variety into new relational properties. The reason is that chemical systems are a special kind of dynamic organization in which the construction of new molecular variety through *dissipative processes* can lead to new, *semiconservative or metastable constraints* (molecular shapes) that can in turn modify the whole organization (and this new organization can bring about new molecules, which may produce new forms of organization, and so on). That is why chemistry opens a practically unlimited combinatorial domain. In contrast, purely physical dynamics (i.e., behavior fundamentally related to the change of position of a body, or of the many bodies that may constitute a system) are not sufficient, as such, to provoke spontaneous phenomena in which complexity can be produced indefinitely.

3. Different authors (Morowitz 1968, 1992; Kauffman 2000, 2003), including ourselves (Moreno and Ruiz-Mirazo 1999; Ruiz-Mirazo and Moreno 2000, 2004), have dealt with this apparent contradiction between thermodynamic (material-energetic) openness and organizational/causal closure.

4. As Wayne Christensen has recently pointed out:

Whilst 'self-organization' is celebrated for its capacity to generate global patterns it has significant limitations as a means of resolving the problems presented by integration pressure. The most important of these are slow action and poor targeting capacity. Precisely because achieving the global state depends on propagating state changes through many local interactions, the time taken to achieve the final state can be long and increases with the size of the system. Moreover, since there is no regulation of global state, the ability of the system to find the appropriate collective pattern depends on the fidelity of these interactions. Here there is a tension: if the self-organization process is robust against variations in specific conditions the process will be reliable, but it will be difficult for the system to generate multiple finely differentiated global states. Alternatively, if the dynamics are sensitive to specific conditions it will be easy for the system to generate multiple finely differentiated global states, but difficult to reliably reach a specific state.... Consequently the most effective means for achieving the type of global coherence required for functional complexity is through regulation, including feedback mechanisms and instructive signals operating at both local and larger scales. (2007, 10–11)

5. Di Paolo mentions the model of the chemoton (Gánti 1975) as a possible minimal example of this, when it is still just the length of the polymer that counts in the control mechanism (not its specific sequence).

6. If we consider carefully which are the specific properties of a polymer in relation to its template and catalytic-regulatory activity, it turns out that they are rather different structurally. The former (template activity) requires a stable, uniform "generic morphology" (a spatial structure with low reactivity—although optimal for copy/replication—and, more importantly, which should be kept unaltered despite variations in the particular sequence of building-blocks), whereas the latter (enzymatic activity) requires precisely the opposite: a very wide range of sequence-dependent "specific morphologies" in order to drive the highly diverse metabolic processes according to the 3D structures of the molecules involved (substrates, products, inhibitors).

7. An important exception is enzymatically active RNA. For reasons not yet well understood, neither DNA nor proteins have completely substituted those template and catalytic functions supported by RNA in an earlier stage of evolution. Thus, RNAs play important roles in the set of processes of transcription and translation of the DNA sequential order into specific strings of aminoacids, including what is called "DNA editing."

8. This causally circular process is not a vicious circle, but is the only way to avoid infinite causal regress. It is a type of complementary interaction that Pattee designated *semantic closure* (1982) in his attempt, since the late 1960s, to

characterize this complex, intertwined set of processes (see, e.g., Pattee 1969, or the more extensive review in Umerez 2001).

9. We are well aware that this is not exactly what Aristotle (1995) meant by formal cause, as for him both formal and material causes are intrinsic, whereas efficient (and final ones) are extrinsic. In our view, however, only when the whole organization is taken into account does formal cause appear as intrinsic (insofar as it is inherently generated in the very system, which becomes an autonomous complex system). Thus, formal cause can also be seen as extrinsic with respect to some levels or subsystems (or even systems), which allows for relatively independent kinds of description.

10. It is also possible to interpret the role played by DNA in terms of *final* causation. For example, when it is said that DNA acts as a program or recipe regulating a developmental process, it is (at least implicitly) considered a final cause, because to regulate something means that certain processes are being constrained according to a norm. Insofar as the norm depends on a long-term evolutionary process, teleology lies outside the organization of the organism. Although this view is consistent with the Aristotelian concept of final cause, there is another view of teleology in biology that places finality within the metabolic organization. This second view has its origins in Kant, who, in his *Critique of Judgment*, considered that the causal circularity of the living organization can be considered as a form of (internal) final causation (1790/1993).

11. We should highlight that, in an RNA-like world, these changes in sequential space would be still bounded, though, by a key restriction: those sequential changes affecting the template capacity would be obviously lethal. Accordingly, in contrast to what happens in present-day life, evolution could not freely explore an unlimited domain of patterns.

12. Current living metabolisms are based not only on two, but on three polymers, as the (supposedly) precedent one polymer-based apparatus, performing both catalytic and template functions, is partly maintained.

13. It is just when we address the question of the DNA proteins from an organizational perspective that the emergence of a referential relation becomes understandable in naturalistic terms: what appears referential is the consequence of the creation of a set of relations giving rise to a new kind of formal causation and the compositionality of the discrete elements constituting the instructing molecule (i.e., the triplets of nucleotides of the DNA). All together, the consequence is the appearance of rewriteable sets of triplets that (within the organization of the cell) *stand for* specific sets of amino acids. Thus, DNA strings in the cell are information, but information in the minimal expression. In DNA strings, the semantic content is nothing other than its instructive function in the synthesis of proteins.

14. This typically happens when RNAs can evolve independently and be recruited—not so much for coding metabolically functional proteins, but for RNA-based processing tasks. For instance, when the organization of the cell allows some of their RNAs to change independently of the DNA and play the role of intron processors, or when the creation of an internal compartment sepa-

rating DNA allows that transcription and translation occur in different places and not simultaneously, or when messenger RNAs are separated from intronic RNA.

References

Aristotle. 1995. *Physics* and *Metaphysics*. In *The complete works of Aristotle*, 2 vols., ed. Jonathan Barnes. Princeton: Princeton University Press.

Benner, Steven A. 1999. How small can a microorganism be? In *Size limits of very small microorganisms*, ed. National Research Council, 126–135. Washington, D.C.: National Academy Press.

Bickhard, Mark H. 2004. The dynamic emergence of representation. In *Representation in mind: New approaches to mental representation*, ed. Hugh Clapin, Phillip Staines, and Peter Slezak, 71–90. Amsterdam: Elsevier.

Christensen, Wayne. 2007. The evolutionary origins of volition. In *Distributed cognition and the will: Individual volition and social context*, ed. David Spurrett Don Ross, Harold Kincaid, and G. Lynn Stephens, 255–287. Cambridge, Mass.: MIT Press.

Collier, John D., and C. A. Hooker. 1999. Complexly organised dynamical systems. *Open Systems & Information Dynamics* 6 (3): 241–302.

Di Paolo, Ezequiel A. 2005. Autopoiesis, adaptivity, teleology, agency. *Phenomenology and the Cognitive Sciences* 4 (4): 429–452.

Dobzhansky, Theodosius. 1973. Nothing in biology makes sense except in the light of evolution. *American Biology Teacher* 35 (3): 125–129.

Eigen, Manfred, and Ruthild Winkler-Oswatitsch. 1992. *Steps towards life: A perspective on evolution*. New York: Oxford University Press.

Fontana, Walter. 1992. Algorithmic chemistry. In *Artificial life II*, ed. Christopher G. Langton, Charles Taylor, J. Doyne Farmer, and Steen Rasmussen, 159–209. Redwood City, Calif.: Addison-Wesley.

Gánti, Tibor. 1975. Organization of chemical reactions into dividing and metabolizing units: The chemotons. *BioSystems* 7 (1): 15–21.

Kant, Immanuel. 1790/1993. Critique of judgment. In *Britannica great books of the Western world*, vol. 42, ed. Robert Maynard Hutchins. Chicago: Encyclopedia Britannica.

Kauffman, Stuart. 1986. Autocatalytic sets of proteins. *Journal of Theoretical Biology* 119 (1). 1–24.

Kauffman, Stuart. 2000. *Investigations*. Oxford: Oxford University Press.

Kauffman, Stuart. 2003. Molecular autonomous agents. *Philosophical Transactions of the Royal Society: Mathematical, Physical and Engineering Sciences* 361 (1807): 1089–1099.

Keller, Evelyn Fox. 2007. The disappearance of function from "self-organizing systems." In *Systems biology: Philosophical foundations*, ed. Fred C. Boogerd,

Frank J. Bruggeman, Jan-Hendrik S. Hofmeyr, and Hans V. Westerhoff, 303–317. Dordrecht: Elsevier.

Kacian, Daniel L., Don R. Mills, Fred R. Kramer, and Sol Spiegelman. 1972. A replicating RNA molecule suitable for a detailed analysis of extracellular evolution and replication. *Proceedings of the National Academy of Sciences of the United States of America* 69 (10): 3038–3042.

Mattick, John S. 2004. The hidden genetic program of complex organisms. *Scientific American* 291 (October): 61–67.

Maturana, Humberto R., and Francisco J. Varela. 1973. *De máquinas y seres Vivos—Una teoría sobre la organización biológica*. Santiago, Chile: Editorial Universitaria S.A.

Maynard Smith, John. 1986. *The problems of biology*. Oxford: Oxford University Press.

Maynard Smith, John, and Eörs Szathmáry. 1995. *The major transitions in evolution*. Oxford: Freeman & Company.

Mayr, Ernst. 1961. Cause and effect in biology. *Science* 134 (3489): 1501–1506.

Moreno, Alvaro. 2000. Closure, identity and the emergence of formal causation. *Annals of the New York Academy of Sciences* 901 (April): 112–121.

Moreno, Alvaro. 2007. A systemic approach to the origin of biological organization. In *Systems biology: Philosophical foundations*, ed. Fred C. Boogerd, Frank J. Bruggeman, Jan-Hendrik Hofmeyr, and Hans V. Westerhoff, 243–268. Dordrecht: Elsevier.

Moreno, Alvaro, and Julio Fernández. 1990. Structural limits for evolutive capacities in molecular complex systems. *Biology Forum* 83 (2–3): 335–347.

Moreno, Alvaro, and Kepa Ruiz-Mirazo. 1999. Metabolism and the problem of its universalization. *BioSystems* 49 (1): 45–61.

Moreno, Alvaro, and Jon Umerez. 2000. Downward causation at the core of living organization. In *Emergence and downward causation*, ed. Peter Bøgh Andersen, Niels Ole Finnemann, Claus Emmeche, and Peder Voetmann Christiansen, 99–117. Aarhus, Denmark: University of Aarhus University Press.

Morowitz, Harold J. 1968. *Energy flow in Biology*. New York: Academic Press.

Morowitz, Harold J. 1992. *Beginnings of cellular life: Metabolism recapitulates biogenesis*. New Haven, Conn.: Yale University Press.

Nicolis, Gregoire, and Ilya Prigogine. 1977. *Self-organization in non-equilibrium systems: From dissipative structures to order through fluctuations*. New York: Wiley.

Pattee, Howard H. 1969. How does a molecule become a message? *Developmental Biology* 3 (Supplement): 1–16.

Pattee, Howard H. 1977. Dynamic and linguistic modes of complex systems. *International Journal of General Systems* 3 (4): 259–266.

Pattee, Howard H. 1982. Cell psychology: An evolutionary approach to the symbol-matter problem. *Cognition and Brain Theory* 5 (4): 325–341.

Pross, Addy. 2004. Causation and the origin of life. Metabolism or replication first? *Origins of Life and Evolution on the Biosphere* 34: 307–321.

Rosen, Robert. 1973. On the dynamical realizations of (M, R)-systems. *Bulletin of Mathematical Biophysics* 35 (1–2): 1–9.

Ruiz-Mirazo, Kepa. 2000. Condiciones físicas para la aparición de sistemas autónomos con capacidades evolutivas abiertas. Ph.D. dissertation, University of the Basque Country.

Ruiz-Mirazo, Kepa, and Alvaro Moreno. 2000. Searching for the roots of autonomy: The natural and artificial paradigms revisited. *Communication and Cognition—Artificial Intelligence* 17 (3–4): 209–228.

Ruiz-Mirazo, Kepa, and Alvaro Moreno. 2004. Basic autonomy as a fundamental step in the synthesis of life. *Artificial Life* 10 (3): 235–259.

Ruiz-Mirazo, Kepa, and Alvaro Moreno. 2006. On the origins of information and its relevance for biological complexity. *Biological Theory* 1 (3): 227–229.

Ruiz-Mirazo, Kepa, Jon Umerez, and Alvaro Moreno. 2008. Enabling conditions for "open-ended evolution." *Biology and Philosophy* 23 (1): 67–85.

Segré, Daniel, Dafna Ben-Eli, and Doron Lancet. 2000. Compositional genomes: Prebiotic information transfer in mutually catalytic non-covalent assemblies. *Proceedings of the National Academy of Sciences of the United States of America* 97 (8): 4112–4117.

Umerez, Jon. 2001. Howard Pattee's theoretical biology: A radical epistemological stance to approach life, evolution and complexity. *BioSystems* 60 (1–3): 159–177.

Van Gulick, Robert. 1993. Who's in charge here? And who's doing all the work? In *Mental causation*, ed. John Heil and Alfred R. Mele, 233–256. Oxford: Clarendon Press.

Varela, Francisco, Humberto R. Maturana, and Ricardo Uribe. 1974. Autopoiesis: The organization of living systems, its characterization and a model. *BioSystems* 5 (4): 187–196.

Wicken, Jeffrey S. 1987. *Evolution, thermodynamics and information. Extending the Darwinian program*. Oxford: Oxford University Press.

7
The Self-construction of a Living Organism

Natalia López-Moratalla and María Cerezo

Recent developments in genetics and developmental biology have shown that the role of genetic information in living organisms is, in one sense, stronger than had been thought and, in another sense, weaker. Insofar as the living organism has certain material constituents—namely DNA, whose sequence of nucleotides (DNA *primary structure*, or *genetic* information) can be interpreted as a kind of sign of something (be this a synthesis of proteins or particular features of the organism)—it seems beyond question that the genome is informative (Barbieri 2003; Maynard Smith 2000). Epigenetics has further amplified and limited this role. It has amplified it by showing that the information on which the life of a living organism depends is richer than that provided by the genome as such; epigenetic processes not only depend on but also give rise to a new kind of information about the state of the genome (DNA *secondary structure,* or *epigenetic* information) and about the correct way to disclose it at each particular set of time and space coordinates. Epigenetics has also limited the role of information in living organisms, insofar as it has stressed the crucial importance of environmental factors in the process of disclosing the genetic information.

When debates in the field of philosophy of biology address recent discoveries in epigenetics, discussion usually centers on the tension between evolution and development, or the so-called *evo-devo* issues: what the actual role of developmental genetic machinery and, in particular, epigenetics is; the relevance of the inheritance of developmental patterns in accounting for evolution; how to relate the idea of self-organization with that of evolution; and so on. In this chapter, we plan to leave aside the evo-devo topic as such. We are not interested here in giving an answer to the question of what the consequences of recent discoveries in epigenetics are for evolutionary biology.

Our concerns are what we ordinarily understand as a living being or organism—specifically, its biological identity and its individuality—and the relation between these aspects. The questions we are interested in pertain to, for example, how a living organism can be individuated, or whether genomes code for organisms and in what sense. We seek to address the question about what makes an organism have a particular biological identity, and what makes it an individual organism distinct from others with whom it shares that biological identity. In particular, we intend to concentrate on the issue of the role of genetic and epigenetic information in accounting for the biological identity of an organism and for the diversity of organisms and their individuality. In order to answer these questions, we offer an interpretation of the self-development of a living organism in which genetic and epigenetic information are mutually coordinated with environmental factors in such a way that each organism has both a particular biological identity and is spatiotemporally individuated.

Our account of the crucial role of epigenetics in development will also attempt to overcome the classical tensions between preformationism and vitalism, or the current version of the tension, between genetic determinism and extreme contingentism or the primacy of environmental plasticity (Maynard Smith 2000; Oyama 1985; Oyama, Griffiths, and Gray 2003). We depart from the premise that either everything is given from the beginning, or the form of the organism emerges as a consequence of an open class of environmental factors that drives the development. In a nutshell, we affirm both horns of the classic dilemma concerning whether a living organism is organized or formed from the beginning, or whether instead organization and form arise over time, for there is a sense in which the form is there from the beginning, and also a sense in which the form arises over time. The key to accounting for this nature of the form is the idea, central to our proposal, that living creatures are entities that are in a continuous process of self-change. Given this fact, the information required to account for such entities is not only the information that codifies features of the result of such a process (or of subresults of smaller phases of the whole process), but also that which codifies features concerning the self-regulation of the process of change itself as development. Of course, the nature of this second type of information must be crucially different from the former, as the information itself must be changing.

Our main thesis is that an adequate conception of the living organism must account for it in dynamic terms, and that epigenetics provides some crucial elements for doing so. We will show that there is sufficient empirical evidence for advancing the hypothesis that, on the assumption that a living organism is a self-developing being, it is possible to identify *a* living organism spatiotemporally for at least a subclass of living organisms. We will be facing issues related to the identity and individuality of living organisms, and we will do so by making a crucial distinction between the constitution of living organisms and their self-construction or self-development.

Four final preliminary comments regarding the methodology and scope of this work are in order. First, in offering our model, we have paid attention mainly to the empirical data observed in mammalian organisms. Any generalization from these data should be undertaken with caution, due to the difficulty of making empirical generalizations about development given the diversity of organisms. In any case, we are not attempting to state the necessary conditions for the simplest forms of life on earth, but rather to state them only for the most complex forms.[1]

Second, and consequently, because most discoveries concerning epigenetic mechanisms are very recent and continue to take place currently,[2] we expect new data to become available in the near future; we thus propose our model with caution, and hope that it will in turn give rise to clues for its future improvement.

Third, it is beyond the scope of our discussion to address issues related to human beings, and thus we do not aim to answer questions related either to the rationality or personhood of the human embryo or human fetus, or to enter into the standard ethical dilemmas concerning those issues. These are different issues from those we are tackling here, but we hope that philosophers interested in those issues will take into account the results of this work.

Finally, this work is strongly interdisciplinary: we have tried to take into consideration the available empirical data as well as to address classical philosophical questions, such as those concerning the biological identity and individuality of living organisms. At the end, we face the question of why it seems that the beginning of a typical individual organism takes place by the emission of the message of its first cell, yet this does not appear to be the case with monozygotic twins. In trying to apply our model to this case, we have come up also with a possible scientific hypothesis that we expect to be empirically explored in the future.

In the following section, we begin by presenting certain notions that are necessary to set the background for the section "The Self-development of a Living Organism," where we deploy our model. The final section applies the model to the case of monozygotic twins.

Life, Organisms, and Epigenesis

The twentieth-century development of thermodynamics and biochemistry has led to a deeper understanding of living systems. Life can be defined as the crucial attribute of certain systems: "open systems *continuously maintained* in steady-states, far from equilibrium, due to matter-energy flows in which informed (genetically) autocatalytic cycles extract energy, build complex internal structures, allowing growth even as they create greater entropy in their environments" (Weber 2003). Thus, one characteristic that is essential to living things is that, in contrast to inert things, these continuous changes depend mostly on intrinsic conditions, as living things possess within themselves the origin and source of such continuous change.

There is a twofold sense in which living creatures can be said to have within themselves the origin and source of their change, and also a twofold sense in which this claim must be restricted. They have the origin and source of their change within themselves in a dynamic and informative sense. On the one hand, while they are alive, they have their own mechanisms to maintain themselves far from equilibrium; they do so by means of vital operations, for example, nutrition, and they react to possible (internal or external) circumstances that endanger their condition of being *alive*. Living organisms have within themselves the dynamic source that allows them to maintain themselves in such a state.

On the other hand, they do so in a particular way, by possessing a specific kind of matter, *genetic matter*: a molecule that is present in all living organisms—that is, DNA—and that accounts for the manner in which structural organization is related to the initial matter from which the organisms are formed. Notice that these two senses in which a living organism possesses the origin of its own change are related, as it is precisely the activated DNA—that is, dynamic genetic matter—that guides the growth and/or development of a living organism in a way such that the system remains in a far-from-equilibrium state. The necessary condition for an organism being alive is not only that it has DNA, a molecule constituted by a sequence of nucleotides arranged in particular ways, but rather that certain parts (genes) of the organism's DNA are expressed at specific space and time coordinates. In this way, living organisms can be continuously changing: insofar as their change is spatiotemporally ordered from within, it is guaranteed that each change will be followed by a further one. The role of *epigenetic information* is precisely to account for such a spatiotemporal order of self-development. In more technical terms, and if we focus on the molecular descriptive level, the relation of both aspects can be put in the following terms: the regulated expression of an organism's genes results in the production of proteins—some of which are transcriptional factors—and of sRNA (small RNA); these proteins and sRNA in turn have an essential role in the expression (activation or silencing) of genes. In this way, the spatiotemporal order of the organism is self-generated.

There are two restrictions on the idea that organisms have within themselves the origin and source of their development. First, because the DNA molecule is inherited in one way or another, the sense in which the origin of change is in *themselves* is not absolute. Second, given the important role of environmental factors, some of which are external to the living organism, the sense in which the source of change is in themselves is not absolute either. Contemporary biology has thus reached a deeper and broader comprehension of life by understanding it as a dynamic cooperation between genes and environment that gives rise to the regulated expression of genes in which the development of a living organism consists.

The scenario so far described allows us to venture an initial, tentative elucidation of what living beings are. A living being is an entity that is capable of *continuous self-development* in time and space on the basis of certain inherited materials (DNA). As we have seen, to possess inherited materials is considered here a necessary but not sufficient condition for development. In order to illustrate this point, consider a conception of a living organism that accounts for its development by recourse only to the genotype-phenotype dyad. It is crucial to realize that this dyad is static. Let us represent the development of a living organism as a temporally ordered sequence of *n*-tuples f_i such as the following:

$$f_0 - f_1 - \ldots f_i - \ldots f_{n-1} - f_n$$

where each f_i is the phenotype of the living organism at time t_i, and where f_0 is the phenotype of the zygote, with its particular configuration (see the following section), and each f_i can be defined as the tuple

$$f_i = (t_i, GI, g_i)$$

where GI is the genetic information (the inherited genome, the sequence of nucleotides) and g_i is the gene or set of genes expressed at t_i. Notice that the sequence of n-tuples can represent the different states of the phenotype along the process of development, and its relation to genetic information, but the sequence cannot account for the dynamic and causal relation between one state and the following. But that is precisely what characterizes a living organism—its self-development potential—that is not captured in the sequence.

Nevertheless, the sequence as it is formulated previously shows that even if the second component, GI, is constant, the third component, g_i, co-varies with time. At each time, there is thus some unchanging information and some changing expression of part of that information, and what is missing is the explanation of the change from one state to the other. But because a living organism is essentially dynamic, it is crucial to account for such a transition. Because GI remains constant, another level of information is required to regulate such changes, and this is precisely the role of epigenetic information. We are now in a position to formulate the *Biological Identity Thesis* and the *Epigenesis Thesis*:

Biological Identity Thesis (BIT) The *biological identity* of a living organism—that is, what that organism biologically is and will be during its whole life—is determined by its inherited GI or genome (the sequence of nucleotides of its DNA), which is constant, and is determined once the GI proper to that organism is formed.

Epigenesis Thesis (ET) Because living beings are continuously self-changing entities in space and time, their individuality is determined by their unitary spatiotemporal phenotypic trajectory, which in turn depends on its epigenetic dynamic information.

Because, in the rest of the chapter, we will focus mainly on ET, three remarks are important in order to clarify the scope of our notion of biological identity. First, as the DNA primary structure or GI includes regulatory regions or sequences that encode features related to the control of the developmental process itself, significant differences between two organisms with two distinct biological identities may depend on small dissimilarities in regulatory sequences that modify the self-construction process at crucial points, and in particular at points that take place in early stages of development.[3] Biological identity thus includes crucial features about the organism's development itself.

Second, biological identity includes characteristics common to all living beings of the same species, as well as variable intraspecific differences in some DNA regions that encode phenotypic features. Roughly speaking, contemporary genetics can explain the former in terms of the number of genes, their loci in the chromosomes, and the presence of particular kinds of regulatory and monomorphic genes that encode for proteins with particular functions. Those features that are common to organisms of the same species are the ones that are vertically transmitted from one generation to the next, and that determine the limits of the species. Intraspecific diversity depends on allelic variability, that is, minor variations in the sequence of a gene that induce individual differences without essentially changing the function of the encoded protein.

Third, the possibility of genetic mutations along the life of the organism does not threaten its biological identity. Mutations that introduce novelty, and are thus at the basis of the evolution of systems, occur *before* the self-construction of the living organism starts, either during the constitution of the gametes out of which the new biological identity arises or else during the process of fertilization itself. Other genetic mutations that happen during the development of the organism are accidental to the biological identity itself, in that such mutations are changes of the genetic information that are not necessary for the development of the living organism. They are better understood as failures in development caused by mistakes in the replication of DNA due to extrinsic casual factors. They, in fact, result in disease or deformity, and this shows that they do not threaten the biological identity of the organism, but rather its normal development as an individual that is alive. Furthermore, the organism has its own repairing mechanisms, and when mutations take place, repairing enzymes tend to restore the mutated sequence to its original state.

Epigenesis as the process or set of sub-processes of self-development must be distinguished from the *epigenetic mechanisms* and *epigenetic information* that induce and regulate self-development. Epigenesis is the process of self-construction of a living being by means of cell differentiation and the formation of tissues and organs beginning with the zygote stage, and continuing through different processes of interaction between genes and environment. Epigenesis, like any other process, has in turn its various phases or subprocesses in which the whole epigenesis of an organism can be divided. The process of division of a zygote into a two-cell or eight-cell organism is a simple example, and the process of construction of the lung is a different one, which is at a higher level of complexity and in turn can be analyzed into its subprocesses. As discussed in the next section, even if the process of shaping and building vital organs of a living organism can be considered to be spatially designed and performed at early stages of its life, epigenetic processes persist until the death of the living organism and account for the maturation of its organs and its aging and death. Obviously, these changes are more notable and faster during the early stages of embryonic development. *Epigenetic mechanisms* are the particular subprocesses of interaction between genes and environment that account for the global process of epigenesis and give rise to new information, that is, epigenetic information. *Epigenetic information* is the particular ordered configuration of matter that results from those mechanisms and that can be interpreted as a sign of something. This information is not in the genome, but it is generated through epigenesis itself.

Any epigenetic process involves the regulated expression of selected genes and the generation of new epigenetic information. Both mechanisms and the interaction between them regulate, in turn, the selective expression of genes in space and time, since different signals occur at different time and space coordinates. Expression of genes takes place by means of a twofold phase: *transcription* of DNA into mRNA and *translation* of mRNA into proteins. Regulation of the process and increasing of information take place by means of various epigenetic mechanisms that can occur during both phases of the process. Two well-characterized examples of epigenetic mechanisms that take place at the phase of transcription are cytosine methylation and histone modification, which induce the silencing or activation of genes by changing the DNA secondary structure, without changing the sequence of nucleotides. Cytosine methylation consists in the addition of a methyl group to a segment of DNA, which can silence the corresponding gene. Demethylation and methylation are thus associated with the expression and repression of gene transcription, respectively. Different methyl groups that can be added and different locations in the DNA of the cytosine to which they are added generate distinct DNA methylation patterns that involve an increase of epigenetic information (Jenuwein and Allis 2001; Jones and Takai 2001; Strahl and Allis 2000; Torres-Padilla et al. 2007; Turner 2000). Among other effects, DNA methylation regulates parental genome imprinting and determines the paternal and maternal specific imprinting, which results in an early developmental asymmetry in the function of parental genomes (Ferguson-Smith and Surani 2001).

For our purposes, the relevant facts are that (1) the DNA methylation patterns change during development, giving rise to methylation patterns proper to cells of specific organs and tissues, and (2) the DNA methylation degree (or number of methyl groups added) increases in already differentiated cells. As a consequence, at each time and space coordinate, those patterns have information about the past life of the corresponding cell and, in particular, temporal information about the amount of time that has elapsed after the organism has been fully constituted (the "molecular clock of life"; see Bird 2002). The DNA methylation pattern and degree of a cell at each stage determines what can and cannot occur in future stages. Similar statements can be made concerning the histone modification mechanism, even if the mechanism is different (Torres-Padilla et al. 2007). The notion of *epigenetic reprogramming* thus arises as the process required to alter the potentialities of natural expression of genes in a cell by changing epigenetic patterns. Normally, this includes, among other things, DNA demethylation processes (Surani 2001). Natural epigenetic reprogramming takes place at the beginning of development in order to regulate the transition from oocyte to embryo. The expression *epigenetic reprogramming* is also used in a different sense to refer to artificial epigenetic reprogramming, which can be conceived as a kind of backward reprogramming, as it consists of restoring potentialities of the past life of a cell by artificially inducing certain epigenetic mechanisms that allow the reproduction of a previous state of the differentiation of the cell (Dean, Santos, and Reik 2003).[4]

Among epigenetic mechanisms that take place at the second phase of gene expression (translation), a well-characterized example is alternative splicing. It consists of the possibility of processing the information of the mRNA in different ways by recombining its exons, thus enabling the synthesis of different proteins. This mechanism allows for an increase of information, as the protein synthesized depends on an alternative splicing that varies according to different signals, which in turn vary according to the space and time coordinates of the trajectory. In this way, at different space and time coordinates, diverse signals are able to translate the same gene into different proteins. This mechanism multiplies the potential of each gene and shows the relevance of environmental factors for the synthesis of proteins.

The epigenetic mechanisms previously described (and other ones) play a crucial regulatory role by determining *which* genes are expressed, *when* they are expressed, and *where* they are expressed. At particular time and space coordinates, the DNA secondary structure and the environmental signals, which both result from the epigenetic processes that occurred in previous stages, determine the genes to be expressed at those time and space coordinates. In this way, the unitary spatiotemporal phenotypic trajectory of the living organism is deployed.

Information about the process itself, epigenetic information, comes in the form of *epigenetic patterns*. Epigenetic patterns are particular configurations of matter that result from epigenetic mechanisms, such as, for example, cytosine methylation patterns and histone modification patterns. Epigenetic mechanisms constitute a variety of processes with either short-term functional effects or long-term functional effects, and, accordingly, the patterns to which they give rise can be considered to be weakly or strongly informative. The patterns of cytosine methylation or histone modification associated with the expression of particular genes can change rapidly in response to external environmental factors. In this case, they have short-term functional effects, and the patterns they give rise to are weakly informative. In order for epigenetic patterns to be strongly informative, they must have long-term functional effects. Functional effects can be characterized as *long-term* when they are permanent in some sense, either through the whole cell cycle or when they are inherited through cell division. In this case, they contribute to generate a particular pattern proper to differentiation of a cellular line, and give information about the kind of cell (Turner 2007).

Permanence of the pattern is a necessary condition for the pattern to be strongly informative, because otherwise, *being informative* could be reduced to meaning "being immediately causal." This weak sense in which a system is informative (when there is a direct causal connection) is frequent in natural processes. Smoke is a sign of fire, and in a less intuitive but permissible way, fire is a sign of smoke. That there are epigenetic processes, mechanisms, and patterns that are informative in this sense is uncontroversial, even if the notion of causality involved in biological systems is particularly complex (Wagner 1999). Cell signaling is also informative in this sense, insofar as it is co-causal for the progress of the process.

Is there a stronger, indirectly causal notion of information that can be applied to epigenetic patterns? In other words, if we grant that there is such a thing as a genetic code—in the restricted sense that genes *code for* the amino acid sequence of protein molecules, or in the broader sense that particular patterns of genetic matter (nucleotide sequence) *give information about* particular phenotypic features[5]—can we apply in a similar way the notion of information, the idea of a "code," to epigenetic patterns?

In what sense is the DNA sequence informative? A is strongly informative if and only if

1. there is some B such that A is about B or conveys B,
2. the correlation between A and B is arbitrary,
3. there is no direct causal connection between A and B.

The correlation between certain genes and phenotypic features of the organism satisfies those three conditions. In a similar manner, certain epigenetic patterns may be correlated with, for example, spatiotemporal coordinates of development, be they past or future. Further, particular cytosine methylation patterns, or methylation degree, may be correlated with the particular spatial location of the cell or the time passed from a particular point in development, and specific histone modifications patterns may be correlated with certain changes in transcription. From these considerations, the *epigenetic code* can be said to describe the way in which the potential for expression of genes in a particular cell type is specified by chromatin modifications put in place at an earlier stage of differentiation (Turner 2007). Crucially, the epigenetic code is generated through the process and changes over the course of the process itself.

Three principles govern epigenetic processes:[6]

(1) The properties of the result of epigenesis are more than the combination of the properties of its previous parts. Epigenesis gives rise to *emergent properties*.[7]

(2) Environmental factors are crucial for epigenetic processes. They are changing factors that are generated by the process itself, have signaling effects regulating gene expression, and give rise to new epigenetic information. Environmental factors or signals are thus constitutive of epigenetic information in a way that they are not constitutive of genetic information, which remains constant. These factors can be intracellular or intercellular (IEF), or external (EEF).

(3) The process has the nature of feedback: the result of each change has both certain phenotypic features and certain epigenetic informative consequences that, in turn, determine future changes. In this way, the feedback nature of epigenetic processes enables a temporally ordered sequence of the messages to be emitted.

Two corollaries of (1) and (2)–(3), respectively, are:

(4) Given two organisms A and B, A is more complex than B if and only if some epigenetic processes in A are richer than those in B.

(5) The phenotype f_i of an organism A at a time t_i is a function of GI, EI at t_{i-1}, IEF at t_{i-1}, and EEF at t_{i-1},

where EI is the epigenetic information (different DNA secondary structure in cells at different locations), IEF is the set of intra- and intercellular factors or signals, and EEF is the set of environmental factors external to the organism.

We will deal with (5) in the next section, but some comments about (4) are in order.

Complexity can be measured in different ways, but there are at least three criteria for determining the degree of complexity: (a) the self-regulation of the processes that allow for the differentiation of cells and parts of the organism; (b) the nonreversible character of these processes; and (c) the degree of developmental autonomy from the external environment:

a. The regulation of differentiation depends on epigenetic information. Simpler organisms have weaker regulation systems in which the development plan is predetermined in large part either spatially or temporally. An example of *spatial regulation* occurs in the case of the fruit fly (*Drosophila melanogaster*), in which segmentation genes expressed locally in the fertilized egg determine the number, size, and polarity of the body segments. Homeotic genes that depend on the previous expression of the segmentation genes determine the process of cell differentiation according to their location in the embryo. An example of *temporal regulation* occurs in the case of the nematode *Caenorhabditis elegans*, in whose development heterochronic genes form a regulatory pathway that controls the temporal sequence of its cell lineage. Heterochronic genes encode proteins that constitute a temporal developmental switch, because the protein levels vary at different times and thereby regulate the expression of different genes. In more complex animals—for example, mammals—regulatory molecular signals are not predetermined spatially or temporally; rather, they appear along with the process itself.

b. The DNA primary structure does not change throughout the whole process of development of a living organism, and thus nonreversibility has to do with epigenetic information; indeed, this is predicted by the fact that the temporal order of messages depends on epigenetic information. The simpler the EI is, the easier it is to reverse the process. This fact is shown in regeneration potentialities in simple organisms, in which changes in DNA second structure are simpler, and cells can dedifferentiate and transdifferentiate more easily as a result of the appearance of signals. On the one hand, these abilities depend on the level of complexity and cell differentiation of the organism in question. Organisms possessing fewer tissues and kinds of cells do not require as many epigenetic processes as more complex ones do. On the other hand, in simple species, development is determined more by intra- and intercellular signals and patterns with short-term functional effects than by long-term patterns. The cnidarian *Hydra vulgaris,* for example, is able to regenerate its head by resetting positional values along the remaining body axis, and it is also able to reform an animal from dissociated cells due to the conservation of signaling pathways that are involved in axial specification (Sánchez Alvarado and Tsonis 2006).

c. Simpler forms of life depend more on their external environment for their development than more complex ones. The complexity of epigenetic processes not only explains the difference between the various stages of development of an organism and its higher degree of regulation, but also accounts for the differences among species. The fact that different animal species (for example, among primates) have very similar genomes, even if they show very different phenotypic features, shows that epigenetic processes have a crucial role in the actual development of each living organism.

Corollary (4) warns about the difficulty of generalizing results from one species to others. This is particularly important if we consider the possibility of artificially reprogramming an individual trajectory, as it is predictable that the more complex an organism is, the more difficult it will be to reprogram its development.

The Self-development of a Living Organism

A living being is an entity that is capable of *continuous self-development* in time and space, based upon certain inherited materials (DNA) that determine its biological identity (BIT), and in such a way that its individuality is determined by its unitary spatiotemporal phenotypic trajectory (ET).

A living organism is thus an entity that changes, and the trajectory of these changes represents the different states of its development; the entire trajectory, from its initial to final point, represents the life of the organism as a temporal sequence of changes. The entity that evolves is constituted as an entity with a specific biological identity (determined by GI) such that the trajectory is determined from the beginning as the trajectory of a particular individual of a particular species. We can now introduce a distinction between the *constitution* of a living organism and the *self-development* of an already constituted organism.

Because BIT establishes that GI is sufficient to establish the biological identity of an organism, no living organism depends on the environment to be *what it is*, but it does depend on the environment *to be* what it is, *to live,* that is, *to spatiotemporally self-develop* according to its own identity.

ET requires that the self-changing entity develops in time and space. The fact that a living organism has a phenotypic trajectory that is spatial involves a twofold component. First, there is a *spatial limiting component*: the organism must be spatially closed, or limited, in order for it to be able to interact with its environment and to maintain itself far from equilibrium. Notice that it is possible for two different living organisms to share part of the space they occupy, as long as each of them has its own biological identity and has a defined boundary.[8] Second, there is a *spatial unity component*: the differentiation, shaping, sizing and distribution of its parts and the emerging relations among them, even if it varies along the trajectory, is ordered and regulated. The particular structure that constitutes the limit of the organism (a membrane or skin, for example) is in turn one of its parts, and its structure and relation to the organism and to the environment is crucial.

If the living organism self-develops spatially, it requires certain mechanisms of *positional regulation* to control the differentiation, shape, size, and distribution of its parts, and its relation to the external environment. Spatial organization takes place by means of the creation of diverse asymmetries during the course of development, as in, for instance, the establishment of different body axes. Cell polarization is a basic and general mechanism that underlies the processes of cell differentiation and spatial organization. The existence of poles in a cell involves an asymmetry in the distribution of components, and thereby its potential division into two different cells. A particular example, relevant for our purposes, is zygote polarization: in mammals, body axes are prepatterned early in development, as a consequence of an initial cell polarization that takes place during fertilization.

There are two distinct mechanisms of positional regulation during development. First, asymmetries in the body according to three axes result from the presence of different concentrations of molecules codified by the *morphogenes*. Morphogenes are molecules that function as signals for the selective expression of other genes, and in this way carry causal positional information during development. Second, the presence of distinct molecules in the cellular membrane allows for the phenomenon of *induction:* a process of regulation of the selective expression of genes in different spatial locations of the organism that takes place by means of specific interactions between two cells, or between a cell and a component of the extracellular matrix.

Because the phenotypic trajectory is *temporal*, self-development also requires temporal regulation. This regulation is shown, for instance, in the fact that the cell multiplication rate is not constant. The rate of development is faster at the beginning of life (during embryonic development) and slower later (during maturation and aging). Indeed, in the processes of differentiation itself, the rate of multiplication of diverse types of cells is different. A particular case, interesting due to its early time in development, is the 3-cell stage of the embryo, which is a consequence of the division of one of the blastomeres faster than the other (Pearson 2002; Piotrowska et al. 2001; Piotrowska and Zernicka-Goetz 2001; Zernicka-Goetz 2002). The best-characterized epigenetic mechanism with a temporal regulation effect is the degree of cytosine methylation.

The spatiotemporal phenotypic trajectory of a living organism has a spatial limit (see earlier) and also a temporal limit: a beginning and an end. The beginning is determined by the conclusion of the *constitution period* that results in the totipotent first cell (*zygote phenotype*) that is capable of developing into the complete organism. The end is established by the death of the organism. Both temporal events are epigenetically regulated. The beginning, for example, cannot be reduced to just the generation of the biological identity by the union of parental chromosomal DNA. The particular context in which a genome exists (its secondary structure or epigenetic patterns, and the zygote phenotype with its particular signals) determines that the spatiotemporal phenotypic stage is actually the beginning of a life, of a phenotypic trajectory. Finally, a corollary of the idea of a living organism as changing according to a trajectory is that any entity that moves forward along a segment of a trajectory from a beginning can be said to have lived, whatever the duration of its life might be.

We can now reformulate our representation of the development of a living organism, taking into account (5):

(5) The phenotype f_i of an organism A at a time t_i is a function of GI, EI at t_{i-1}, IEF at t_{i-1}, and EEF at t_{i-1}.

An individual living organism can thus be represented as a temporally ordered sequence of n-tuples f_i:

$$f_0 - f_1 - \ldots f_i - \ldots f_{n-1} - f_n$$

where each f_i is the phenotype of the living organism at t_i, and where f_0 is the phenotype of the zygote, and each f_i can be defined as the tuple

$$f_i = (t_i, \text{GI}, \text{EI at } t_i, \text{IEF at } t_i)$$

and where EI at t_i and IEF at t_i are a consequence of EI at t_{i-1}, IEF at t_{i-1}, and EEF at t_{i-1}.

The model includes four necessary conditions for development: (1) constant genetic information (GI); (2) changing epigenetic information (EI); (3) environmental signals, intra- and intercellular factors (IEF) and external factors (EEF); (4) the sequential and regulative character of the trajectory that accounts for the spatio-temporal continuity of the process and thus for its unity.

To understand a living organism as a self-continuously-developing entity is to realize that (1), (2), and (3) are *co-causal* in such a way that each of the stages of the trajectory induces the following step, and that the feedback nature of the process causes epigenetic information to change and to regulate the process together with the signals. Condition (4) is then, in fact, enabled by (1)–(3). Because the regulative character of the process accounts for the unity of the organism, and in turn that character is enabled by (1)–(3), this unity can also be expressed in different forms according to the phenotypic features of the organism at each time. At the beginning of development, the unity of the trajectory is shown in particular component gradients and cellular interactions. Later, as new structures and properties emerge, new forms of coordination also emerge, and the unity of the trajectory is then shown in the role of particular tissues or systems, in particular that of the nervous system. Crucially, because each phenotype stage f_i at t_i is enabled by f_{i-1} at t_{i-1}, the organism is the same along the trajectory.[9]

We can now turn to the tension between preformationism and extreme developmentalism, and see how the two levels of information GI and EI allow for a solution to the difficulty. On the one hand, there is a sense in which the form is present at the beginning, as GI is present from the beginning, but this does not imply preformationism, as there is a sense in which the form is not present, but rather is generated, because EI is generated during the process. If we appeal to Aristotelian causal terminology to explain this point, we could rephrase our conception as follows:

the form of a living organism is the form of an essentially continuously self-changing entity with a biological identity; it is therefore a form in which the formal and efficient causes are completely interrelated with the material cause in such a way that the form can be said to be an *increasing form*, with a direction given by the biological identity.[10]

Because the spatiality of the trajectory has a limiting component and a unity component, both are constitutive of the individual trajectory. The unitary character of the trajectory that is expressed by (4) does not consist only of the fact that all changes take place in a limited or closed space, but also involves the fact that the development of each part is intrinsically related to the development of the limited whole, and vice versa. The development of each cell, tissue, and organ is dependent on the development of the whole, because their development depends on epigenetic memory, which comprises the previous states of the organism at other times and spaces, and on signals of the whole coming from other parts. Crucially, intra- and intercellular signals play a role that is essentially different from the one played by external signals. The development of the whole is dependent on the development of the parts in both the more trivial part-whole sense and the functional sense: the whole space is constituted by the spaces of the parts, and the life of the whole at each phenotypic stage depends on the functions of the parts at such phenotypic stages.[11]

We can now propose the following condition of individuality for living organisms:

Condition of Individuality (CI) If x and y are living organisms, $x \neq y$ if and only if the trajectory of $x \neq$ the trajectory of y.

As we have seen, self-development requires a phase of constitution, prior to self-development itself, which in turn is preceded by a phase of preparation of the male and female gametes. Once the new organism is constituted, its self-development can be divided into different stages, but in general four phases can be distinguished: (a) from zygote to gastrula; (b) from gastrula to birth; (c) growth and maturation; and (d) aging and death. The general trajectory of self-development is well characterized. For our purposes, in what follows we will describe the four phases with a broad brush, and we will go into details only about the constitution period and the first stages of the first phase, which are crucial for the application of the model to extreme cases such as twinning.

The limits of each spatiotemporal phase of the trajectory of an organism are defined by the emergence of particularly relevant phenotypic features. The first phase starts with the zygote that results from constitution, and finishes with the phenotype gastrula, in which three types of tissue are already present, formed of differentiated cells out of which organs will be constructed. During this initial phase, as we will see, the cells have already started their differentiation, because they have significantly different relational properties: they relate to other cells in different ways by means of different signals. Thus there is a notion of relational differentiation that already applies: cells differ from each other according to the diverse relations that they have to other cells.

The second phase is the construction of different organs, tissues, and systems, during which the organism evolves from the gastrula state to the fetal state, and carries on this process until birth. During this phase, those groups of cells that have locally migrated and have given rise to specific interactions with other cells initiate the building of particular tissues, organs, and systems. Because different functional parts of the organism appear at different times, it is possible to distinguish early phenotypes of the fetus (the origin of the heart and nervous system), intermediate phenotypes (for example, the construction of extremities, the liver and pancreas, etc.), and later phenotypes (the formation and maturation of the lungs). The longest phase in mammals is the third, that of growth and maturation, where the organism grows in size, concludes the period of its brain plasticity, and achieves the maturity of its sexual organs in order to reproduce. The last phase is the aging of differentiated cells, in which epigenetic mechanisms bring about a decline in the multiplication rate of cells (senescence) and the weakening of the DNA repair systems. Finally, the repression of genes of cellular survival gives rise to the *apoptosis* or death of the cells of vital organs.

Constitution of a Living Organism by Fertilization
The constitution of the living organism takes place by means of the process of fertilization, which results in the zygote phenotype. Early fertilization takes place in a series of stages that are interrelated and ordered: mutual activation of the gametes by means of specific interactions, membrane fusion and penetration of the male pronucleus, and rearrangement of the maternal and paternal chromosomes. Each stage consists of various events synchronized in time. Late fertilization ends with the asymmetric distribution of cellular components and the creation of a plane that drives the first cell division.

Among the multiple events that take place during the fertilization process, two are particularly important: *epigenetic parental imprinting reprogramming* and *cell polarization*. Because male and female gametes have their own methylation patterns that repress the expression of genes, the fertilization and constitution of a new organism requires not only the constitution of the new GI and thus the biological identity of the organism, but also the specific DNA secondary structure that makes possible the beginning of the expression of genes specific to the new organism. Once it is constituted, the zygote not only has a new sequence of nucleotides proper to itself, but it also starts its own epigenetic activity precisely by modifying the parental imprinting, through specific demethylation and methylation and by the change of the spatial configuration of the chromosomes.

Cell polarization during fertilization takes place as the result of the diffusion of calcium from the sperm entry point in the membrane toward the opposite pole, created by polar bodies that remain attached to the external surface of the early embryo. The diffusion of calcium ions originates a calcium concentration gradient that regulates the events in fertilization in a synchronized form. The result of fertilization is thus a cell with (1) a new and distinct GI, (2) a particular epigenetic state different from the parental one, and (3) an irregular concentration of distinct molecules, in particular of calcium, and of cellular organelles, which makes the zygote an asymmetric or polarized cell. As a consequence, there is in the zygote phenotype a first cleavage plane that will allow for its division into an organism of two different cells.

Early Development
The early development of the mammalian embryo consists of a series of critical events after fertilization: the first and asymmetric cell division, the establishment of specific cellular contacts among blastomeres in different stages, the initial lineage of differentiation, and all the early steps required to set up a future body plan. The initial phase of embryonic patterning in mammals takes place by a developmental signaling pathway that induces a selective patterning of expression of genes.

The blastomeres arising from the first cleavage division have distinguishable fates, and are thus responsible for the beginning of cell lineage specification and the patterning of the embryo (Piotrowska et al. 2001). At this early stage, spatial and temporal regulation processes take place. The two-cell blastomere that inherits the sperm entry point tends to divide first to produce cells that predominantly populate the embryonic part of the blastocyst, and there is a stereotyped cleavage pattern up to the four-cell stage (Piotrowska-Nitsche and Zernicka-Goetz 2005). The specification of embryonic axes thus begins before cleavage. The first cleavage plane present in the zygote comes into view in the blastocyst as the boundary between the embryonic and abembryonic parts (Zernicka-Goetz 2002). This is verified by the fact that the structural patterning of the blastocyst is not established if the period of constitution does not correctly resemble fertilization, as with the parthenotes produced by the activation of a nonfertilized mature oocyte, and the pseudo-embryos derived from a zygote from which the cortical cytoplasm of sperm entry point has been removed.

The zygote phenotype of the organism is followed by the 2-cell, 3-cell, 4-cell, 8-cell, morula, blastocyst, and gastrula stages. At the 8-cell stage, the embryo acquires a characteristic phenotype as a consequence of the process of compaction: four internal blastomeres increase their contact, creating a compact group of cells that relate to each other by gap interactions, and that relate to the external blastomeres by different interactions. We see here a phenomenon of relational differentiation of cells. The internal cells will develop into embryonic tissues, and the external ones will give rise to extraembryonic tissues.

From the morula stage on, cells multiply at different rates according to their signals and state, and the organism passes through states of even and odd numbers of cells. The process of differentiation of cells into embryonic and extraembryonic tissues continues, and at the blastocyst stage two different parts in the organism can be distinguished, resulting from both lines respectively: the internal cellular mass (ICM) and the trophectoderm. At this stage, the process of implantation in the uterus begins and the differentiation of embryonic cells starts in the internal cellular mass, giving rise to the embryonic disk, composed of the endoderm and the ectoderm. Gastrulation is the process in which the disk grows and migration and differentiation of cells occur, producing the three cellular types from which all the cells of the organism will be formed.

Explanatory Scope of the Model: The Case of Monozygotic Twins (MZTs)

We have been defending the idea that the self-development of a living organism originates at the zygote state and progresses through a process that is spatiotemporally regulated. In particular, we have claimed that by means of CI, the individuation of the living organism is determined by the individuation of the spatiotemporal phenotypic trajectory. Our aim in this final part of the chapter is to apply our model to the case of monozygotic twins (MZTs; two genetically identical individuals). Normally, the existence of MZTs is seen as a difficulty for the individuality of the early embryo, because some sort of fission of the embryo has been the commonly accepted hypothesis for explaining MZTs.[12] We intend to show that our model allows an alternative explanation of what until now has been accounted for by means of the fission hypothesis. We will first briefly summarize the commonly accepted hypotheses (*received view*) of explaining MZTs. Second, we will propose a new explanation, based on the model of self-development as a unitary spatiotemporal trajectory (the *alternative view*). Third, we will provide empirical and theoretical evidence in favor of this new hypothesis. Finally, we will explore the way in which our model could be adjusted to the received view, if it happened to be empirically confirmed as the true mechanism of natural twinning. In any case, this exploration can account for the empirical data derived from artificial in vitro mammal twinning.

The Received View

MZTs result from a certain kind of fission of the developing organism prior to implantation, or at the beginning of implantation. Fission is the process by means of which one or more pluripotent cells of the early embryo may separate, giving rise to a second individual that is genetically identical to the one already developing. This hypothesis has been based on a morphologic explanation of the fact that some MZTs seem to share extraembryonic tissues (chorions and amnios). In humans, a third of MZTs are *dichorionic* (i.e., they have two complete and separate chorions). This could indicate that their origin takes place by the separation of the blastomeres prior to the formation of the blastocyst. Almost two thirds of MZTs are *monochorionic* and *diamniotic*. This might suggest that separation is produced in the ICM after the formation of the blastocyst. Finally, very few MZTs seem to be *monochorionic* and *monoamniotic*, so that separation might occur after the formation of the *amnios* during implantation.

Twinning as the Constitution of Two Zygotes through One Longer Fertilization

Recent research in epigenetics and human MZTs points to the fact that phenotypic diversity between them might be due to epigenetic differences (Fraga et al. 2005; Petronis 2006). This indicates the crucial role of epigenetic information in the posterior developmental trajectory of genetically identical individuals. These results pay attention mainly to the presence of different epigenetic patterns of DNA methylation in adult MZTs. As we have seen, these patterns arise during development, accounting for the expression or repression of different genes and thus for phenotypic differences. However, in addition to EI, intra- and intercellular signaling (IEF) plays a crucial role in development. As opposed to long-term epigenetic patterns, IEF appear and disappear at different times in development, so that they do not remain in the organism during its life. Nevertheless, such IEF are causally constitutive of the individual trajectory of an organism: they are variables on which different stages defining the trajectory depend.

The *alternative view* claims that MZTs are due to differences in the constitution process (*fertilization*) and not in the self-construction process. The standard constitution process concludes in the formation of *one* totipotent cell with a zygote phenotype: a polarized cell with a new biological identity, from which the organism self-develops. A plausible explanation of MZTs is that the constitution process results in *two* genetically identical zygotes out of a longer fertilization process in which the first mitosis of the new cell takes place before polarization by calcium ions. We would then have a long fertilization of one oocyte and sperm resulting in two totipotent cells (zygote phenotype). Such fertilization would include an intermediate state in which a nonpolarized cell would be formed with the new biological identity (a new GI), but still without the phenotype proper to a zygote. The division of this cell and simultaneous polarization of the two resulting cells would generate two zygotes. The constitution process of MZTs would give rise to two unitary spatiotemporal trajectories. The alternative view is in agreement with the empirical observation of lower levels of calcium in mothers at the time of the natural conception of MZTs (Steinman and Valderrama 2001), as concentration of calcium temporally regulates the process of fertilization (Ciemerych, Maro, and Kubiak 1999; Day, Johnson, and Cook 1998).

Note also that because the first stages of development depend on the zygote state, different variants of possible MZT trajectories are possible, according to the level of calcium that determines the strength of the intercellular interactions in the early embryo. What is crucial is to realize that identical GI, highly similar EI, and crucially different IEF give rise to two trajectories. This is in fact shown by phenotypic features in early stages of MZT development that are different from the standard trajectories.

Casting Doubts on the Received View and Motivating the Alternative View
The empirical data on which the received view is based is not as conclusive as is generally assumed, because there is no necessary connection between the fact that most MZTs share extraembryonic tissues and the explanation of twinning as fission. On the one hand, this explanation does not take into account the data about *fusion* of parts of the embryos as a mechanism for explaining different kinds of Siamese twins. In some cases, the fusion of the trophectoderms of two early embryos by the interaction of their cells inside the same zona pellucida can explain the fact that the two embryos share certain extraembryonic tissues. Given the small space inside the zona pellucida, it is easier that fusion take place than that fission would. On the other hand, *dizygotic* twins have developed with a common placenta, showing that two different individuals constituted by two fertilizations can share extraembryonic tissues. Actually, recent reflection on these and other empirical data show that MZTs and DZTs (dizygotic twinning) are not as different as is usually assumed (Boklage 2006).[13]

In addition, the received view assumes that the establishment of the polarization of the mammal embryo is induced by interaction with the mother during implantation. However, the discovery of the polarized phenotype of the zygote that induces asymmetric consecutive cell division, and hints at the later set up of axes, indicates that the spatial organization of the embryo begins at the zygote state: even if the axes cannot be morphologically observed at the blastocyst stage, they are preestablished due to molecular mechanisms set up during previous phases, which take effect with the first cellular division (Gardner 2007). In a nutshell, the existence of asymmetries from the beginning of development makes it difficult to explain MZTs in terms of accidental fission, unless the anomalies in polarization timing indicate the existence of two trajectories from the beginning.[14]

Finally, the alternative view can be easily adjusted to account for uncommon cases such as the existence of MZTs of different sexes. These MZTs have discordant phenotypes and seem to be a consequence of the nonidentity of the zygote they come from. The alternative view accounts for these cases by hypothesizing that the first intermediate cell formed during fertilization and previous to the two zygotes would be XXY, so that one of the MZTs could be male (an XY zygote) and the other could be female (an XO zygote).

Plausible Extensions of the Model

Current research on multiple artificial twinning (the artificial separation of blastomeres in the morula phase) and cloning (nuclear somatic transfer to oocytes, zygotes or blastomeres) shows that processes different from fertilization might be responsible for the constitution of a new organism. These facts do not threaten the validity of our trajectory model. In such cases, certain processes of constitution resemble that of fertilization in two crucial ways: (1) the constitution process always gives rise to *one* cell out of which the trajectory starts; and (2) in order to constitute an actual entity able to develop, essential characteristics of the zygote phenotype must be reproduced in order to obtain totipotency (Simonsson and Gurdon 2004). Finally, if the received view were to be true, and MZT took place during early self-construction, a new process of constitution either of one or of two new living organisms out of the existing one would be necessary, always including epigenetic reprogramming to totipotency.[15]

Acknowledgments

We would like to thank the Spanish and the Navarre Governments for the research funds that made possible the development of this article (MEC Research Project ref. HUM2005–05910/FISO and Navarre Government Research Project ref. Resolución 67/2006, 29 de marzo). In addition, discussion of these issues with colleagues of the Departments of Biochemistry and Philosophy of the University of Navarra, conversations with Angel d'Ors, and comments by the editors of the volume helped us to improve our work. We are grateful to all of them.

Notes

1. The molecular developmental biologist seeks to make comparisons and to accommodate results to similar organisms (Rosenberg 1997, 462). In doing philosophy based on the scientific data, we think that this procedure must be respected.

2. Even if epigenetic discoveries can be traced back to the 1980s and 1990s, 2001 can be considered as the year in which the field was consolidated by the

publication of volume 293 of *Science* on August 10, where important results in epigenetics came to light. In 2002, the *Annals of the New York Academy of Sciences* published a volume on the philosophy of biology on epigenetics.

3. This fact can explain why organisms of different species sometimes have very similar genomes.

4. The only natural backward reprogramming that takes place in mammals is carcinogenesis, and is called *cellular transformation*.

5. Even if the correlation of each gene is with a multiplicity of possible features, this correlation is in some sense informative.

6. Notice that these principles govern epigenesis as a process or subprocess of development, but they do not necessarily govern epigenetic mechanisms.

7. The sense in which these properties are emergent is weaker than the standard sense. Obviously, we do not think that epigenetic processes give rise to substances or properties of a different kind: all that emerges is *biological*, so to speak. We speak of *emergent* in the sense that substances and properties resulting from epigenetic processes are irreducible to those from which they arise, because they require a novel kind of organization or unity that is necessary to account for the corresponding new properties. The new structures are capable of a new kind of activity or agency of which the more fundamental structures were not. Note, in addition, that *different kinds* of epigenetic processes might give rise to emergent properties that are *emergent* in *different senses*, and this would require a detailed analysis of each of the processes. For example, what we will call later *constitution process* (fertilization in the case of mammals) is an epigenetic process, and the result of such a process might be considered to be emergent in a substantial sense. The development process and its subprocesses seem to be emergent in an ontological dynamic-causal sense, but not in a substantial sense (there are *new irreducible* properties, but fundamental properties and prior emergent properties play a role in determining the properties that emerge at each time). In any case, we think that an accurate analysis of this issue exceeds the aim and scope of this chapter, and our views here are not definitive.

8. Note that the idea of the relation of niche/tenant is seen as different from the part/whole relation in Smith and Varzi 1999. This is in turn confirmed by the existence of different immunological systems and the development of tolerance systems during pregnancy.

9. Notice that our proposal indirectly solves the problems posed by Olson (1997) and Hershenov (2002). The end of a particular trajectory (death) has to do with the loss of the unity proper to the particular stage at which the phenotype is at that time.

10. For Aristotle's notion of cause and its divisions, see *Physics* 2.3 and *Metaphysics* 5.2 (Aristotle 1995).

11. Our unitary trajectory model is not questioned by the existence of natural chimeras. An early embryo in blastocyst state admits in its ICM the incorporation of an embryonic cell with a different genotype. The incorporation of certain cells into the embryo is similar to what happens in organ transplants. Usually those

phenomena happen early in development, when the organism has not yet developed its immune system.

12. For the controversy in philosophy about whether MZT is a difficulty for explanations of individuality, see Ford 1988. For more recent debate, see Smith and Brogaard 2003 and Damschen, Gómez-Lobo, and Schönecker 2006.

13. Boklage (2006) intends to show that characteristics specific to both MZTs and DZTs might arise also as a consequence of partaking a history of deriving two body symmetries (and, thus, two trajectories) out of an embryonic mass.

14. Notice also that recent research in mouse cloning shows that the zygote phenotype can be a suitable recipient for nuclear transfer (Egli et al. 2007).

15. In this case, there are two possible ways in which MZT can arise. First, one of the trajectories can develop normally, and a new constitution takes place by reprogramming. Our model predicts then that there would be a trajectory that starts earlier in time than the other, and that no organism dies in twinning. Second, if fission occurs and totipotency is restored to two of the cells or cell groups, then two new trajectories start, and the initial trajectory comes to an end: the initial individual can be said to have died. In this case, the constitution process is a peculiar one, in which the start of the trajectory does not take place by means of fertilization, and in which the first stage of the trajectory can be constituted by a group of cells. See also Damschen, Gómez-Lobo, and Schönecker 2006.

References

Aristotle. 1995. *Physics* and *Metaphysics*. In *The complete works of Aristotle*, 2 vols., ed. Jonathan Barnes. Princeton: Princeton University Press.

Barbieri, Marcello. 2003. *The organic codes: An introduction to semantic biology*. Cambridge: Cambridge University Press.

Bird, Adrian. 2002. DNA methylation patterns and epigenetic memory. *Genes & Development* 16 (22): 6–21.

Boklage, Charles E. 2006. Embryogenesis of chimeras, twins and anterior midline asymmetries. *Human Reproduction* 21 (3): 579–591.

Ciemerych, Maria A., Bernard Maro, and Jacek Z. Kubiak. 1999. Control of duration of the first two mitoses in a mouse embryo. *Zygote* 7 (4): 293–300.

Damschen, Gregor, Alfonso Gómez-Lobo, and Dieter Schönecker. 2006. Sixteen days? A reply to B. Smith and B. Brogaard on the beginning of human individuals. *Journal of Medicine and Philosophy* 31 (2): 165–175.

Day, Margot L., Martin H. Johnson, and David I. Cook. 1998. Cell cycle regulation of a T-type calcium current in early mouse embryos. *European Journal of Physiology* 436 (6): 834–842.

Dean, Wendy, Fatima Santos, and Wolf Reik. 2003. Epigenetic reprogramming in early mammalian development and following somatic nuclear transfer. *Seminars in Cell & Developmental Biology* 14 (1): 93–100.

Egli, Dieter, Jacqueline Rosains, Garrett Birkhoff, and Kevin Eggan. 2007. Developmental reprogramming after chromosome transfer into mitotic mouse zygotes. *Nature* 447 (7145): 679–685.

Ferguson-Smith, Anne C., and M. Azim Surani. 2001. Imprinting and the epigenetic asymmetry between parental genomes. *Science* 293 (5532): 1086–1089.

Ford, Norman M. 1988. *When did I begin? The conception of the human individual in history, philosophy and science.* New York: Cambridge University Press.

Fraga, Mario F., Esteban Ballestar, Maria F. Paz, Santiago Ropero, Fernando Setien, Maria L. Ballestar, Damia Heine-Suñer, et al. 2005. Epigenetic differences arise during the lifetime of monozygotic twins. *Proceedings of the National Academy of Sciences of the United States of America* 102 (30): 10604–10609.

Gardner, Robert L. 2007. The axis of polarity of the mouse blastocyst is specified before blastulation and independently of the zona pellucida. *Human Reproduction (Oxford, England)* 22 (3): 798–806.

Hershenov, David B. 2002. Olson's embryo problem. *Australasian Journal of Philosophy* 80 (4): 502–511.

Jenuwein, Thomas, and C. David Allis. 2001. Translating the histone code. *Science* 293 (5532): 1074–1080.

Jones, Peter A., and Daiya Takai. 2001. The role of DNA methylation in mammalian epigenetics. *Science* 293 (5532): 1068–1070.

Maynard Smith, John. 2000. The concept of information in biology. *Philosophy of Science* 67 (2): 177–194.

Olson, Eric T. 1997. Was I ever a fetus? *Philosophy and Phenomenological Research* 57 (1): 95–110.

Oyama, Susan. 1985. *The ontogeny of information: Developmental systems and evolution.* Cambridge: Cambridge University Press.

Oyama, Susan, Paul E. Griffiths, and Russell D. Gray, eds. 2003. *Cycles of contingency: Developmental systems and evolution.* Cambridge, Mass.: MIT Press.

Pearson, Helen. 2002. Your destiny from day one. *Nature* 418 (6893): 14–15.

Petronis, Arturas. 2006. Epigenetics and twins: Three variations on the theme. *Trends in Genetics* 22 (7): 347–350.

Piotrowska, Karolina, Florence Wianny, Roger A. Pedersen, and Magdalena Zernicka-Goetz. 2001. Blastomeres arising from the first cleavage division have distinguishable fates in normal mouse development. *Development* 128 (19): 3739–3748.

Piotrowska, Karolina, and Magdalena Zernicka-Goetz. 2001. Role for sperm in spatial patterning of the early mouse embryo. *Nature* 409 (6819): 517–521.

Piotrowska-Nitsche, Karolina, and Magdalena Zernicka-Goetz. 2005. Spatial arrangement of individual 4-cell stage blastomeres and the order in which they are generated correlate with blastocyst pattern in the mouse embryo. *Mechanisms of Development* 122 (4): 487–500.

Rosenberg, Alex. 1997. Reductionism redux: Computing the embryo. *Biology and Philosophy* 12 (4): 445–470.

Sánchez Alvarado, Alejandro, and Panagiotis A. Tsonis. 2006. Bridging the regeneration gap: genetic insights from diverse animal models. *Nature Reviews: Genetics* 7 (11): 873–884.

Strahl, Brian D., and C. David Allis. 2000. The language of covalent histone modifications. *Nature* 403 (6765): 41–45.

Smith, Barry, and Berit Brogaard. 2003. Sixteen days. *Journal of Medicine and Philosophy* 28 (1): 45–78.

Smith, Barry, and Achille C. Varzi. 1999. The niche. *Nous* 33 (2): 214–238.

Steinman, Gary, and Elsa Valderrama. 2001. Mechanisms of twinning III: Placentation, calcium reduction and modified compaction. *Journal of Reproductive Medicine* 46 (11): 995–1002.

Simonsson, Stina, and John Gurdon. 2004. DNA demethylation is necessary for the epigenetic reprogramming of somatic cell nuclei. *Nature Cell Biology* 6 (10): 984–990.

Surani, M. Azim. 2001. Reprogramming of genome function through epigenetic inheritance. *Nature* 414 (6859): 122–128.

Torres-Padilla, Maria-Elena, David-Emlyn Parfitt, Tony Kouzarides, and Magdalena Zernicka-Goetz. 2007. Histone arginine methylation regulates pluripotency in the early mouse embryo. *Nature* 445 (7124): 214–218.

Turner, Bryan M. 2000. Histone acetylation and an epigenetic code. *BioEssays* 22 (9): 836–845.

Turner, Bryan M. 2007. Defining an epigenetic code. *Nature Cell Biology* 9 (1): 2–6.

Wagner, Andreas. 1999. Causality in complex systems. *Biology and Philosophy* 14 (1): 83–101.

Weber, Bruce H. 2003. Life. *Stanford encyclopedia of philosophy*. Available at http://plato.stanford.edu/entries/life/.

Zernicka-Goetz, Magdalena. 2002. Patterning of the embryo: The first spatial decisions in the life of mouse. *Development* 129 (4): 815–829.

8

Plasticity and Complexity in Biology: Topological Organization, Regulatory Protein Networks, and Mechanisms of Genetic Expression

Luciano Boi

This discussion is aimed first at studying some important aspects of the plasticity and complexity of biological systems and their links. We shall further investigate the relationship between the topological organization and dynamics of chromatin and chromosome, the regulatory proteins networks, and the mechanisms of genetic expression. Our final goal is to show the need for new scientific and epistemological approaches to the life sciences. In this respect, we think that in the near future, research in biology has to shift drastically from a genetic and molecular approach to an epigenetic and organismal approach, particularly by studying the network of interactions among gene pathways, the formation and dynamics of chromatin structures, and how environmental (intra- and extracellular) conditions may affect the response and evolution of cells and living systems.

The comprehension of the connection between genetic expression, cell differentiation, and embryo development is one of the more difficult scientific problems of the life sciences and perhaps one of the great intellectual adventures of our times. A better characterization of these deeply related biological events could provide a key to our understanding of the growth and evolution of higher organisms, as well as of several mechanisms responsible for serious health diseases. This task is complex and challenging, and it seems that, in this connection, a profound change in biological thinking is required and a significantly more organismal and integrative approach needs to be carried out. This change entails the working out of a process of relational unification in biology, which should be based on a multiscale method for studying the properties and behaviors of biological systems. This method should be aimed at exploring the different and to some extent irreducible (in the sense that, for instance, epigenetic properties of the genome cannot described or understood solely in terms of genetic properties of DNA sequences) levels of organization of living organisms. Yet the question is whether it can provide a formal systematic basis for the biological sciences as a whole. According to our perspective, a systematic and integrative approach to the biological sciences might allow us to understand a meaningful ontological and epistemological reality, namely, that living beings display a plurality of different ontogenetic, morphological, and functional levels, which—although it is the outcome of genuine biological processes—can affect changes in organisms and also influence the course of evolution.

Let us now describe the state of the field and the main problems we will address in the following discussion. In the last several years, it has become increasingly evident that the linear sequence map of the human genome is an incomplete description of our genetic information. This is because information on genome function and gene regulation is also encoded in the way the DNA sequence is folded up with proteins to form chromatin structures, which then compact through different fundamental steps into chromosomes inside the nucleus. This means that biological information and organization pertaining to living organisms cannot be portrayed in the DNA sequence alone. In a postgenomic era, the importance of chromatin-chromosome/epigenetic interface has become increasingly apparent, and the role of proteins in the regulation and modulation of gene expression and cellular activity appear henceforth to be very fundamental.

In fact, the eukaryotic genome is a highly complex system that is regulated at three major hierarchical levels, as follows: (1) The DNA sequence level that supports, locally and globally, different kinds of elastic deformations of the molecule such as bending, twisting, and knotting and unknotting. These geometrical and topological transformations carry an amount of important information concerning the emergence of certain genetic functions like transcription, replication, recombination, and repair. (2) The chromatin level, which involves three major remodeling processes needing the action of different ensembles of regulatory factors and cofactors, namely, the folding and packing of complex DNA-histones, histone modifications, and the methylation of the DNA molecule. Both the remodeling and compaction of chromatin result from the inherently controlled flexible character of those macromolecular complexes that take part in the formation of the chromatin. (3) The nuclear level, which includes the dynamic and three-dimensional spatial organization of the chromosome within the cell nucleus. It must be stressed at this point that chromatin remodeling and chromosome organization constitute two novel layers of biological information that enrich and complete that carried by the genetic DNA sequence.

There is increasing evidence that such a higher-order organization of chromatin arrangement contributes essentially to the regulation of gene expression and other nuclear functions. Furthermore, in eukaryotes, DNA topology and chromatin remodeling may have allowed the evolution of specific molecular mechanisms to set the default state of DNA functions in response to external and/or internal signals in differentiated cells. Epigenetic aspects of hierarchical DNA/protein complexes have begun to be elucidated in different model systems, including *Drosophila* and yeast. The epigenetic mechanisms might thus constitute the molecular memory of the expression state of genes or gene sets that must be transmitted to progeny.

The length and chromatin organization of the genetic material imposes topological constraints on DNA during fundamental nuclear processes such as DNA replication, gene expression, DNA recombination and repair, and modulation of chromatin/chromosome organization. DNA topology, which rests upon the ideas of deformability (change and adaptability of forms) and plasticity (reversible transformations) of molecular and macromolecular structures, has become a unifying topic for a variety of different fields that deal with DNA dealings such as replication, recombination, and transcription. More than a simple packaging solution of DNA in the cell, chromatin organization has recently emerged as an active gene regulation complex structure, and recent studies (see, e.g., Almouzni and Kaufman 2000). Cremer et al. (2004) and Richmond and Widom (2000) have suggested that the enzymes that modify chromatin generate local as well as global changes in DNA topology that drive the formation of multiple, remodeled nucleosomal states. Thus, DNA topology itself is a possible way of regulating dynamic gene expression. Moreover, DNA topology likely has a crucial role in chromatin and chromosome organization. The organization and the localization of the chromosomes inside the nucleus seem to play a fundamental role in the regulation of gene expression and epigenetic memory. The architecture of the nucleus itself and the relative position of specific chromosomal domains in different and specific nuclear territories might play a role in controlling gene expression and cellular functions.

We think it is worthwhile to emphasize the enormous impact of chromatin organization and dynamics on epigenetic phenomena and cell metabolism. In a postgenomic era, the importance of epigenetics has become increasingly apparent. The definition of epigenetics is constantly evolving to encompass the many processes that cannot be accounted for by the simple genetic (DNA) code, and the term now refers to extra layers of instructions and information (especially cellular, organismal, and environmental) that influence gene activity without altering the DNA sequence. In this context, the chromatin/epigenetics interface is one of the foremost frontiers of recent research in biology. Theoretically, we are thus in need of a deep and global rethinking of some fundamental concepts in biology like "gene code," "molecular mechanisms," and "genetic information." At the very least, they need to be supplemented by the concepts, respectively, of "chromatin code," "multilevel regulatory mechanisms," and "epigenetic information." In fact, contrary to the prevailing dogma in biology during the second half of the twentieth century, according to which the nucleic acids and particularly DNA were the exclusive carrier of genetic information, in recent years our understanding of epigenetic phenomena has progressed to the point where it appears that the true carrier of genetic information is the chromosome rather than just DNA. Indeed, the chromatin substrate appears to harbor metastable key features determining the recognition and the reading of genetic information. This information layer should be deciphered and integrated with the information layer of the genome sequence to model genetic networks properly. In this way, we will acquire a global understanding of the way the information contained in the genome is interpreted by the cell.

This study addresses the correlation between geometrical structure, topological organization, complex dynamics, and biological functions of the cell nucleus and its components, as well as their multilevel regulatory systems. We will focus on the spatial organization of DNA, chromatin, and chromosome, as well as their effects on the global regulation of genome functions and cell activities. Throughout this discussion, we shall suggest a multilevel and integrative approach to the study of some fundamental biological processes and shall deal with a broad spectrum of questions ranging from mathematical ideas, biological implications, and theoretical issues.

Our discussion is divided into four main sections: first, we start with some remarks on the geometry and topology of the genome, its compaction into the chromosome, and the biological meaning of such a process; second, we address, more specifically, the structure and dynamics of the chromatin and the chromosome, and the role of epigenetic phenomena in cell regulation; third, we briefly examine the way in which epigenetic phenomena influence cell differentiation and embryonic development, as well as different types of pathological diseases; fourth, we consider the fundamental relation between form and function in biological systems.

The elucidation of these four central issues may help in understanding some key open-ended theoretical questions, such as the role and meaning of plasticity and complexity in living systems, the nature and interpretation of biological information at the cell and organism levels, and new aspects of the interaction between genotype and phenotype. Our principal goal is to demonstrate that certain geometrical and topological objects and transformations are responsible for the formation of many structures and patterns at the mesoscopic and macroscopic levels of living organisms, and that they carry an enormous amount of important information on the emergence of new biological forms and the developmental paths of organisms. This last remark deserves attention in two respects:

1. The use of the notion of information in biology, in order to avoid semantic ambiguities and to go beyond a reductive (borrowed from cybernetic and the engineering theory of information) translation in biology, should be linked to the concept of topological (dynamic) form, on the one hand, and to that of system complexity, on the other. In fact, (a) what is transmitted in living systems during reproduction, development, and evolution is not only the genetic code carried by the DNA molecule or the chemical instructions of genes, but also at the same time the form and the meaning of these codes and instructions, which are not from the outset completely contained in the DNA-code itself; and (b) the informational content of biological processes can less be measured or captured in terms of discrete units (*bits*) of information, than by determining the degree of complexity of the structural modifications and the functional mechanisms needed to carry out these processes and thus to realize living forms.

2. The expression, regulation, and modulation of a single gene or of sets of genes within the chromatin, the chromosome, and inside the cell could not be performed solely through genetic information, because other more dynamic and complex physical principles and morphogenetic mechanisms are required for promoting and enhancing epigenetic events and cell activity.

From a Genomic to an Epigenomic Approach to Living Systems

Most molecular biologists believe that the molecular building blocks, which form the genetic material of organisms, encapsulate and determine the entire process of life and its evolution on earth. Yet it has become increasingly apparent that we hardly understand in any detail the links between the molecular substrate and the nature of organisms. It is at present widely recognized that the most striking questions in biology are (1) how to explain the organization of chromatin in the cell and its influence on the replication of cells and on the entire metabolism of eukaryotic organisms, and (2) why the different modes of expression of genes essentially rely on different interrelated epigenetic processes and on the existence of sets of regulatory networks that act on the dynamics of organismal evolution.

Before we examine these questions in detail, we would like to stress a few points that show the novelty and the far-reaching significance of these issues for developing new approaches to biological processes and forms:

1. There is now strong evidence for the belief that certain domains of the chromatin and some epigenetic processes control gene expression and regulate chromosome and cell behavior. Eukaryotic DNA is organized into structurally distinct domains that regulate gene expression and chromosome behavior. Epigenetically heritable domains of heterochromatin control the structure and expression of large chromosome domains and are required for proper chromosome segregation. Studies have identified many of the enzymes and structural proteins that work together to assemble heterochromatin (Belmont 2002; Cozzarelli and Holmes 2000; Cremer and Cremer 2001; Grewal and Moazed 2003; Nakayama et al. 2001). The assembly process appears to occur in a stepwise manner involving sequential rounds of histone modification by silencing complexes that spread along the chromatin fiber by self-oligomerization, as well as by association with specifically modified histone amino-terminal tails. Finally, an unexpected role for noncoding RNAs (polymerase) and RNA interference in the formation of epigenetic chromatin domains has been uncovered.

2. The rule governing physiological regulation and higher levels of cellular organization are located not in the genome, but rather in interactive epigenetic networks that themselves organize genomic responses to environmental signals.

3. It is now well known that a genome is prepared to respond in a programmed manner to "shocks," like the "heat shock" response in eukaryotic organisms and the "SOS" response in bacteria. However, in contrast to these programmed responses, there are genome responses to unanticipated challenges that are not so precisely programmed. The genome is unprepared for these shocks. Nevertheless, they are sensed, and the genome responds in a discernible but initially unforeseen manner. Familiar examples are the production of mutation by X-rays and by some mutagenic agents.

4. For a long time, it was believed that the close mapping between genotype and morphological phenotype in many contemporary metazoans shows that the evolution of organismal form is a direct consequence of evolving genetic programs. However, recently it has been proposed that the present relationship between genes and form is a highly derived condition—a product of evolution rather than its precondition. Prior to the biochemical canalization of developmental pathways and the stabilization of phenotypes, interaction of multicellular organisms with their physicochemical environments dictated a many-to-many mapping between genome and forms. These forms would have been generated by epigenetic mechanisms: initially physical processes characteristic of condensed, chemically active materials, and later conditional, inductive interactions among the organism's constituent tissues. This concept—that epigenetic mechanisms are the generative agents of morphological character origination—helps to explain findings that are difficult to reconcile with the standard neo-Darwinian model, such as the burst of body plans in the early Cambrian period, the origins of morphological innovation, homology, and rapid change of form. This concept entails a new interpretation of the relationship between genes and biological form.

Epigenetics

Epigenetics and Change in Living Beings

Loosely defined, epigenetics studies how certain patterns of gene expression may change and become stably inherited, without affecting the actual base sequence of the DNA. We know today that epigenetics plays an important role in gene expression, cell regulation, embryonic development, and also in tumorigenesis. In humans, the main epigenetic events are protein-histone modification and DNA methylation. These events correlate with age and lifestyle. The largest twin study on epigenetic profiles reveals the extent to which lifestyle and age can impact gene expression (Fraga et al. 2005). Its aim was to quantify how genetically identical individuals could differ in gene expression on a global level due to epigenetics. It has been found that 35 percent of twin pairs had significant differences in DNA methylation and histone modification profiles. The study revealed that twins who reported having spent less time together during their lives, or who had different medical histories, had the greatest epigenetic differences. Gene expression microarray analysis revealed that in the two twin pairs most epigenetically distinct from each other—the three- and fifty-year-olds—there were four times as many differentially expressed genes in the older pair than in the younger pair, confirming that the epigenetic differences the researchers saw in twins could lead to increased phenotypic differences. These findings help show how environmental factors can change one's gene expression and susceptibility to disease by affecting epigenetics.

In the nucleus of eukaryotic cells, the three-dimensional organization of the genome takes the form of a nucleoprotein complex: chromatin. As we have already discussed, this organization not only compacts the DNA but also plays a critical role in regulating interactions with the DNA during its metabolism. This packaging of our genome, the basic building block of which is the nucleosome, provides a whole repertoire of information in addition to that furnished by the genetic code. This mitotically stable information is not inherited genetically and is termed "epigenetics." One of the challenges in chromatin research is to understand how epigenetic states are established, inherited, controlled, and modified so as to guarantee that their integrity is maintained while preserving the possibility of plasticity. This plasticity works through different degrees and at different levels, depending on the dynamics of the biological process concerned and on the complexity of the task required to perform these processes. In other words, the aim is to understand the temporal and spatial dynamics of chromatin organization, during the cell cycle, in response to different stimuli and in different cell types.

Chromatin dynamics are affected by alterations to the nucleosome, the basic repeating unit formed from DNA wrapped around an octamer of histone proteins. Such alterations can be controlled by three classes of compounds: (1) histone chaperones principally implicated in the transfer of histones to DNA to form the nucleosome; (2) nucleosome remodeling factors or possibly disassembly factors that enable DNA sequences within nucleosomal structures to become accessible to protein complexes, allowing replication, transcription, repair or recombination; and (3) factors involved in posttranslational modifications of histones. The large repertoire of these epigenetic modifications is at the heart of the chromatin code hypothesis.

Histones are major protein components of chromatin and are subject to numerous posttranslational modifications like acetylation, methylation, phosphorylation, polyADP-ribosylation, and ubiquitination. The combination of these various potential modifications could generate a great diversity of chromatin states in the nucleus. This repertoire of modifications is at the center of the histone code hypothesis, which is considered to establish two types of epigenetic markers: heritable (or epigenetic), which contribute to the maintenance of gene activity during cell divisions, and labile, for rapid response to the environment. This code would be decoded by proteins or protein complexes able to recognize specific modifications. The modification by methylation of lysine 9 on histone H3 recognized by proteins of the HP1 family provides an example that supports this hypothesis.

The structural rearrangements within chromatin, which occur during repair of damage produced by UV radiation, show certain parallels with those occurring during replication. To explain the dynamics of these rearrangements, researchers have put forward a three-step model: access, repair, and restoration of structures. The de novo assembly of nucleosomes takes place during replication and possibly during repair of DNA in the restoration phase. The construction of the nucleosome is facilitated by assembly factors or histone chaperones. The best characterized of these factors to date is CAF-1 (chromatin assembly factor-1), which was discovered in 1986 and which stimulates the formation of nucleosomes on DNA newly replicated in vitro. It comprises three subunits (p150, p60, and p48) and interacts with acetylated histones H3 and H4. It has been subsequently shown that CAF-1 is also able to promote the assembly of nucleosomes specifically coupled to the repair of DNA by nucleotide excision repair, or NER (which involves DNA synthesis) in vitro systems. It has been finally demonstrated (Ehrenhofer-Murray 2004; Fraga et al. 2005) that the in vitro recruitment of a phosphorylated form of CAF-1 (p60) to chromatin occurs in response to UV irradiation of human cells. Such in vitro coupling between chromatin assembly and repair would ensure restoration of chromatin organization immediately after repair of the lesion.

The coupling between repair and assembly in chromatin was reproduced in *Drosophila* embryo extracts to show that the nucleosome assembly commences from the lesion site. Single-strand breaks and gaps are the most effective lesions in stimulating nucleosome assembly via

CAF-1. The search for two-hybrid partners of CAF-1, using the large subunit p150 as bait, has allowed identification of proliferating cellular nuclear antigen (PCNA), a marker of proliferation, whose involvement in replication and repair of DNA is well known. The interaction of CAF-1 with PCNA provides the direct molecular link between chromatin assembly and replication or repair. Finally, protein Asf1 (anti-silencing factor 1) interacts with CAF-1 and can stimulate assembly. A chromatin assembly chain centered on PCNA is therefore gradually coming to light.

After these studies in extracts and in cell cultures, Almouzni et al. (1994) and Scott and O'Farrell (1986) investigated the importance of CAF-1 in a whole organism, *Xenopus*. They identified the *Xenopus* homolog of the p150 subunit of CAF-1. Using a dominant-negative strategy that takes advantage of the dimerization properties of xp150, they have revealed the critical role of CAF-1 in the rapid cell divisions characteristic of the embryonic development of Xenopus. Thus, CAF-1 plays a vital role in early vertebrate development.

Beyond the level of the nucleosomes, the chromatin is compacted into higher structures that delimit specialized nuclear domains such as regions of heterochromatin and euchromatin. Heterochromatin is defined as the regions of chromatin that do not change their condensation during the cell cycle and represents the majority of the genome of higher eukaryotes. Heterochromatin principally comprises repeated noncoding DNA sequences; its characteristics generally contrast with those of euchromatin. One essential characteristic of the heterochromatin regions, which has been highly conserved during evolution, is the presence of hypoacetylated histones (H3 and H4). Apart from its repression of transcription, heterochromatin's function remains largely unknown.

The current evidence suggests that chromatin remodeling factors use the energy of ATP hydrolysis to generate superhelical torsion in DNA, to alter local DNA topology, and to disrupt histone-DNA interactions, perhaps by a mechanism that involves ATP-driven translocation along the DNA. Many chromatin remodeling factors have been found to affect transcription regulation, but it also appears that chromatin remodeling is important for processes other than transcription. Moreover, proteins in the SNF2-like family of ATPases have been found to participate in diverse processes such as homologous recombination (RAD54), transcription-coupled DBA repair (ERCC6/CSB), mitotic sister chromatid segregation (lodestar; Hrp1), histone deacetylation (Mi-2/CHD3/CHD4), and maintenance of DNA methylation states (ATRX). It thus appears that there is a broad range of nontranscriptional functions of chromatin remodeling proteins. In addition, the recent sequencing has led to the identification of many novel SWI2/SNF2-related putative ATPases. For example, there are al least seventeen SWI2/SNF2-related open reading frames in the *Drosophila* genome, and only six of the corresponding proteins have been analyzed. In their native state, SWI2/SNF2-related proteins have been generally found to exist as subunits of multiprotein complexes. Thus, the purification of the native forms of the novel SWI2/SNF2-like proteins would likely reveal many new chromatin remodeling complexes. It will be an interesting and important challenge to identify the functions of these new factors.

Nontranscriptional Chromatin Remodeling Factors

Let us now very schematically indicate the different processes and functions in which chromatin remodeling factors and cofactors seem to be involved:

1. *Chromatin structure is an important component of eukaryotic DNA replication.* Nucleosomes appear to be generally inhibitory to replication. For instance, the positioning of a nucleosome over a yeast autonomously replicating sequence (ARS) inhibits plasmid DNA replication in vivo, and the packaging of DNA into chromatin represses SV40 DNA replication in vitro. It has been found that specific DNA binding factors, such as the yeast origin recognition complex, can establish an arrangement of nucleosomes that allows the initiation of DNA replication. Hence, in the cell, it is possible that replication-competent chromatin structures are generated by the coordinate action of the DNA replication machinery and ATP-dependent chromatin remodeling factors.

2. *Chromatin assembly is a fundamental biological process by which nuclear DNA is packaged into nucleosomes.* ACF (ATP-utilizing chromatin assembly and remodeling factor) was identified and purified on the basis of its ability to mediate the ATP-dependent assembly of periodic nucleosome arrays, and it consists of two subunits: ISWI and a polypeptide termed Acf1. The Acf1 subunit functions cooperatively with the ISWI subunit for the assembly of chromatin. ACF-mediated chromatin assembly can be carried out with purified recombinant ACF, purified recombinant NAP-1 (a core histone chaperone), purified core histones, DNA (either linear or circular), and ATP. ACF requires the hydrolysis of ATP for both the deposition of histones onto DNA as well as the establishment of periodic nucleosome arrays. In addition to its function in the assembly of chromatin, ACF can catalyze the ATP-dependent

mobilization of nucleosomes, and is therefore a chromatin remodeling factor. Thus, ACF provides an example of a chromatin remodeling factor that has been purified and characterized based on its function in a non-transcriptional process.

3. *Chromatin structure is an important component of DNA repair in eukaryotes.* The DNA repair machinery must have access to the DNA lesions in chromatin, and the newly repaired DNA must also be reassembled into chromatin. A biochemical study of chromatin structure and nucleotide excision repair by Almouzni and Kaufman (2000) revealed that ACF is able to facilitate the excision of pyrimidine (6–4) pyrimidonne photoproducts in a dinuclesome. In that study, chromatin-mediated repression of nucleotide excision was only partially relieved by ACF, but it nevertheless appears that ACF can increase the efficiency of nucleotide excision in chromatin. Different SWI2/SNF2-like ATPase seems to contribute to DNA repair. For example, a SWI2/SNF2-related protein with a function in DNA repair is Cockayne syndrome B protein (CSB).

As we have just seen, there are a handful of examples of ATP-driven chromatin remodeling-reorganizing factors in processes other than transcription. These studies are likely to be the proverbial tip of the iceberg of an exciting and important area of chromatin research (see Benecke 2006; Ehrenhofer-Murray 2004; Li 2002). One of the key challenges for the future will be to devise chromatin remodeling assays that accurately reflect the specific functions of the factors in the cell.

DNA Methylation, Silencing, and Gene Expression: From Regulation to Biological Information

Let us now consider the other most important epigenetic event: genomic DNA methylation. Recent studies have illuminated the role of DNA methylation in controlling gene expression and have strengthened its links with histone modification and chromatin remodeling. DNA methylation is found in the genomes of diverse organisms including both prokaryotes and eukaryotes. In prokaryotes, DNA methylation occurs on both cytosine and adenine bases and encompasses part of the host restriction system. In multicellular eukaryotes, however, methylation seems to be confined to cytosine bases and is associated with a repressed chromatin state and inhibition of gene expression. DNA methylation is essential for viability in mice, because targeted disruption of the DNA methyltransferase enzymes results in lethality. There are two general mechanisms by which DNA methylation inhibits gene expression: first, modification of cytosine bases can inhibit the association of some DNA-binding factors with their cognate DNA recognition sequences, and second, proteins that recognize methyl-CpG can elicit the repressive potential of methylated DNA. Methyl-CpG-binding proteins (MBPs) use transcriptional co-repressor molecules to silence transcription and to modify surrounding chromatin, providing a link between DNA methylation and chromatin remodeling and modification.

Recently, there have been significant advances in our understanding of the mechanisms by which DNA methylation is targeted for transcriptional repression and the role of MBPs in interpreting the methyl-CpG signal and silencing gene expression. We emphasize examples from mammalian systems, including studies on animal models, because several recent reviews have covered topics of DNA methylation and silencing in plants and fungi. Mammalian cytosine DNA methyltransferase enzymes fit into two general classes based on their preferred DNA substrate. The de novo methyltransferases DNMT3a and DNMT3b are mainly responsible for introducing cytosine methylation at previously unmethylated CpG sites, whereas the maintenance methyltransferase DNMT1 copies preexisting methylation patterns onto the new DNA strand during DNA replication. A fourth DNA methyltransferase, DNMT2, shows weak DNA methyltransferases activity in vitro, but targeted deletion of the DNMT2 gene in embryonic stem cells causes no detectable effect on global DNA methylation, suggesting that this enzyme has little involvement in setting DNA methylation patterns. Examples of global de novo methylation have been well documented during germ-cell development and early embryogenesis, when many DNA methylation marks are reestablished after phases of genome demethylation. Recent studies based on cell-culture model systems have suggested at least three possible means by which de novo methylation might be targeted: first, DNMT3 enzymes themselves might recognize DNA or chromatin via specific domains; second, DNMT3a and DNMT3b might be recruited through protein-protein interactions with transcriptional repressors or other factors; third, the RNA-mediated interference (RNAi) system might target de novo methylation to specific DNA sequences (Ballestar and Esteller 2005; Esteller 2004, 2005; Jenuwein 2006).

Clearly, DNA methylation is a multilevel regulating process and a multilayer informational mechanism. And there are indeed different conformational and functional complexities associated with DNA methylation in cell activity.

Thus, epigenetic modification of DNA is coupled with gene expression silencing. DNA methylation is linked with transcriptional silencing of associated genes, and much effort has been invested in studying the mechanisms that underpin this relationship. Two basic models have evolved: in the first, DNA methylation can directly repress transcription by blocking transcriptional activators from binding to cognate DNA sequences; in the second, MBPs recognize methylated DNA and recruit co-repressors to silence gene expression directly. For some promoters, repression mediated by DNA methylation is most efficient in a chromatin context, indicating that the "active" component of this repression system might rely on chromatin modification. In keeping with this observation, MBPs associate with chromatin remodeling co-repressor complexes. Two unexpected facets of the DNA-methylation-mediated silencing system have recently become apparent: first, DNA methyltransferase enzymes themselves might be involved in setting up the silenced state in addition to their catalytic activities, and second, DNA methylation can affect transcriptional elongation in addition to its characterized role in inhibiting transcriptional activation.

How Epigenetic Factors Shape Development, Health, and Disease

Epigenetics: A New Relationship between Genetics and Environment

As we mentioned previously, the massive sequencing of our genome leaves unanswered key questions concerning the organization and functioning of cells and organisms. Epigenetics seems to be the most promising line of research able to unfold this postgenomic era. The epigenetic processes discussed earlier are natural and essential to many organism functions, but if they occur improperly, there can be major adverse health and behavioral effects. There is a need for a large-scale epigenetic mapping of our cells that moves biology into this new century with a new vision. Epigenetics can be understood as the process that initiates and maintains heritable patterns of gene expression and gene function in an inheritable manner without changing the sequence of the genome. Epigenetics can also be understood as the interplay between environment and genetics. Regarding this last issue, epigenetics provides the best explanation of how the same genotype can be translated into different phenotypes. Perhaps still more important: epigenetics supplies organisms with new layers of biological information that result from specific rules of regulation and organization inherent in chromosomes, cells, and organisms. This higher-order biological information can, in turn, retroact in many different ways on the genome profile and functioning according to cellular and extracellular contexts.

One essential epigenetic mechanism for repressing transcription is that in which methyltransferases first attach methyl groups (CH_3) to cytosine bases of DNA, and then protein complexes, recruited to methylated DNA, remove acetyl groups and repress transcription. Repression of transcription—the transfer of genetic information from DNA to RNA—is one route by which epigenetic mechanisms can adversely impact health. Examples of the powerful modulator effects of epigenetics in this scenario are beginning to emerge in an exponentially increasing number: strains of Agouti mice can undergo changes of DNA methylation status of an inserted IAP (intracisternal A-particle) element that changes the animal's coat color; cloned animals demonstrate an inefficient epigenetic reprogramming of the transplanted nucleus that is associated with aberrations in imprinting, aberrant growth, and lethality beyond a threshold of faulty epigenetic control; and monozygotic twins who thus share the same DNA sequence can present anthropomorphic difference and distinct disease susceptibility related to epigenetic differences such as DNA methylation and histone modifications. The goal of epigenetics research is to identify all the organizational and chemical changes and relationships among chromatin constituents that functionally contribute to the genetic code, which will allow a better understanding of normal development, aging, abnormal gene control in cancer, and other diseases, as well as the role of environment in human health.

Chromosome Organization in Topological Territories
Gene silencing in mammalian cells may be mediated by the positioning of a gene in proximity to the heterochromatic territory in interphase nuclei, suggesting that the eukaryotic nucleus is divided into heterochromatin territories that repress transcription, and territories in which transcription is favored. It is currently believed that this spatial organization of the nucleus in discrete territories may help to establish a tissue-specific pattern of gene expression required for the onset and progression of cellular differentiation. During erythroid maturation, the entire genome is progressively silenced and packaged into heterochromatin, whereas the b-globin locus is among the last to be silenced. The tissue-specific activation and maintenance of its expression in the repressive environment of a terminally differentiating red cell are due, at least in part, to the Locus Control Region (LCR), composed of several DNaseI

hypersensitive sites (HS) that contain numerous binding sites for erythroid and ubiquitous transcription factors. It has been previously demonstrated that stable gene expression and open chromatin configuration require both a functional enhancer and positioning away from centromeric heterochromatin, and revealed that enhancers can mediate the localization of genes to nuclear compartments that favor gene activation, away from the repressive compartment of heterochromatin. This led to the hypothesis that activators bound to tissue-specific LCRs/enhancers may act to establish and maintain gene expression in differentiated cells by ensuring that a linked gene resides in a nuclear compartment permissive for transcription, thereby preventing its inclusion in facultative heterochromatin that forms during cell differentiation, and permitting it to be active in the appropriate lineage. Interestingly, as red cells mature, the majority of DNA is heterochromatized at noncentromeric sites in the nucleus. Thus, this observation that repressors move from centromers during differentiation provides a possible basis for this maturation-associated increase in heterochromatin formation.

The main point to be stressed here is that the way in which the genome is organized within the nuclear space—that is, how chromatin fibers are folded across the entire human genome—both within normal and diseased cells influences gene regulation and chromosome function. This kind of organization entails a layer of biological information that stands at a level beyond that carried by DNA sequence, and that is essential for processing the genetic information itself. The spatial organization of human chromosomes and genes in the nucleus is changed, for example, during development and in certain diseases.

Let us describe the spatial organization of chromosomes in greater detail. The nucleoplasm is territorially organized into a number of mobile subnuclear organelles in which certain protein and nucleic acid components with specific biological activities are concentrated. The most prominent example is the nucleolus, which contains regions with ribosomal RNA genes from several chromosomes, and also contains the machinery for the assembly of ribosomal subunits. Other nuclear substructures include the SC35 domains, Cajal and promyelocytic leukemia (PML) bodies. The localization, organization, dynamics, and biological activities of these suborganelles appear to be closely related to gene expression. Using whole chromosome painting probes and fluorescence in situ hybridization (FISH), a territorial organization of interphase chromosomes has been demonstrated (Cremer et al. 2004). Chromosome territories have irregular shapes and occupy discrete nuclear positions with little overlap. In general, gene-rich chromosomes are located more in the nuclear interior, and gene-poor chromosome territories are located at the nuclear periphery. In agreement with this, nontranscribed sequences were predominantly found at the nuclear periphery or perinucleolar, and active genes and gene-rich regions tended to localize on chromosome surfaces exposed to the nuclear interior or on loops extending from the territories.

Recent experimental findings support the concept of a functional nuclear space, the interchromosomal domain (ICD) compartment. According to the ICD model, the interface between chromosome territories is more easily accessible to large nuclear complexes than regions within the territory. More recently, it has been proposed that chromosomes territories are further organized into 1-Mb domains, extending the more accessible space to open intrachromosomal regions surrounded by denser chromatin domains. Using high-resolution light microscopy, an apparent bead-like structure of chromatin can be visualized in which approximately 1-Mb domains of chromatin are more densely packed into an approximately spherical subcompartment structure with dimensions of 300–400 nm. These domains are thought to be formed by a specific folding of the 30-nm chromatin fiber to which the chain of nucleosomes associates under physiological salt concentrations. Other models have been proposed. The radial-loop models propose small loops of roughly 100 kb arranged in rosettes, and the random-walk/giant-loop (RW/GL) model proposes large loops of chromatin back-folded to an underlying structure. In the chromonema model, the compaction of the 30-nm fiber is achieved by its folding into 60- to 80-nm fibers that undergo additional folding to 100- to 130-nm chromonema fibers.

Chromatin Compaction and Biological Processes: How Form Determines Function

Depending on the degree of compaction, chromatin regions have different accessibility, and this accessibility is related to the biological function of chromosomes. The organizational, topological, and dynamical properties of the chromatin environment determine the mobility of Cajal and PML bodies and other supramolecular complexes. Large particles with sizes of around 100 nm (100-nm diameter nanospheres, 2.5-MDa dextrans) are completely excluded from dense chromatin regions. The nuclear Cajal and PML bodies with a size of about 1µm will therefore have access to only a subspace of the nucleus. Any movement of these bodies over distances above a few hundred nanometers will require a chromatin reorganization that allows the separation of chromatin subdomains to create accessible regions within and throughout the chromatin network. The interface between chromosome territories would provide such a subcompartment. More generally, the nuclear subspace accessible for nuclear bodies is likely to include the interface between chromosome territories that allows a movement of the bodies by a transient separation of chromatin domains. By random movements, the nuclear bodies explore this accessible space and are expected to be localized more frequently in these more open chromatin regions. Inasmuch as they coincide with regions of active gene transcription, they would also constitute possible biological targets of PML and Cajal bodies. This hypothesis can be tested by analyzing the intranuclear mobility and localization of multiple nuclear components simultaneously. In this respect, however, we need to take an ambitious step to develop a more integrative and global approach by studying nuclear bodies, RNA, and chromatin loci in parallel within the same living cell in order to identify their functional relations.

Finally, we should not forget that DNA methylation and histone modifications occur in the context of a higher-order chromatin structure. Nucleosomes, formed by the wrapping of 147 bp of DNA segment around a histone octamer core organized into the central $(H3-H4)_2$ tetramer and two peripheral H2A-H2B dimers, are the champions of that league. The nucleosome is the first level of DNA compaction in the nucleus.[1] A second level of compaction consists of a solenoid structure formed by the nucleosomal array and stabilized by the linker histone H1 (Boi 2008). Multisubunit complexes, such as those constituted by the SWI/SNF proteins, use the energy of ATP to mobilize nucleosomes and allow the access of the transcriptional machinery; or massive repressive complexes counteract SWI/SNF functions, as does the polycomb group gene family. In the end, the gene expression and function, and the overall genome activity, of the healthy cell is the result of the balance between these massive forces shaping our human epigenome.

The organization of DNA into chromatin is a highly dynamic process that reveals different degrees of plasticity of the many nuclear complexes involved in the remodeling of chromatin. Moreover, this plasticity plays a dynamic regulatory role in the gene transcription process and in the DNA repair process. The organization of DNA must be compatible with access of those DNA binding factors that regulate genome replication: the transcription of genes, recombination of chromosomes, and repair of damaged DNA. The modulation and the extent of chromatin folding by histone acetylation above the 30-nm fiber level and by other enzymatic processes constitute an important, early regulatory principle of gene expression and cell differentiation (Widom 1998). Recent years have witnessed a rapid discovery of enzymes that modify the structure of chromatin in response to cell internal and external cues, rendering chromatin a highly dynamic structure. Chromatin "plasticity" is mainly brought about by ATP-dependent chromatin "remodeling" factors, multiprotein complexes containing nucleic acid-stimulated DEAD/H ATPases of the Swi2/Snf2 subfamily. These enzymes couple ATP hydrolysis to alterations of the chromatin structure at the level of the nucleosomal array, which generally facilitates the access of DNA binding proteins to their cognate sites. Energy-dependent modifications of chromatin structure revealed distinct phenomena for individual classes of remodeling factors, such as the modification of the path of DNA supercoiling around the histone octamer, the generation of accessibility of nucleosomal DNA to DNA-binding proteins, and the stable distortion of nucleosome structure, including formation of particle with "dinucleosome" characteristics.

Not all enzymes are equally active in all assays, but all are capable of inducing ATP-dependent relocation of histone octamers, their "sliding," on DNA. It seems that the diversity of nucleosome "remodeling" phenomena brought about by the enzymes of the different classes may result from variations of one basic plastic theme, reminiscent of the action of DNA translocases. Conceivably, nucleosome remodeling enzymes simply enhance the intrinsic dynamic properties of nucleosomes by lowering the energy barrier due to the destabilization of histone-DNA interactions. Accordingly, "twisting" and "looping" models have also been invoked to explain catalyzed nucleosome mobility. Taken together, the available data do not support models invoking DNA twisting as the main driving force for nucleosome movements. Rather, the data favor a "loop recapture" model, in which the distortion of DNA into a loop at the nucleosome border initiates nucleosome sliding. Overall, the current phenomenology is consistent with chromatin remodelers working as anchored DNA translocases. Most, if not all, experimental results can be explained by one general mechanism. DNA translocation against a fixed histone body leads to detachment of DNA segments from the edge of the nucleosome, that is, to their bending and recapture by the histones to form a loop. Depending on the step length of the remodeling cycle, which corresponds to the length of the DNA segment detached from the nucleosomal edge and the extent of inclusion of nucleosomal linker DNA into a loop, variable-sized DNA loops may be generated on the nucleosome surface. The speed, directionality and processivity with which such loops are propagated will determine the predominant result of a remodeling reaction, such as the detection of an "altered path" of the DNA or nucleosome translocation. Thus, the rich phenomenology of nucleosome necessarily involves quantitative differences in certain kinetic and geometric parameters. However, it is clear that all types of remodeling enzymes are able to increase the dynamic properties of nucleosomes in arrays and to generate accessible sites. Therefore, the looping model mentioned previously will need to be refined as one considers nucleosome dynamics in a folded nucleosomal fiber.

Epigenetics, Diseases, and Aberrant Chromatin Alterations
In order to highlight the fundamental fact that the organizational properties of chromatin influence genome activity, in the sense that they are the principal carrier and activator of the multilevel genetic and epigenetic information, let us now consider the epigenome of a sick cell. Most human diseases have an epigenetic cause. The perfect control of our cells by DNA methylation, histone modifications, chromatin-remodeling and microRNAs becomes dramatically distorted in the sick cell. In other words, severe alterations of nuclear forms and especially of the chromatin and the chromosome may provoke different types of damage to the cell's activity, thus suggesting that the topological forms of living systems are one of the most fundamental determinants of the unfolding of biological functions during development and evolution. The groundbreaking discoveries have been initially made in cancer cells, but it is just the beginning of the characterization of the aberrant epigenomes underlying neurological, cardiovascular, and immunological pathologies.

In human cancer, the DNA methylation aberrations observed can be considered as falling into one of two categories: transcriptional silencing of tumor suppressor genes by CpG island promoter hypermethylation in the context of a massive global genomic hypermethylation, and a massive global genomic hypomethylation. CpG islands become hypermethylated, with the result that the expression of the contiguous gene is shut down. If this aberration affects a tumor suppressor gene, it confers a selective advantage on that cell and is selected generation after generation. Recently, researchers have contributed to the identification of a long list of hypermethylated genes in human neoplasias, and this epigenetic alteration is now considered to be a common hallmark of all human cancers affecting all cellular pathways. At the same time that the aforementioned

CpG islands become hypermethylated, the genome of the cancer cell undergoes global hypomethylation. The malignant cell can have 20–60 percent less genomic 5mC than its normal counterpart. The loss of methyl groups is accomplished mainly by hypomethylation of the "body" (coding regions and introns) of genes and through demethylation of repetitive DNA sequences, which account for 20–30 percent of the human genome.

How does global DNA hypomethylation contribute to carcinogenesis? Three mechanisms can be invoked as follows: chromosomal instability, reactivation of transposable elements, and loss of imprinting. Undermethylation of DNA may favor mitotic recombination, leading to loss of herezygosity as well as promoting karyotypically detectable rearrangements. Additionally, extensive demethylation in centromeric sequences is common in human tumors and may play a role in aneuploidy. As evidence of this, patients with germline mutations in DNA methyltransferase 3b (*DNMT3b*) are known to have numerous chromosome aberrations. Hypomethylation of malignant cell DNA can also reactivate intragenomic parasitic DNA, such as L1 (Long Interspersed Nuclear Elements, LINEs) and Alu (recombinogenic sequence) repeats. These and other previously silent transposons may now be transcribed and even "moved" to other genomic regions, where they can disrupt normal cellular genes. Finally, the loss of methyl groups can affect imprinted genes and genes from the methylated-X chromosome of women. The best-studied case is of the effects of the H19/IGF-2 locus on chromosome 11p15 in certain childhood tumors (Ballestar and Esteller 2005; Esteller 2005). DNA methylation also occupies a place at the crossroads of many pathways in immunology, providing us with a clearer understanding of the molecular network of the immune system. Besides, aberrant DNA methylation patterns go beyond the fields of oncology and immunology to touch a wide range of fields of biomedical and scientific knowledge.

The most important theme of the previous remarks on the human epigenome, which is mainly related to a methodological and epistemological revolution in epigenetics, may be summarized as follows. Cells of a multicellular organism are genetically homogeneous but structurally and functionally heterogeneous, owing to the differential expression of genes. Many of these differences in gene expression arise during development and are subsequently retained through mitosis. Stable modifications of this kind are said to be "epigenetic," because they are heritable in the short term but do not involve mutations of the DNA itself. The two most important nuclear processes that mediate epigenetic phenomena are

DNA methylation and histone modifications. Epigenetic effects by means of DNA methylation have an important role in development but can also arise stochastically as humans and animals age. Identification of proteins that mediate these effects has provided insight into this complex process, as well as into the diseases that occur when it is perturbed. External influences on epigenetic processes are seen in the effects of diet on long-term diseases such as cancer. Thus, epigenetic mechanisms seem to allow an organism to respond to the environment through changes in gene expression. The extent to which environmental effects can provoke epigenetic response is a crucial question that is still largely unanswered.

New Paradigms for Explaining Gene Expression and Cell Activity

The development in recent years of epigenetics entails the emergence of a more integrative and global approach to the study of biological forms and functions. To tackle the whole human epigenome and to deal with the entire organism, it is necessary to elucidate the relationship between the different levels of plasticity of protein complexes associated with chromatin remodeling and gene regulation, and the various levels of complexity exhibited by the phenotypic patterns during embryogenesis. The landscape of genetic expression revealed by epigenetics studies appears to be much more complex than that showed by DNA sequencing alone, and it clearly results from diverse layers of biological information (DNA folding, histone modifications, the complex regulatory roles of DNA methylation, chromatin remodeling complexes, the spatial organization of chromosomes, the architecture of nuclear bodies, cell morphology, and mobility), which intervene at different stages of the spatial and temporal development and evolution of a living human organism.

In fact, the most unusual genetic phenomena have very little to do with the genes themselves. True, as the units of DNA that encode proteins needed for life, genes have been at biology's center stage for decades. But work over the past ten years suggests that they are little more than puppets (Abbott 1999; Coen 1999; Cornish-Bowden and Cárdenas 2001; Jaenisch and Bird 2003; Misteli 2001). An assortment of proteins and sometimes RNAs pull the strings, telling the genes when and where to turn on or off. These findings are helping researchers understand long-standing puzzles. Why, for example, are some genes from one parent "silenced" in the embryo, so that certain traits are determined only by the other parent's genes? Or how are some tumor suppressor genes inactivated—without any mutation—increasing the propensity for cancer? Such phenomena are clues suggesting that gene expression is not determined solely by the DNA code itself. Instead, as we have shown in the previous sections, that cellular and organismic activity also depends on a host of so-called epigenetic phenomena—defined as any gene-regulating activity that doesn't involve changes in the DNA code and that can persist through one or more generations. Over the past ten years, mainly thanks to the development of a more integrative and global approach, cell and molecular biologists have been able to show the four fundamental facts that, taken together, represent a major conceptual and experimental breakthrough in the life sciences:

1. Gene activity is influenced by the proteins that package the DNA into chromatin, the protein-DNA complex that helps the genome fit nicely into the nucleus, by enzymes that modify both those proteins and the DNA itself, and even by RNAs. Chromatin structure affects the binding of transcription factors, proteins that control gene activity, to the DNA. Protein histones in chromatin are modified in different ways to modulate gene expression. The chromatin-modifying enzymes are now considered the "master puppeteers" of gene expression. During embryonic development, they orchestrate the many changes through which a single fertilized egg cell turns into a complex organism. And, throughout life, epigenetic changes enable cells to respond to environmental signals conveyed by hormones, growth factors, and other regulatory macromolecules without altering the DNA itself. In other words, epigenetic effects provide a mechanism by which the environment can very stably change living beings. The most important point that needs to be stressed here is that chromatin is not just a way to package the DNA to keep it stable. All the recent work on acetylation, methylation, phosphorylation, and histone modifications and their direct correlation with gene expression show that chromatin's proteins are much more than static scaffolding (Vogelauer et al. 2000; Lutter, Judis, and Paretti 1992; Fraga et al. 2005; Almouzni et al. 1994). Instead, they form an interface between DNA and the rest of the organism. The topological and dynamical modifications of chromatin structure play a crucial role, sometimes for clearing the way for transcription and other times for blocking it. The exact nature of these modifications remains largely mysterious. One may think that the different modifications mean different things, because they recruit different kinds of proteins and prevent other kinds of modifications.

2. These proteins and RNAs control patterns of gene expression that are passed on to successive generations. A variety of RNAs can interfere with

gene expression at multiple points along the road from DNA to protein. More than a decade ago, plant biologists recognized a phenomenon called posttranscriptional gene silencing, in which RNA causes structurally similar mRNAs to be degraded before their messages can be translated into proteins (Costa and Shaw 2006; Haswell and Meyerowitz 2006). In 1998, a similar phenomenon was found in nematodes, and it has since turned up in a wide range of other organisms, including mammals (Widom 1998). RNAs can also act directly on chromatin, binding to specific regions to shut down gene expression. Sometimes an RNA can even shut down an entire chromosome. Furthermore, newly formed female embryos solve the so-called dosage compensation problem—female mammals have two X chromosomes, and if both were active, their cells would be making twice as much of the X-encoded proteins as males' cells do—with the aid of an RNA called XIST, translated from an X chromosome gene. By binding to one copy of the X chromosome, XIST somehow sets in motion a series of modifications of its chromatin that shuts the chromosome down permanently. Thus, regulatory noncoding RNAs could be widespread in the genome, and influence gene function. The unit of inheritance—that is, a gene—now extends beyond the sequence to epigenetic modifications of that sequence. Moreover, the various epigenetic profiles that generate phenotypic differences may retroact on the arrangement of gene sequences and thus influence genome integrity.

3. In 1957, Waddington proposed an epigenetic hypothesis according to which patterns of gene expression, not genes themselves, define each cell type (Waddington 1957). Moreover, many biologists thought that the genome changes all the time as cells differentiate. Liver cells, for instance, became liver cells by losing unnecessary genes, such as those involved in making kidney or muscle cells. In other words, certain genes would be lost during development. One of the best clues for this phenomenon came from the realization that the addition of methyl groups to DNA plays some role in silencing genes—and that somehow the methylation pattern carries biological information over from one generation to the next. Besides, since the 1970s, cancer biologists have observed that the DNA in cancer cells tends to be more heavily methylated than DNA in healthy cells. So methylation might contribute to cancer development by altering gene expression. Indeed, a demonstration of this claim was recently established (Ballestar and Esteller 2005). The combined observations that DNA methylation can result in repression of gene expression, and that promoters of tumor suppressor genes are often methylated in human

cancers, provided an alternative mechanism for the inactivation of these genes that does not involve genetic mutations. Thus, the changes in methylation in tumors are in fact the cause, and not merely a consequence, of tumor formation.

4. Many observational data concerning anatomic and morphological differences in the phenotypic lineage made researchers aware that there could be parent-specific effects in the offspring. Other observations made through the centuries suggested that the genes passed on by each parent had somehow been permanently marked—or imprinted—so that expression patterns of the maternal and paternal genes differ in their progeny. These so-called imprints have since been found in angiosperms, mammals, and some protozoa. Over the past few years, several genes have been identified that are active only when inherited from the mother, and others only when inherited from the father. Many imprinted genes have been found; about half are expressed when they come from the father and half when they come from the mother. Among these are a number of disease genes, including the necdin and UBE3A genes on chromosome 15 that are involved in Prader-Willi and Angelman syndromes, and possibly p73, a tumor suppressor gene involved in the brain cancer neuroblastoma. Several others, including Peg3 and Igf2, affect embryonic growth or are expressed in the placenta.

A Few Fundamental Themes in the Life Sciences

In this section, we would like to make several remarks on some new research roads in biology to which we should pay much more attention in the future. The crucial point is that there are, as we have tried to show throughout this discussion, different layers of biological information depending on the level of organization one considers for studying the properties and behaviors of any living organism at the various stages of its embryogenetic development and overall growth. Moreover, these different layers are interconnected and may all be involved simultaneously and in a coordinated way in cell activity and an organism's development. Let us point to a few important aspects of this multilayer organization of biological information.

A first worthwhile theoretical remark is that biologists are striving to move beyond a "parts list" to more fully understand the ways in which network components interact with one another to influence complex processes. Thus attention has turned to the analysis of networks that operate at many levels. At the scale of networks of interacting proteins that govern cellular function, the flagellated bacterium *Caulobacter crescentus* has been a model system for cell cycle regulation for at least twenty-five years. This example shows clearly that the transcriptional regulatory circuits provide only a fraction of the signaling pathways and regulatory mechanisms that control the cell. Rates of gene expression acting in cells are modulated through posttranscriptional mechanisms that affect mRNA half-lives and translation initiation and progression, as well as DNA structural and chemical state modifications that affect transcription initiation rates. Phosphotransfer cascades provide fast point-to-point signaling and conditional signaling mechanisms to integrate internal and external status signals, activate regulatory molecules, and coordinate the progress of diverse asynchronous pathways. As if this were not complex enough, we are now finding that the interior of bacterial cells is highly spatially structured, with the cellular position of many regulatory proteins as tightly controlled at each moment in the cell cycle as are their concentrations.

The second remark relates to the proteomics challenge. Learning to read patterns of protein synthesis could provide new insights into the working of the cell and thereby a better understanding of how organisms, including humans, develop and function. By identifying proteins on the scale of the proteome—which can involve tens or even hundreds of thousands of proteins, depending on the state of the cells being analyzed—proteomics can answer fundamental questions about biological mechanisms at a much faster pace than the single-protein approach. The "global" picture painted by proteomics can, for example, allow cell biologists to start building a complex map of cell function by discovering how changes in one signaling pathway—the cascade of molecular events sparked by a signal such as a hormone or neurotransmitter—affect other pathways, or how proteins within one signaling pathway interact with each other. The "global" picture also allows medical researchers to look at the multiplicity of factors involved in diseases, very few of which are caused by a single gene. Proteomics is very likely one of the most important of the "postgenomic" approaches to understanding gene function, because it is the proteins encoded by genes that are ultimately responsible for all processes that take place within the cell. But although proteins may yield the most important clues to cellular function, they are also the most difficult of the cell's components to detect on a large scale.

A second, complementary postgenomic approach is expression profiling, also known as transcriptomics. When a gene is expressed in a cell, its code is first transcribed to an intermediary "messenger RNA" (mRNA), which is then translated into a protein. Transcriptomics involves identifying the mRNAs expressed by the genome at a given time. This provides a snapshot of the genome's plans for protein synthesis under the cellular conditions at that moment. Transcriptomics can, specifically, yield important biological information about which genes are turned on, and when. But it has the disadvantage that although the snapshot it provides reflects the genome's plans for protein synthesis, it does not represent the realization of those plans. The correlation between mRNA and protein levels is poor—generally lower than 0.5—because the rates of degradation of individual mRNAs and proteins differ, and because many proteins are modified after they have been translated, so that one mRNA can give rise to more than one protein. In even the simplest self-replicating organism, *Mycoplasma genitalium*, there are 24 percent more proteins than genes, and in humans there could be at least three times more. Posttranslational modification of proteins is important for biological processes, particularly in the propagation of cellular signals, for example, where the attachment of a phosphate group to a protein can trigger either activation or inactivation of a signaling cascade. So measuring proteins might directly provide a more accurate picture of the biological information involved in a cell's activity.

The development of a proteomics program has led in recent years to a significant elucidation of the relationship between structure and function in biomolecules and to an important revision of the prevailing paradigm that *(rigid) structure (linearly) determines function*. Several studies on the role played by proteins and protein interaction in biological phenomena have uncovered several misconceptions regarding the nature of the relation between the structure and function of biomolecules:

1. Because the overall three-dimensional structure of proteins is always much better conserved than their sequence, it is not uncommon for members of a protein family that possess no more than 10 to 30 percent sequence identity to have structures that are practically superimposable. Residues critical for maintaining the protein-fold and those involved in functional activity tend to be highly conserved. However, because proteins during evolution gradually lose some functions and acquire new ones, the residues implicated in the function will not necessarily be retained even when the protein-fold remains the same. Conservation of protein-fold will not, then, be correlated with retention of function, because a link between structure and function would be expected only

if attention were restricted to the functional binding site region instead of the whole protein.

2. Another difficulty in analyzing correlations between structure and function lies in the fact that individual proteins usually have several functions. It has been estimated that proteins are able, on average, to interact with as many as five partners through a variety of binding sites.

3. A further ambiguity lies in the term "function" itself. This term is used in different ways, and a possible correlation with structure will depend on which aspect of function and which level of biological organization are being considered. Biochemists tend to focus on the molecular level and consider mainly activities like binding, catalysis, or signaling. In many instances, the only activity that is discussed is binding activity, and so function is taken as synonymous with binding. However, functions can also be defined at the cellular and organismic level, in which case they acquire a meaning only with respect to the biological system as a whole, for instance, by contributing to its health, performance, survival or reproduction.

4. Protein functions can also be distinguished in terms of the biological roles they play at the organismic level, and this has led to a classification of functions that corresponds to the classification of energy-, information-, and communication-associated proteins. The link between such biological roles and protein structure is less direct than between binding activity and structure, as these functions tend to result from the integrated interactions of many individual proteins or macromolecular assemblies.

5. The prevailing paradigm that structure determines function is often interpreted to mean that there is a causal relation between structure and function. Although a biological activity always depends on an underlying physical structure, the structure in fact does not possess causal efficacy in bringing about a certain activity. Causal relations are dynamic relations between successive events, not between two material objects or between a geometrical static structure and a physicochemical event. Thus a biological event such as a binding reaction cannot be caused by something that is not an event, like the structure of one or both interacting partners. It is also impossible to deduce binding activity from the structure of one of the interacting molecules if a particular relationship with a specific partner has not first been identified. This is because a binding site is essentially a relational entity defined by the interacting partner and not merely by structural features that are identifiable independently of the relational nexus with a particular ligand.

The structure of a binding site, as opposed to the structure of a molecule, cannot be described without considering the binding partner. Because the static geometrical structure of a protein is not the only and most important cause of its function, attempts to analyze structure-function relationships should consist of uncovering correlations rather than (linear) causal relations. A conception shift is thus needed in proteomics, because there is not a unique, necessary, and sufficient relation between the three-dimensional structure of a protein and its biological activity, but a nexus of dynamical relationships between protein complexes and their interactions and activities. There is definitely a direct and fundamental link between the topological folding of proteins, the tertiary forms that result from this folding, and their dynamics in the context of a cell's activity. However, the biological information associated with proteins does not derive only from structural information, but also from the complex functional networks that connect specific binding sites at the molecular level to the cell's activity and to the more global organismic level of organization and working.

Systems Biology

The Need for a Systems Biology Approach

The third remark, which relates to the previous one, is aimed at highlighting the importance of a systems biology approach. Systems biology is about *interactions* rather than about constituents, although knowing the constituents of the system under study may be a prerequisite for starting description and modeling. Interactions often bring about properties sometimes called "emergent properties." For example, a system may start oscillating, although its individual constituents do not. For example, evolutionary biologists have for a long time wondered about how jump-like transitions can occur in evolution. From the viewpoint of systems theory, the answer arises from bifurcations. In a nonlinear system, at certain points in parameter space (called *critical points*), bifurcations occur; that is, a small change in a parameter leads to a qualitative change in system behavior (e.g., a switch from steady state to oscillation). It is clear that the number of potential interactions within a system is far greater than the number of constituents. If only pairwise interactions were allowed, the former number would be n^2 if the latter number were denoted by n. The number of interactions is even larger if interactions within triples and larger sets are allowed, as is the case in multiprotein complexes. In systems biology, a biological object or being is a *system* if emergent properties result from it. Genomics has certainly been a very important and fruitful undertaking and has given us many new insights into molecular biology. However, much of molecular biology is based on reductionism and simple determinism. It is an extreme exaggeration to say that the human genome has been "deciphered." Besides the fact that functions have not yet been assigned to all ORFs, it should be acknowledged that even if all functions were known, we would be far from understanding the phenomenon of life, because knowledge of all the individual gene products does not say much about the interactions between them. According to a systems view of life, the study of the dynamics and interaction networks is essential for understanding the ways in which living organisms regulate their cellular activity and organize their physiological growth. One of the major goals of systems biology is to find appropriate ways of diagramming and mathematically describing the specific, complex interactions within and between living cells. Because complex systems have emergent properties, their behavior cannot be understood or predicted simply by analyzing the structure of their components. The constituents of a complex system interact in many ways, including negative feedback and feed-forward control, which lead to dynamic features that cannot be captured satisfactorily by linear mathematical models that disregard cooperativity and nonadditive effects. In view of the complexity of informational pathways and networks, new types of mathematics are required for modeling these systems.

It is worth noticing that the specificity of a complex biological activity does not arise from the specificity of the individual molecules that are involved, as these components frequently function in many different processes. For instance, genes that affect memory formation in the fruit fly encode proteins in the cyclic AMP (camp) signaling pathway that are not specific to memory. It is the particular cellular compartment and environment in which a second messenger, such as camp, is released that allow a gene product to have a unique effect. Biological specificity results from the way in which these components assemble and function together. Interactions between the parts, as well as influences from the environment, give rise to new features, such as network behavior, which are absent in the isolated components. Consequently, "emergence" has appeared as a new concept that complements "reduction" when reduction fails. Emergent properties resist any attempt at being predicted or deduced by explicit calculation or any other means. In this regard, *emergent properties* differ from *resultant properties*, which can be predicted from lower-level information. For instance, the resultant mass of a multicomponent protein assembly is simply equal to the sum of the masses of each individual component. However, the way in which we taste the saltiness of sodium chloride is not reducible to the properties of sodium and chlorine gas. An important aspect of emergent properties is that they have their own causal powers, which are not reducible to the powers of their constituents. According to the principles of emergence, the natural world is organized into stages that have evolved over evolutionary time through continuous and discontinuous processes. Reductionists advocate the idea of "upward causation," by which molecular states generally bring about higher-level phenomena, whereas proponents of emergence admit "downward" causation, by which higher-level systems may influence lower-level configurations.

Chromatin Code: Complex Regulatory Principle of Cell Activity
The last remark relates to the dynamic reorganization of chromatin during the cell cycle. Chromatin remodeling faces major questions concerning the intricate and multilevel interplay between the topological plasticity of nuclear structures involved in genome regulation and cell activity and the ever-increasing complexity of gene regulatory networks. The experimental evidence suggests that chromatin form and its modifications play a critical role in gene regulatory coding (gene activation or gene silencing), in the emergence of cellular differentiation and in development. Even if genomic DNA is the ultimate template of our heredity, clearly DNA is far from being the exclusive entity responsible for generating the full range of information that ultimately results in a complex eukaryotic organism, such as a human. We favor the view that epigenetics, imposed at the level of DNA-packaging proteins (histones and nonhistones), is a critical feature of a genome-wide mechanism of information storage and retrieval that is only beginning to be understood. All the theoretical and experimental work we considered in this discussion suggests that a "chromatin code" exists that may considerably extend the information potential of the genetic (DNA) code. Chromatin coding is a second layer of coding implemented by histone tail posttranslational modifications outside the nucleosome. This second-level code is required in eukaryotic cells to provide the additional information necessary to process their long genome (compared to prokaryote ones). There is more and more evidence that histone proteins and their associated covalent modifications contribute to a mechanism that can alter chromatin

structure, thereby leading to inherited differences in transcriptional "on-off" states or to the stable propagation of chromosomes by defining a specialized higher-order structure at centromers. Differences in "on-off" transcriptional states are reflected by differences in histone modifications that are either "euchromatic" (on) or "heterochromatic" (off). The "chromatin code" is read out by co-regulators analog and prior to the reading out of the primary coding (the nucleotide sequence) by transcription factors, and then translated by means of different transcriptional and posttranscriptional steps into biological functions.

Furthermore, it has been shown that control of chromatin packaging into condensed or decondensed fibers plays a major role in the regulation of gene expression. Several complex events are associated with the decondensation of chromatin, which is characteristic of the "open" or active state. Genes in open regions of chromatin can be expressed efficiently, whereas genes in condensed or closed regions are silent. Recent research suggests that large regions of chromatin are literally "ploughed" open by the RNA polymerase II complex in a process known as intergenic transcription (Cremer and Cremer 2001; Ehrenhofer-Murray 2004; Santoro and Grummt 2005). The RNA polymerase complex has been shown to contact a number of factors capable of modifying the structure of the chromatin fiber by adding chemical side groups to nucleosomes and other chromatin proteins. These modifications are thought to increase the accessibility of the genes within these regions or domains, resulting in augmented binding of transcription factors to the gene. However, this alone is not enough for efficient gene expression. Many genes require additional regulatory regions of DNA, known as enhancers, that are often located at considerable distances from the gene along the chromatin fiber. It has been shown that distant enhancers actually physically contact their target genes in the nucleus by looping out the intervening DNA (Cozzarelli and Holmes 2000; Calladine et al. 2004; Boi 2007; Belmont 2002). Such long-range interactions between enhancers and genes are powerful switches that turn on transcription of individual genes resulting in high levels of expression. Recent work suggests that these regulatory interactions between enhancers and genes can occur only if the chromatin containing them is first remodeled to the open state by intergenic transcription (Costa and Shaw 2006; Cremer et al. 2004; Ehrenhofer-Murray 2004; Gilbert and Bickmore 2006). These are essential processes in the chain of events that control gene expression.

The previous discussion clearly indicates that there is a strong correlation between the different local plastic modifications and the overall topological reorganization of chromatin structure and the series of events that lead to high levels of gene expression. Chromatin remodeling and dynamic chromatin reorganization is a key regulatory principle whose processing is essential to open the way to gene activity, but also to control cell differentiation and to orchestrate embryonic development. Overall chromosome stability and identity seem to be influenced by epigenetic alterations of the underlying chromatin structure. In keeping with the distinct qualities of accessible and inaccessible nucleosomal states, it could be that "open" (euchromatic) chromatin represents the underlying principle that is required for inheritance of progenitor character and young cell division. Conversely, "closed" (heterochromatic) chromatin is possibly the reflection of a developmental "memory" that stabilizes lineage commitment and gradually restricts the self-renewal potential of our somatic cells. Whatever it may be, epigenetics imparts a fundamental regulatory system beyond the sequence information of our genetic code and emphasizes that Mendel's gene is much more than just a DNA moiety.

Research in the Biological Sciences and Concluding Remarks

What we have tried to show in the previous analysis is, first, that the phenomena of epigenetics clearly reveal the existence of a cryptic code—that is, a systemic web of regulated processes—of physicochemical (the DNA sequence is chemically altered) and topological (the structure of chromatin is spatially modified into new forms) nature which is written (or unfolds) over our genome's DNA sequence. Second, and even more fundamental, not only is the DNA sequence important, but so is gene activity that is regulated in response to the environment. In other words, epigenetics gives greater place to the interactions of genes with their environment, which bring the phenotype into being. Thus, the epigenetic phenomena refer to extra layers of nuclear plasticity and information processing that influences in an essential way gene activity and cell functioning without altering the DNA sequence. Furthermore, recent research shows that the epigenetic code is "read" out by co-regulator complexes prior to the reading out of the primary coding (the nucleotide sequence) by transcription factors (Georgel 2002; Li 2002). These co-regulators interact with other genome maintenance and regulation pathways for permitting the cellular transcription machinery to "interpret" properly the gene regulatory code. Chromatin is indeed highly dynamic rather than static, and this dynamism represents another important mechanism of gene regulation in eukaryotes.

Biologists now understand that the cell must continually "remodel" the local chromatin structure to give regulatory molecules access to the DNA substrate. This remodeling rests on the action of multiprotein complexes, and generally involves either the hydrolysis of ATP to alter histone-DNA interactions, or the posttranslational modification of histone tails. In contrast to transcription factors, co-regulator complexes, though not interacting directly with DNA, frequently generate appreciable "synergies" between the actions of transcription factors, with changes in concentration having disproportionate (nonadditive) consequences on the rate of transcription. For example, studies show that the nuclear receptor transcription intermediate factor-2 (TIF2) promotes synergy between two transcription activators (Cremer et al. 2004; Benecke 2003, 2006). Researchers have also found that co-regulators can act to switch transcription factors from being repressors to being activators, as well as link gene regulation to other molecular processes such as DNA maintenance and replication (Görisch et al. 2004; Benecke 2003; Jin et al. 2005). All this suggests that transcription regulation can be predicted accurately only if co-regulators' activity is taken into account.

The sequence of the human genome is the same in all our cells, whereas the epigenome differs from tissue to tissue, from organ to organ, and from organism to organism, and changes in response to the cell's environment. Epigenetic codes are much more subject to environmental influences than the DNA sequence. This could also help to explain how lifestyle and toxic chemicals affect susceptibility to diseases. In fact, up to 70 percent of the contribution to a particular disease can be nongenetic. The challenge today is to pin down this vast, complex, and ever-changing code in a meaningful way. The diversity of epigenomes in different cell types means that it may not make sense to restrict our study to one single tissue, or to a particular time in a tissue's development, but the epigenome of all tissues and the overall organism's growth processes have to be mapped out. If the DNA sequence is like the musical score of a symphony, the epigenome is like the key signatures, phrasing, and dynamics that show how the notes of the melody should be played. Thus, although necessary, this "musical score" is not at all sufficient for cells to start and develop their multilevel activity.

When thinking at the systems level, it becomes clear that genes matter only because they are one of many cellular and organismic codes, each of which contributes to the construction of an organism by conveying a specific form of biological information. No single gene is more or less important than any other, and the loss of function of gene x causing phenotype X is not itself an interesting observation. It is interesting only if we can begin to qualitatively and quantitatively explain how gene x interacting with genes w, y, and z together produce phenotype X in context A but not in context B, and what predictive value this interaction has on the system. Stated another way, the complex processes of reading and interpreting the genome can take place only in the context of embryonic development (during which, besides, the genome is reprogrammed many times), the action of all proteins, all lipids, and other cellular mechanisms inherited from one parent. There are at least one hundred different proteins involved in the cellular machinery, and without their action the genome couldn't be expressed, which means that the biological informational content of DNA sequences would be very poor without this role of orchestration and construction of organisms played by proteins.

Another fundamental fact is the role at almost every level of organization and communication in living cells of protein-protein interaction. Before the proteomics revolution, it was known that proteins were capable of interacting with each other and that protein function was regulated by interacting partners. However, the extent and degree of the protein-protein interaction network was not realized. It is now believed that not only are the majority of proteins in a eukaryotic cell involved in complex formation at some point in the life of the cell, but also that each protein might have, on average, six to eight interacting partners (Black 2000; Misteli 2001; Remaut and Waksman 2006). The ways in which proteins interact range from direct apposition of extensive complementary surfaces to the association of specialized, often modular domains, to short, unstructured peptide stretches ("linear motifs"). With the availability of complete genome sequences, it has emerged that up to 30 percent of the proteomes in higher organisms consists of natively disordered elements. This material is now understood to often have a major role in the formation of large regulatory protein complexes by incorporating linear binding motifs or acting as spacers between protein-binding and activity-bearing modules (Collier 1998; Davidson et al. 2002; Georgel 2002). Association of such a sequence with modular domains affords greater flexibility, because both the peptide and peptide-binding domain can more readily be separated from the functional core of the binding partners (such as the active site of an enzyme), and thus do not usually impose considerable evolutionary constraints on the domains that support activity. Peptides or linear motifs can bind to proteins in a variety of ways. They are frequently held in an extended conformation, but recognition motifs can also consist of ß-turns, ß-strands, or α-helical structures. One particularly interesting case is the protein-protein interactions that involve association of a ß-strand from the ligand with a strand or a ß-sheet in the binding partner. The point to be stressed is that different kinds of ß-strand additions mediate protein-protein interactions in important cell processes, such as cell signaling or host-pathogen interaction.

We have to consider that a fully detailed image of a complex organism requires knowledge of all of the proteins and RNAs produced from its genome. Due to the production of multiple mRNAs through alternative RNA-processing pathways, human proteins often come in multiple variant forms. Only a global view of splicing regulation combined with a detailed understanding of its mechanisms will allow us to paint a picture of an organism's total complement of proteins and of how this complement changes with development and the environment. From the very beginning of a living organism—that is, the fertilized egg and embryogenesis—these proteins and cellular systems promote and control transcription and posttranscriptional modifications. And it is this whole system that enables cellular machinery to translate the lower layers of chemical information stored in the DNA sequence and packaged in the nucleus in viable and significant biological information and also to interpret genes properly in the framework of cell differentiation and an organism's growth.

The understanding that has recently emerged is that there are many crucial biological questions that cannot be resolved only by means of genetic sequencing and local molecular mechanism analysis. Development and evolution, and the formation and the function of the neural networks in the brain, are processes that are not easily broken down into elements corresponding to the effects of individual genes, individual biochemical components, or even individual cells. A holistic or systems approach seems to be required, and this is a challenge for theoretical as well as for molecular biologists: in particular, if development as such is to be understood, we need to uncover—presumably in a topological-combinatorial and dynamical manner—patterns of the activation of different sets of genes. In this respect, it has been demonstrated that a genomic regulatory network can explain the early development of the sea urchin embryo (Davidson et al. 2002; De Witt, Greil, and van

Steensel 2005). These regulatory circuits prescribe the ordered expression of genes that determine the fates of developing cells and move those cells together down a one-way path to yield a functional organism.

The definition of systems biology requires a theoretical integration of mathematics, physics, and biology for a better understanding, for instance, of a range of complex biological regulatory systems. It is very likely that the answer to many interesting biological issues lies on the frontier of mathematical patterns, physical constraints, and biological processes. In the previous section, we tried to show two very significant examples in which mathematics, physics, and biology interact very deeply—namely, the topological manipulations of topoisomerases on DNA, and the spatial folding of DNA into chromatin structure within the nucleus. In both cases, it is impossible to separate the following three levels of activity and organization of cells and organisms: the extremely accurate conformational flexibility of macromolecules (which seems to be a fundamental property of living matter acting locally and globally), the tendency of biological systems to work cooperatively in a wealthy variety of complex regulatory networks for ensuring the overall physiological integrity of organisms, and the multilevel dynamics whose action is essential for sustaining biochemical metabolism, cell activity, and the growth of any embryo into an adult organism. Among the different dynamical principles acting in living systems, two seem to be very pervasive: the continuous remodeling of the macromolecular structures that constitute the cell's nucleus (especially chromatin and chromosome), and the role of self-organization in the formation, maintenance, and organization of cellular structures. The most important processes related to the remodeling of macromolecular structures, like chromatin and chromosome, are those of folding (which leads to their condensation) and unfolding (which leads to their decondensation). These processes are connected, respectively, to conformational constraints (large-scale interactions and binding sites connectivity) and to organizational regulatory functions (expression of gene and cell activity)

A large part of this article has been dedicated to the first principle. But let us make a few brief remarks on the second principle. In contrast to the mechanism of self-assembly, which involves the physical association of molecules into an equilibrium structure (e.g., virus and phage proteins self-assemble to true equilibrium and form stable, static structures), the concept of self-organization is based on observations of chemical reactions far from equilibrium, and the associated processes involve the physical interactions of molecules in a steady-state structure.

This concept is well established in chemistry, physics, ecology, and sociobiology. Self-organization in the context of cell biology can be defined as the capacity of a macromolecular complex or organelle to determine its own structure based on the functional interactions of its components (Misteli 2001; Karsenti 2007). In a self-organizing system, the interactions of its molecular parts determine its architectural and functional features. The processes that occur within a self-organized structure are not underpinned by a rigid architectural framework; rather, they determine its organization. For self-organization to act on macroscopic cellular structures, three requirements must be fulfilled: a cellular structure must be dynamic, material must be continuously exchanged, and an overall stable configuration must be generated from dynamic components. Observations from recent studies on the dynamic properties of cellular organelles indicate that many macroscopic cellular structures, such as cytoskeleton, the cell nucleus, and the Golgi complex, fulfill the requirements for self-organization. These structures are characterized by two apparently contradictory properties.

On one hand, they must be architecturally stable; on the other hand, they must be flexible and prepared for change. Self-organization ensures structural stability without loss of plasticity. Fluctuations in the interaction properties of its components do not have deleterious effects on the structure as a whole. However, global and persistent changes rapidly result in morphological changes. The basis for the responsiveness of self-organized structures is the transient nature of the interactions among their components. The dynamic interplay of components generates frequent windows of opportunity, during which proteins can change their interaction patterns or be modified. The effective availability of components is controlled by posttranslational modifications via signal transduction pathways. Another important point that applies to many large biological systems (proteins networks, cell components) is the principle of specificity, which enables the different constituent molecules or macromolecules to recognize each other and to exclude others that do not belong, so that no external instructions are necessary to form the assembly. In other words, the pattern of an ordered structure is built into the bonding properties of its constituents, so that the system "assembles itself" without the need for a scaffold, which means that the system is capable of self-organization.

It has to be stressed that a systems approach may benefit strongly from currently discussed large-scale programs of "transcriptomics" and "proteomics." These include systematic studies of the expression of messenger RNA and proteins within cells of one and the same organism under different conditions of development. Further aspects of such postgenomic or epigenomics programs are systematic comparative analyses of structures, modules and functions of proteins and regulatory sequences outside of genes, as well as their posttranslational modifications and associations. For the understanding of cell differentiation, the regulatory functions of noncoding sequences are of particular importance. Many different fundamental issues are connected with such programs. For example, for the developmental biologist, it is hoped that in this way the internal order of the network of gene regulation—namely, its relation to morphogenetic processes—may be revealed. Comparison of different organisms may allow us to reconstruct pathways of evolution with respect to protein structure and function and the genomic organization of the regulation of gene activities.

One of the many differences between biology and the physical sciences lies in the uniqueness of biological entities and the fact that these are the product of a long history (Jacob 1970; Nicolis and Prigogine 1987). Living beings are truly historical structures. It can be said that all biological order results from structural and geometrical constraints, biological robustness and adaptation, epigenetic flexibility and variability, and historical contingencies (the possible pathways of evolution). This simultaneously controlled and contingent natural history unfolds along different scales in time and space and follows different possible paths, so living systems may encounter bifurcations, singularities, and criticalities during their development. In fact, time and space are highly dynamic as geometrical and topological parameters, and also in the sense that they are the very outcome of the action of intrinsic physical dynamics and of their interactions with external factors or constraints, such as environmental changes. This problem may also be characterized by saying that there are not absolute phenomena in biology. Another outstanding feature of all organisms is their unlimited organizational and dynamical complexity. Every biological system is so involved in multiple interactions and pathways, so rich in feedback devices, and so plentiful in retroactions and unforeseen effects that one wonders whether a complete description is possible. As one goes to higher levels of organization, not all the properties of the new entity are knowable consequences of the properties of the components—no more than chemistry is, in practice, predictable from physics. We mention a last significant characteristic of the complex structures that one finds in biology, namely, the difference between contingency (variation) and necessity (sameness). In the inorganic world, this distinction may be illustrated by the problem of the shape of snow crystal (Stewart 1998). The fairly correct explanation, known since the time of Kepler and Descartes, is that the hexagonal form of the crystals is produced by the close packing in a plane of spherical water globules; in more precise terms, the internal structure involves puckered hexagonal layers. The hexagonal symmetry is thus a necessity and follows from what Kepler called the "demands of matter" (1611). (Necessity does not mean, however, that the crystals fulfill a universal law, as some of them may present erratic features or even other kind of symmetries.) But what of the external (or, more exactly, dynamic) shape in the living world? Many different individual shapes are found, and each is contingent on the particular history of its formation. How the symmetry of the dynamic shape is maintained during growth remains an unsolved problem.

A few ideas are at the core of this chapter. To conclude, it might be useful to rephrase them as brief statements:

1. The DNA sequence does not contain *all* the information for producing an organism. In other words, the genomic DNA is not the sole purveyor of biological information, because, first of all—as we have thoroughly shown—in most relevant cases, genes are expressed in an epigenetic framework and activated within specific cellular regulatory activities. Even the so-called central dogma of molecular biology—namely, that DNA sequences define protein sequences in a way in which the latter do not define the former (or, stated differently, that a DNA sequence contains the necessary and complete information for a protein)—is only partly true, and it needs to be deeply revised.

2. In addition, the concept of information itself is misleading and reductionist and so must take into account the highly complex processes responsible for gene expression and regulation and cell activity, as research in epigenetics plainly demonstrates (Li 2002; Jaenisch and Bird 2003; Callinan and Feinberg 2006). For biologists, reductionism means that a particular characteristic of a living organism can be explained in terms of chemistry and physics. This reductionism would eliminate the need for biology as a science. The problem with biology, unlike physics, is that its objects of interest are extremely complex. Exploring the limits of reductionism in biology is important, because there is ample evidence that many fields of biological studies are nonreductionist in nature; in other words, much of biology cannot be reduced to physicochemical properties.

3. Many relevant biological mechanisms involved in the development of the embryo and in morphogenesis reside less in the genomic DNA than in those epigenetic phenomena such as chromatin dynamics and chromosome organization, which control cell fate and embryo development. For example, it has been shown that organization and cell fate switch respond to positional information in certain plants; in other words, cells are sensitive to (and are controlled by) the spatial location of gradients of morphogenetic substances or morphogens in tissues in the developing embryo (Crick et al. 1961; Wolpert 1971).

4. Genes are not the whole of life, and genetic coding is unable to explain many fundamental biological processes of living systems, such as chromatin structures and chromosome spatial organization, cytoskeleton dynamics and mobility, cell signaling and communication, the mechanisms of pattern formation, and embryogenesis development. Moreover, the role of the non–genetically encoded properties of elasticity and of deformability of biological structures at the macromolecular, cellular, or multicellular level are now acknowledged to contribute to the regulation and generation of active physiological processes. Two very significant examples are: first, the motor role of biological membrane elasticity in the driving force of vesiculation-initiating plasma membrane endocytosis, and second, the role of geometrical strains and deformations of embryonic tissues in the regulation of developmental genes expression during the early steps of *Drosophila* embryo development at gastrulation (Farge 2003; Boi 2008).

5. In fact, the concept of creativity (which includes the ideas of mobility, action, and emergence) comes closer to describing the process of development, rather than the prevalent notion of simply following a set of instructions (i.e., a mechanical code). Biological development of an organism is not merely a read-out and implementation of a set of genetic instructions. Development is a continuing interaction between the "painter genotype" and the "canvas phenotype" that finally produces a living organism (Coen 1999). The generation of an individual entity through developmental processes is indeed a more creative act than the purely mechanical concept of molecular copying and reproducing could explain (the previously explained property of self-organization serves as a beautiful example that illustrates clearly certain striking features of creativity in biological structures and patterns).

6. The formation of patterns during development (cell differentiation, tissue shaping, organogenesis) cannot be explained solely by genetic coding and molecular mechanisms. We need much more than the

metaphors of coding and machines. For example, the principles of topological flexibility and dynamic organization allow (at least in part) us to explain how such patterns are generated by functionally driven remodeling processes (such as in chromatin folding or in chromosome reconfiguration), by geometric and mechanical constraints (see point 4), and by principles of self-reorganization under the action both of intrinsic dynamical factors (chemical, kinetic, and catalytic parameters and reactions) and the influence of external or environmental factors (energetic, metabolic, and so on).

We would like to conclude by stressing the fact that a new theory of living organisms requires not only new mathematics, new physics, and new epistemology, but also that some striking new ideas and methods borrowed from each of these disciplines be worked out together and merged into a meaningful interdisciplinary and global explanation of biological systems. Such an approach might contribute to restore to our fragmented biological sciences the kind of integration and unity they strongly need. The plethora of recent biological observations and theoretical models may pave the way to discover new mathematical objects and concepts and to open new frontier problems, and conversely, biology may benefit from these mathematical structures in organizing experimental data and clarifying their descriptions. This gathering of deep mathematical concepts, physical principles, and epistemological analysis is necessary for constructing a kind of relational biology, which will enable us to explore relationships among the systems, properties, and behaviors of living organisms. An attempt should be made to construct a theory—say, a semiobiology—in which the informational language (code, program, computation) so characteristic of molecular biology be completed by (and somehow translated into) the language of dynamical systems (phase space, bifurcations, trajectories) and the language of topology (deformations, plasticity, forms). These considerations lead ultimately to the suggestion that our traditional modes of system representation, involving fixed sets of sequential states, together with imposed mechanical laws, strictly pertain to an extremely limited class of systems that can be called simple (static) systems or mechanisms. Biological systems are not in this class, and they must be called complex or dynamic. Complex systems can be in some sense only approximated, locally and temporally, by simple ones, and so require a new set of mathematical ideas to be described. Such a fundamental change of viewpoint leads to a number of theoretical and experimental consequences, some of which were described in this discussion.

Notes

1. The genomic DNA of eukaryotes is very long (about 2 m in humans) compared to the diameter of the cell's nucleus (about 10^{-5} m). Packaging of the genome involves coiling of the DNA in a left-handed spiral around molecular spools, made of histone octamers, to form nucleosomes. About 80 percent of the genomic DNA is organized as nucleosomes. Nucleosome assembly is initiated by wrapping a 121 bp DNA segment around a tetramer of histones $(H3/H4)_2$. Association of H2A/H2B dimmers at either side of the tetramer organizes 147 bp of DNA. DNA is a moderately flexible polymer with a persistence length of about 150 pb. In the absence of exogenous forces, 150 pb of DNA follows essentially a straight path, but in a nucleosome, it coils in 1.65 toroidal superhelical turns around the octamer and is thus severely distorted. This means that DNA bending around the nucleosome is expected to happen at high energy costs. This energy cost is compensated by DNA-histone interactions occurring approximately every 10 bp on each DNA strand, generating 7 histone-DNA interaction clusters per DNA coil (superhelical locations (SHL) 0.5, 1.5, 2.5, . . . , 6.5). The DNA-histone interactions are stabilized by more than 116 direct and 358 water-bridged interactions, rendering the nucleosome a stable particle in the absence of additional factors.

References

Abbott, Alison. 1999. A post-genomic challenge: Learning to read patterns of protein synthesis. *Nature* 402 (6763): 715–720.

Almouzni, Geneviéve, and Paul Kaufman. 2000. DNA replication, nucleotide excision repair, and nucleosome assembly. In *Chromatin structure and gene expression*, ed. Sarah Elgin and Jerry Workman, 24–48. Oxford: Oxford University Press.

Almouzni, Geneviéve, Saadi Khochbin, Stefan Dimitrov, and Adolfus Wolffe. 1994. Histone acetylation influences both gene expression and development of *Xenopus laevis*. *Developmental Biology* 165 (2): 654–669.

Ballestar, Esteban, and Manel Esteller. 2005. The epigenetic breakdown of cancer cells: From DNA methylation to histone modification. In *Progress in molecular and subcellular biology, vol. 38: Epigenetics and chromatin*, ed. Philippe Jeanteur, 169–181. Berlin: Springer.

Belmont, Andrew. 2002. Mitotic chromosome scaffold structure: new approaches to an old controversy. *Proceedings of the National Academy of Sciences of the United States of America* 99 (25): 15855–15857.

Benecke, Arndt. 2006. Chromatin code, local non-equilibrium dynamics, and the emergence of transcription regulatory programs. *The European Physical Journal E. Soft Matter and Biological Physics* 19 (3): 379–384.

Benecke, Arndt. 2003. Genomic plasticity and information processing by transcriptional coregulators. *Complexus* 1 (2): 65–76.

Black, Douglas. 2000. Protein diversity from alternative slicing: A challenge for bioinformatics and post-genome biology. *Cell* 103 (3): 367–370.

Boi, Lucino. 2007. Geometrical and topological modeling of supercoiling in supramolecular structures. *Biophysical Reviews and Letters* 2 (3–4): 1–13.

Boi, Luciano. 2008. Interfaces between geometry, dynamics and biology: from molecular topology to the chromosome organization. In *New trends in geometry, and its role in the natural and living sciences*, ed. Luciano Boi and Carlo Bartocci, 220–248. London: Elsevier.

Calladine, Chris, and Horace Drew, Ben Luisi, and Andrew Travers. 2004. *Understanding DNA. The molecule and how it works*. London: Elsevier.

Callinan, Pauline A., and Andrew P. Feinberg. 2006. The emerging science of epigenomics. *Human Molecular Genetics* 15 (R1): R95–R101.

Coen, Enrico. 1999. *The art of genes: How organisms make themselves*. Oxford: Oxford University Press.

Collier, John. 1998. Information increase in biological systems: How does adaptation fit? In *Evolutionary systems: Biological and epistemological perspectives on selection and self-organization*, ed. Gertrudis Van de Vijver, Stanley N. Salthe, and Manuela Delpos, 129–140. Dordrecht: Kluwer.

Cornish-Bowden, Andreas, and Miguel Cárdenas. 2001. Complex networks of interactions connect genes to phenotypes. *Trends in Biochemical Sciences* 26 (8): 463–465.

Costa, Silvia, and Peter Shaw. 2006. Chromatin organization and cell fate switch respond to positional information in *Arabidopsis*. *Nature* 439 (7075): 493–496.

Cozzarelli, Nathan R., and Victor F. Holmes. 2000. Closing the ring: Links between SMC proteins and chromosome portioning, condensation, and supercoiling. *Proceedings of the National Academy of Sciences of the United States of America* 97 (4): 1322–1324.

Cremer, Thomas, Katrin Küpper, Steffen Dietzel, and Stanislav Fakan. 2004. Higher order chromatin architecture in the cell nucleus: on the way from structure to function. *Biology of the Cell* 96 (8): 555–567.

Cremer, Thomas, and Christoph Cremer. 2001. Chromosome territories, nuclear architecture and gene regulation in mammalian cells. *Nature Reviews: Genetics* 2 (4): 292–301.

Crick, Francis, Leslie Barnett, Sydney Brenner, and Richard J. Watts-Tobin. 1961. General nature of the genetic code in proteins. *Nature* 192 (4809): 1227–1232.

Davidson, Eric H., Jonathan P. Rast, Paola Oliveri, Andrew Ransick, Cristina Calestani, Chiou Hwa Yuh, and Takuya Minokawa, et al. 2002. A genomic regulatory network for development. *Science* 295 (5560): 1669–1678.

De Witt, Elzo, Frauke Greil, and Bas van Steensel. 2005. Genome-wide HP1 binding in *Drosophila*: Developmental plasticity and genomic targeting signals. *Genome Research* 15 (9): 1265–1273.

Ehrenhofer-Murray, Ann E. 2004. Chromatin dynamics at DNA replication, transcription and repair. *European Journal of Biochemistry* 271 (12): 2335–2349.

Esteller, Manel, ed. 2004. *DNA methylation: Approaches, methods and applications*. New York: CRC Press.

Esteller, Manel. 2005. Aberrant DNA methylation as a cancer-inducing mechanism. *Annual Review of Pharmacology and Toxicology* 45: 629–656.

Farge, Emmanuel. 2003. Mechanical induction of twist in the *Drosophila* foregut/stomodeal primordium. *Current Biology* 13 (16):1365–1377.

Fraga, Mario, Esteban Ballestar, Maria Paz, Santiago Ropero, Fernando Setien, Maria L. Ballestar, and Damia Heine-Suñer, et al., 2005. Epigenetic differences arise during the lifetime of monozygotic twins. *Proceedings of the National Academy of Sciences of the United States of America* 102 (30): 10604–10609.

Georgel, Philippe. 2002. Chromatin structure of eukaryotic promoters: A changing perspective. *Biochemistry and Cell Biology* 80 (3): 295–300.

Gilbert, Nick, and Wendy A. Bickmore. 2006. The relationship between higher-order chromatin structure and transcription. *Biochemical Society Symposium* 73: 59–66.

Görisch, Sabine M., Malte Wachsmuth, Carina Ittrich, Christian P. Bacher, Karsten Rippe, and Peter Lichter. 2004. Nuclear body movement is determined by chromatin accessibility and dynamics. *Proceedings of the National Academy of Sciences of the United States of America* 101 (36): 13221–13226.

Grewal, Shiv I. S., and Danesh Moazed. 2003. Heterochromatin and epigenetic control of gene expression. *Science* 301 (5634): 798–802.

Haswell, Elisabeth S., and Elliot M. Meyerowitz. 2006. MscS-like proteins control plastid size and shape in *Arabidopsis thaliana*. *Current Biology* 16 (1): 1–11.

Jacob, François. 1970. *La logique du vivant: Une histoire de l'hérédité*. Paris: Gallimard.

Jaenisch, Rudolf, and Adrian Bird. 2003. Epigenetic regulation of gene expression: How the genome integrates intrinsic and environmental signals. *Nature Genetics* 33 (March): 245–254.

Jenuwein, Thomas. 2006. The epigenetic magic of histone lysine methylation. *FEBS Journal* 273 (14): 3121–3135.

Jin, Jingji, Yong Cai, Bing Li, Ronald C. Conaway, Jerry L. Workman, Joan Weliky Conaway, and Thomas Kusch. 2005. In and out: Histone variant exchange in chromatin. *Trends in Biochemical Sciences* 30 (12): 680–687.

Karsenti, Eric. 2007. Self-organisation processes in living matter. *Interdisciplinary Science Reviews* 32 (2): 163–175.

Kepler, Johannes. 1611. *De Niue Sexangula*, ed. Godfrey Tampach. Frankfurt am Main. Republished 1966. *The six-cornered snowflake*. Oxford: Clarendon Press.

Li, En. 2002. Chromatin modification and epigenetic reprogramming in mammalian development. *Nature Reviews: Genetics* 3 (9): 662–673.

Lutter, Leonard, Luann Judis, and Robert Paretti. 1992. Effects of histone acetylation on chromatin topology in vivo. *Molecular and Cellular Biology* 12 (11): 5004–5014.

Misteli, Tom. 2001. Protein dynamics: implications for nuclear architecture and gene expression. *Science* 291 (5505): 843–847.

Nakayama, Jun-ichi, Judd C. Rice, Brian D. Strahl, C. David Allis, and Shiv I. S. Grewal. 2001. Role of histone H3 lysine 9 methylation in epigenetic control of heterochromatin assembly. *Science* 292 (5514): 110–113.

Nicolis, Gregoire, and Ilya Prigogine. 1987. *Exploring complexity: An introduction*. Munich: Piper.

Remaut, Henri, and Gerald Waksman. 2006. Protein-protein interaction through ß-strand addition. *Trends in Biochemical Sciences* 31 (8): 436–444.

Richmond, Timothy J., and Jonathan Widom. 2000. Nucleosome and chromatin structure. In *Chromatin structure and gene expression*, ed. Sarah C. R. Elgin and Jerry L. Workman, 1–23. Oxford: Oxford University Press.

Santoro, Rafaella, and Ingrid Grummt. 2005. Epigenetic mechanism of rRNA gene silencing: temporal order of NoRC-mediated histone modification, chromatin remodeling, and DNA methylation. *Molecular and Cellular Biology* 25 (7): 2539–2546.

Scott, Matthew P., and Patrick H. O'Farrell. 1986. Spatial programming of gene expression in early Drosophila embryogenesis. *Annual Review of Cell Biology* 2: 49–80.

Stewart, Ian. 1998. *Life's other secret: The new mathematics of the living world*. London: Allen Lane.

Vogelauer, Maria, Jiansheng Wu, Noriyuki Suka, and Michael Grunstein. 2000. Global histone acetylation and deacetylation in yeast. *Nature* 408 (6811): 495–498.

Waddington, Conrad. 1957. *The strategy of the genes: A discussion of some aspects of theoretical biology*. London: Allen & Unwin.

Widom, Jonathan. 1998. Structure, dynamics, and function of chromatin in vitro. *Annual Review of Biophysics and Biomolecular Structure* 27: 285–327.

Wolpert, Lewis. 1971. Positional information and pattern formation. *Current Topics in Developmental Biology* 6: 183–224.

III
Information and the Biology of Cognition, Value, and Language

9

Decision Making in the Economy of Nature: Value as Information

Benoit Hardy-Vallée

All organic beings are striving to seize on each place in the economy of nature
—Darwin 1859/2003, 90

Cognition, to use Reuven Dukas's formula, is the set of "neuronal processes concerned with the acquisition, retention, and use of information" (Dukas 2004, 347). Decision making is one of the principal *uses* of that information. Classically, decision making is not a topic of discussion in biology and philosophy of biology. The analysis and study of decision making is usually left to philosophy of mind, economics and psychology (see Hardy-Vallée 2007, forthcoming). Philosophers are mostly concerned with the normative features of decisions, that is, what makes a decision rational or not. In philosophy of mind, the standard conception of decision making equates *deciding* and *forming an intention before an action* (Audi 2001; Davidson 1980; Searle 2001). According to a different analysis, this intention can be equivalent to, inferred from, or accompanied by desires and beliefs. If desires are represented as utilities, and beliefs as probabilities, then decisions can be represented by rational-choice theory (RCT) models. The two branches of RCT, decision theory and game theory, formalize the logical relationships between reasons. RCT specifies the formal constraints on optimal decision making in individual and interactive contexts. Rational agents select actions that have the higher subjective expected utility (obtained by multiplying probabilities and utilities) and select equilibrium strategies, that is, n-tuples of states where no player has an advantage to deviate from (see Baron 2000 for an introduction). RCT is also a framework for building predictive models of choice behavior: which lottery an agent would select, whether an agent would cooperate or not in a prisoner's dilemma, and so on. Experimental economics, behavioral economics, cognitive science, and psychology (I will refer to these disciplines broadly as "psychology") use this model to study how subjects make decisions and which mechanisms they rely on for choosing. These patterns of inference can then be compared with RCT (Camerer 2000; Kahneman and Tversky 1979, 1991, 2000; Thaler 1980).

Standard philosophical, economic, and psychological analyses of decision making implicitly or explicitly adopt what could be called a "cogitative" conception. On this account, decision making is a high-level, explicit, and deliberative process analogous to reasoning. Philosophers explain decisions as inferential transitions between propositional attitudes. These transitions can—at least in theory—be made explicit. As Davidson explains, "If someone acts with an intention then he must have attitudes and beliefs from which, had he been aware of them and had he the time, he could have reasoned that his act was desirable" (Davidson 1980, 85).

Economists and rational-choice theorists represent agents as *Homo economicus*: all their preferences are transitive and they select actions by computing probabilities and utilities. In a game-theoretic context, thanks to common knowledge of rationality and backward induction, they infer the equilibrium strategy and act upon it. Psychological research also clearly assumes that deciding is an explicit, "high-level process" (Johnson-Laird and Shafir 1993, 1); decisions are "reached by focusing on reasons that justify the selection of one option over another" (Shafir, Simonson, and Tversky 1993, 34). The fact that it is studied mostly by multiple-choice tests using paper and pen illustrates well how decisions are conceived in psychology: subjects' decision-making competence is supposedly revealed by questionnaires on probabilistic reasoning. Consequently, it is not surprising that decision making does not stimulate many debates or research in biology and philosophy of biology: it is not construed as a biologically significant capacity. It compares to chess playing: an occasional activity made possible by uniquely human, high-level cognitive capacities.

I would like to suggest here that contrary to common wisdom, decision making is not specifically human, but rather a behavioral control scheme typically found in animals endowed with sensory, motor, and control apparatuses, and, more specifically, in brainy animals (craniates, arthropods, and cephalopods). There are of course some decisions that will fall outside the scope of this analysis: some because they involve multiagent coordination (e.g., jury decision making, institutional processes), and others (e.g., in ethical or scientific contexts) because they appeal to our *theoretical* rationality. Humans have, for instance, to ponder the justification of a moral claim or the soundness of an inference in interpreting an experiment; in doing so, they don't—or don't only—make use of their *practical* rationality (what to do) but their *theoretical* rationality (what to believe). I am not therefore making the bold claim that *all* types of decision are present in humans and animals, but that humans and animals share many decision-making mechanisms. Consequently, some mechanisms are not shared: for instance, language, and all the cognitive enhancements it affords (recursivity, communication, abstraction, etc.) is a properly human mechanism. Language recruits sensorimotor faculties common in primates, but the complex organization of these faculties is uniquely human (Hauser, Chomsky, and Fitch 2002). I will be thus be interested in *natural* rationality, that is, the practical competences shared by human and animals, and not theoretical rationality, afforded by language and human complex sociality.

Neurobiologists interested in the neural basis of decision making recently began to use a continuist concept of decision making. They labeled "biological decision making" their object of study (Glimcher 2003a; Montague, King-Casas, and Cohen 2006; Montague and Quartz 1999; see also Gintis 2007 for a similar construal of decision making). Without explicitly defining the term, their research and methodology clearly show that they attempt to identify the structures and mechanisms that animals employ in the valuation, selection, and attainment of certain goals. Glimcher, for instance, argues that "fundamental features of decision making are common to many species" (Glimcher and Rustichini 2004). Risk aversion, for instance, can be found in humans and birds; given that birds and humans share a common reptilian ancestor, a risk-averse utility function might be "an efficient and evolved feature of vertebrate choice." Thus the epithet "biological" indicates that they inquire into a *natural* phenomenon—as opposed to the ideal-agent models of rational-choice theorists—and that they analyze primarily its biological substrate, not its logicolinguistic structure. Belief-desire-intention reasoning would not then be core of decision making, but a uniquely human implementation of it.

My aim in this chapter is to analyze the concept of *biological decision making* along these lines, to show its theoretical ramifications with neuro- and evolutionary biology and how it should affect our understanding of a variety of biological information, that is, *neural information*. I define biological decision making as *goal-oriented*, *value-based* information-processing.[1] This concept, I suggest, should be understood in a wider theoretical framework initiated by Darwin, that of the "economy-of-nature." This framework (discussed in the following section) applies the concepts and tools of economics to the living world. It states that living beings try to make optimal choices and that evolution endowed them with a certain proficiency in selecting, valuing, and achieving goals. I suggest that these considerations make the case for a value-based account of neural information (see section entitled "The Natural Economics of Information"). Although most conceptions of intentional ("cognitive") information are usually value-neutral, I suggest that neural information-processing is primarily evaluative; that is, neural information-processing is, at its basis, concerned with cost-benefits computations. The economy-of-nature is, among others, a methodological principle for building models of animals as maximizers of scarce resources and models of their brains as economic decision-making organs. Biological decision making requires goals (see section entitled "Goal Orientation: Behavior and Mechanisms") and values (see "Value and Valuation Mechanisms"). I describe how behavioral ecology and neuroeconomics support the idea that biological agents are goal-oriented, value-based information processors. I conclude with remarks on rationality and biological decision making (see "Conclusion: On Natural Rationality").

The Economy of Nature

A Darwinian Economy

Needless to say, Darwin's theory of descent with modification was a true conceptual revolution (Darwin 1859/2003). It provided a general mechanism to explain the diversity and adaptivity of living beings—natural selection—and a guiding principle for organizing the mass of facts about them, the tree of life (see Dennett 1995; Gayon 2003; Thagard 1992, chap. 6). In recalling how this idea had come to his mind, Darwin wrote that he was trying to solve one problem: why plants and animals sharing a common ancestry end up to be so different? Because "the modified offspring of all dominant and increasing forms tend to become adapted to many and highly diversified places in the economy of nature" (Darwin 1887, 84).

Darwin repeatedly uses the expression "economy of nature" in *The Origin of Species* and other writings. He was not the first to consider nature as an economy, although he was among the first to suggest an explicit similarity between natural and political economy. Before Darwin, the idea of nature as an economy had no particular application to human economic practices. In *The Sacred Theory of the Earth*, theologian Thomas Burnet referred to the "Oeconomy of nature" as the "well ordering of the great Family of living Creatures" an order of divine origin (Burnet ca. 1692/1965, II:x). Swedish naturalist Carl Linnaeus, in his *Specimen Academicum de Oeconomia Naturae*, construed this divine order as being self-organized, exhibiting a balance of births and deaths, a complementarity between the function and purpose of life forms (Hestmark 2000; Linnaeus 1751). Adam Smith recognized the unity of this economy, where all living forms strive for "self-preservation, and the propagation of the species" (Smith 1759/2002, 90). Lyell, in his *Principle of Geology*, describes how the involuntary agency of human and other animals "contribute to extend or limit the geographical range and numbers of certain species, in obedience to general rules in the economy of nature" (Lyell 1853, 664). Where Linnaeus saw a clockwork organization, Lyell saw a dynamic equilibrium. Thus, from natural theology to geology, the economy of nature referred to the complex organization of the universe (Bowler 1976; Ghiselin 1978, 1995, 1999; Hammerstein and Hagen 2005; Hodgson 2001; Schabas 2005). With Darwin, *natural* economy began to be understood using the conceptual tools of *political* economy. The division of labor, competition ("struggle" in Darwin's words), trading, cost, the accumulation of innovations, the emergence of complex order from unintentional individual actions, the scarcity of resources, and the geometric growth of populations are ideas borrowed from Adam Smith, Thomas Malthus, David Hume, and other founders of modern economics. Thus, the economy of nature ceased to be an abstract representation of the universe and became a depiction of the complex web of interactions between biological individuals, species and their environment—in short, the subject matter of ecology. Consequently, Darwin's main contributions are his transforming biology into a *historical* science—like geology—and into an *economic* science (Ghiselin 1999, 7; see also Hirshleifer 1978). I will here be interested in the second contribution, and how it should shape our conception of information processing.

In this perspective, the representation of nature as an economy is not a new theory of nature, for it is not a body of structured knowledge about phenomena and their cause. Neither is it an a priori truth, an unfalsifiable proposition, a law of nature or a mechanism. More precisely, I suggest, it is a *principle*—in the sense employed by model-based science theoreticians—for building models (Cartwright 1989; Giere

2004; Morgan and Morrison 1999). Principles are background assumptions, or "templates" for building models. Scientific models, as the idea is employed here, are simplified representations of complex phenomena used as mediators between principles on one side and data on the other. The principle of the economy-of-nature is, in biology, analogous to the principles of thermodynamic in physics: these general statements cum specific conditions allow scientists to work out a representation of a physical process such as heat dissipation in a gasoline engine. Similarly, the general principle of an economy of nature cum specific details about an animal (ecology, needs, physical constitution) allows biologists to build simplified models of animal behavior.

Biological Decision Making
I take the economy-of-nature principle to be a refinement of the natural selection principle: although it describes general features of the biosphere, it puts emphasis on the intersection between individual biographies and natural selection, especially in regard to decision making. On the one hand, the decisions biological individuals make increase or decrease their fitness; therefore, good decision makers are more likely to propagate their genes. On the other hand, natural selection is likely to favor good decision makers and to get rid of bad decision makers. Thus biological decision making is a central concept for models based on the economy-of-nature principle. It is not, however, *equivalent* to fitness maximization. It is rather one of the means by which biological agents attempt—but may fail—to maximize their fitness. They maximize their fitness when they generate copies of their own genes or through their life-history strategies, not when they catch a fish or climb a tree.

Biological agents do not choose to become sexually mature at a particular age or to invest a large part of their caloric intake in reproduction. Most of these traits are chosen by natural selection and unfold in an entire lifetime. What agents *do* choose, however, is to mate with *this* particular partner, run from *this* predator, eat *this* particular prey, and so on. Even when partner preferences are fixed by natural selection, the explanation of the fact that they choose *this* particular one refers to the individual's own internal mechanisms. Natural selection may be able to choose which *type* of food or partner a biological agent will seek, but not which *token*. As Richard Dawkins puts it, "Genes are the primary policy-makers; brains are the executives" (Dawkins 1976, 59). It is therefore more appropriate to see organisms as "adaptation executors," not fitness maximizers (Tooby and Cosmides 2005, 14)—although it is likely that executing adaptations, in the appropriate environment, usually leads to fitness maximization.

In this framework, a biological decision maker is any agent who can control its behavior. More precisely, in order to have genuine control over its behavior, an agent must possess *control mechanisms*, that is, internal structures that process sensory information and motor commands. It is meaningful to talk of individual decision making when information-processing mechanisms are among the proximal causes of choice behavior, for otherwise it cannot be considered as a decision. A flame or a water drop does not *decide* to go up or down, because their "behavior" is not driven by some information it might have about its internal or external environment.

Moreover, we do not consider that a flame or a water drop decides because it lacks two important features: goal orientation and valuation. In the two following sections, I argue that these two concepts are fundamental to the understanding of biological decision making and that taking brainy animals as implementing *goal-oriented, value-based* information-processing was—and still is—successful for revealing their decision-making mechanisms. This suggests that one of the most basic ways to be informed about X is to know the value of X for the attainment of a goal. I shall therefore present first my account of value-based information (see the following section), and then show in the two sections that follow that goal-orientation and valuation mechanisms shared by humans and animals shed light on the nature of biological decision making and natural rationality (concluding section).

The Natural Economics of Information

Biological agents have two fundamental imperatives: survival and reproduction. They need not be aware of these necessities, but if their decisions never achieve some degree of success in these two matters, natural selection will easily discard them. Biological decision making is therefore oriented toward certain goals. The goals can be survival and reproduction per se, instrumental goals that allow individuals to survive and reproduce, instrumental goals that once served survival and reproduction, or goals that are achieved thanks to mechanisms that once served survival and reproduction. (Of course, many goals can be irrelevant for survival, but I will focus here on adaptive ones.) There are at least three ways of being goal-oriented: decision policies can be *goal-achieving*, *goal-seeking*, or *goal-directed* (McFarland and Bösser 1993, although I follow their definitions loosely). A system implements *goal-achieving* control when it can either recognize the goal or change its behavior when the goal is achieved (whether the goal is represented or not, and sought or not), *goal-seeking* control when it is able to reach a goal (without necessarily having a representation of the goal), and *goal-directed* control when it entertains an explicit representation of the goal. These three control schemes can be thought of as a nested hierarchy of functions. An extremely simple system may have a goal built in, without any representation of it and without actively seeking it: *when variable V reaches a certain limit, do A*. Thus the system does not represent anything and does not seek anything, but is able to recognize a goal. Another, more complex system may have control and monitoring mechanisms by which it actively performs and assesses its actions so that it can see to it that variable V reaches a certain limit. Thus the system seeks, but need not have a representation of what it seeks: servomechanisms may be enough. Finally, an even more complex system may entertain a representation (either a linguistic representation or a sensorimotor simulation) of V reaching a certain limit. Note that some functions might be dissociable: it is always theoretically possible that a nonrepresentational system seeks a goal without modifying its behavior once the goal is reached (pure goal seeking) or that representational system does not seek anything and does not modify its behavior once the goal is reached (pure goal direction). I do not take any stance on these matters, but will simply say that these functions tend to co-occur in brainy animals, and that their performances are nonexclusive. These three functions correspond, however, to significant aspects of goal orientation. In every case, information-processing mechanisms are sensitive to the attainment of goals. Without goals, there is no need to make decisions since there are no priorities.

Because biological agents have goals, they cannot be systematically indifferent between different courses of actions. Moreover, certain goals are universal and nonnegotiable, such as finding nutrients. Certain actions facilitate the acquisition of resources, predator avoidance, reproduction, and so on, and others impede it. In order to choose the proper courses of action, and given that energy and information are not free and unlimited, biological agents need to be able to "care." In less anthropomorphic terms, it is fair to say that biological agents that strive to survive, reproduce, or achieve other instrumental goals must ground the selection of actions on devices that rank preferences, that is, on valuation mechanisms. *Value*, as I will use the term, is the currency by which decisions and their performance, outcomes, or expected payoffs are compared. It is not synonymous with reward: the latter refers to an immediate advantage of the outcome of a decision, and the former is an "estimate about how much reward (or punishment) will result from a decision, both now and into the future" (Montague, King-Casas, and Cohen 2006, 419). Valuation mechanisms are usually designed to assign a value or a rank to internal states and behaviors with respect to their effectiveness in goal achieving or seeking. They assess "how good" the acquisition of a resource is relative to a goal, or how good an action is relative to the attainment of a goal. Neural and behavioral models of decision making should therefore specify the goals sought by animals (human or not) and the valuation mechanisms by which internal states and behaviors are assessed. This suggests that the primary means of information processing are cost-benefit computations and that information in its basic form is inherently value-based—at a certain level of organization. I will clarify this immediately.

Let's distinguish first different theories and conceptions of information. Although information is a complex notion, a recent effort—known as the philosophy of information—to clarify its nature began to identify the different understandings and their relationships (Floridi 2004a, 2004b). First, information can be said of external (i.e., noncognitive) structures that embody organization. In this sense, information is that which can "fight," temporally and locally, entropy. This is *physical* information, either classical (as a pure state) or quantum (as a wave function). Second, information can be a message that reduces uncertainty: information is the vehicle of syntactic communication (Shannon and Weaver 1949). This is *syntactic* information. Because a Turing machine is a Shannon-style channel of information (Bohan Broderick 2004), any computational device is also a syntactic information-processing device. Individual neurons, for instance, implement information-processing capacities (Koch and Segev 2000; Eliasmith and Anderson 2003). Thus one can talk of *neural* information as a "wet" form of syntactic information. A related concept, specific to living beings, is *genetic* information (Maynard Smith 2000; Mameli 2005). Our DNA stores syntactic information that, once interpreted and modulated by developmental processes, environmental pressures, and epistatic interactions (between genes), actively participate in the construction and maintenance of our bodies (this information was once referred to as a "blueprint," but since the advent of developmental biology, it is more adequate to speak of genes as an important contributor, together with the environment, in gene expression, rather than the "master code"). Information can also be *intentional*, in the philosophical sense of being *about* something: assertions, thoughts, and other mental and nonmental representations (pictures, maps, etc.) provide nonsyntactic information (description or indication) about their referent (Dretske 1981; Millikan 1984). This information can be true or false, and that which makes it true is what it is about: this is semantic information. (As will be apparent, I use the words *intentional* and *semantics* as nonsynonymous, because I want to make room for another kind of intentional information, thus making it clear that there might be intentional information whose meaning is not primarily descriptive or indexical.) Thus, in living beings, there is genetic information at the replicator level (the genes), syntactic information at the neuronal level, and intentional information at the vehicle level (the whole, situated agent).

These forms of information are not completely independent. External patterns (physical information) excite sensors that convert energy fluctuations in electric influx (syntactic information). These signals are then normalized, amplified, filtered, and so on by neurons (neural information/computation), that is, cells whose structure is in large part determined by DNA (genetic information). Given certain conditions (left to the theory of representation one advocates), a neural structure—that is, a set of neurons—their connections and spike behaviors can conjointly track external structures and thus be *about* them (Thagard 2007). Hence neural computation, although it is a syntactic process at the neural level, can, at the system level, be semantic (the "system level" being the whole brain, inserted in a body, situated in an environment). For instance, a research team identified hippocampal cells in the mouse brain that selectively fire when the mouse perceives a nest. However the shape, color, odor, location or any other physical property of the nest is, these cells are activated only by external structures that may fill the role of a nest. Moreover, these nest-responsive cells do not respond to similar stimuli, suggesting that the mouse brain can store invariant representations of functional categories (Lin et al. 2007). Of course, the complete representation will probably involve other cells, but the point is that once all the elements are identified, it is possible to think of a semantic representation as a distributed pattern or neural information processing that tracks environmental features (functional or not). This intentional information is thought to be primarily semantic, that is, *representational* (it provides a proxy of an external feature) and *truth-functional* (its object is what makes it true; if something activates the complex of nest-responsive cells without being a nest, then it *misrepresents*). I would like to suggest, however, another mode of intentional information crucial for biological decision makers, where the informational content is not primarily representational and truth-functional, but *evaluative*. Evaluative information is not primarily about providing the decision maker with an indication of, or a better description of, an external structure but a better assessment of (1) the worth (cost-benefits) of that structure or (2) the worth of an action targeted at a structure. Evaluative information is intentional because it is ultimately about something else, but it is not semantics *stricto sensu* because it is not primarily descriptive or representational. It is more about the "fit" of a structure or an action in the attainment of a goal than about the structure itself. Information construed as such corresponds to *utility* in its most basic sense: the agent-relative appraising of an action or a structure.

The gist of a value-based account of intentional information is that neural systems filter the flux of stimuli (extero- or proprioceptive) primarily in reference with their own interests. Efficiency requirements strongly constrain syntactic information processing: no Buridan's ass would live for a long time in nature. Certain behavioral options ought to be preferred. A sine qua non of survival and reproduction is a preference for behaviors that lead to energetic resources acquisition—life runs on batteries, to use Montague's expression (2006). Energy, however, is costly and scarce: one needs to spend some in order to get some. Hence no efficiency, no survival, no reproduction. Consequently, evaluative information processing is the adaptation of computational means to survival ends through valuation mechanisms: semantic information thus complements, refines, or improves evaluative information. This suggests also a reversal in the order of explanation.

Usually, philosophers assume that semantic information is first, and evaluative second. Agents allegedly represent their environment (e.g., through perception), then add a particular value to some representation: the perceived object is perceived *as* desirable or likable. As Millikan says, "representations that tell what to do have no utility unless they can combine with representations of facts" (Millikan 1996, 152). What I suggest here is that, evolutionarily and conceptually, evaluative information is anterior to semantic information. Brainy animals may be able to synchronize their neural processing so as to assess the utility of a resource (e.g., food) before being able to have any other semantic information about the nature of this food. A sugary taste in the mouth does not count as a representation or description of a sugary object, but the firing of reward centers in the brain elicited by sugar constitutes intentional, nonrepresentational, nonsemantics information *about* the sugared objects: the whole agent is informed that something may be helpful in survival. It is possible that the taste is elicited by another product that mimics sugar, but because this information is not semantic and truth-functional, there this no alethic criteria (e.g., truth) constitutive of the intentional nature of evaluative information. Evaluative information can be wrong, but only in complex animals able to mispredict their future valuation, or to deceive themselves.

Hence I am not saying that undifferentiated evaluative and semantic information may be prior, but that evaluative information, in the world of biological decision makers, is primary. We should explain evaluative information first, then show how semantic information (when there is some) figure in this account. For instance, we should not explain first the content of the mouse representations of nest (semantic information) and then look for its use in decision making, but rather start with explaining nest-related decision making (goals-orientation and valuation mechanisms) and then see how semantic information about nests figures in this explanation. Starting with the representational to explain the intentional is what Brandom called the Platonist order of explanation: explaining the use of representations in terms of their content (2000). What I put forth here is a pragmatic order of explanation, a naturalistic version of Brandom's: starting with the use (construed as primary intentionality) to explain the content, with the evaluative to explain the semantics.

To sum up: genetic information, together with other factors in development, contributes to building and maintaining living beings. Those endowed with brains also process *syntactic* information, in the form of signal transformation. Properly organized, distributed patterns of neurobiological computation can process intentional information, i.e., providing information about something else. Intentional information can be either semantics (descriptive or indexical) or evaluative (about the utility of a structure or of an action). Evaluative information processing (goal-oriented and value-based) is a basic requirement for survival and reproduction, because it is the basis of biological decision making, a fundamental feature of agents in the economy of nature.

Goal Orientation: Behavior and Mechanisms

Economic Goals

Behavioral ecology (Krebs and Davies 1997; Pianka 2000) models animals as economic agents that achieve ultimate goals (survival and reproduction) through instrumental ones (partner selection, food acquisition and consumption, and so on). Even though behavioral ecologists claim to study fitness-maximizing behaviors, as pointed out as by White, Dill, and Crawford, they "almost always ignore the number of offspring produced and study, instead, how a particular adaptation contributes to some fitness proxy, for example, net energy intake rate" (2007, 276). Optimal foraging theory thus represents foraging as a maximization of net caloric intake. With general principles derived from microeconomics, optimization theory, and control theory, coupled with information about the physical constitution and ecological niche of the predator, it is possible to predict what kind of prey and patch an animal will favor, given certain costs such as search, identification, procurement, and handling costs. Optimal foraging theory (OFT), as their founders suggested, tries to determine "which patches a species would feed on and which items would form its diet if the species acted in the most economical fashion" (MacArthur and Pianka 1966, 603). OFT primarily models animals as efficient *goal seekers* and *goal achievers*: they engage and succeed in searching for nutrients.

OFT modeling thus incorporates agents, their choices, the currency to be maximized (most of the time a caloric gain), and a set of constraints. Most researches study where to forage (patch choice), what to forage for (prey choice), and for how long (optimal time allocation). It is supposed that the individual animal makes a series of decisions in order to solve a problem of sequential optimization. An animal looking for nutrients must maximize its caloric intake while taking into account those spent in seeking and capturing its prey. To this problem one must also add, among others, the frequency of prey encounter, the time devoted to research, and the calories each prey type affords. All these parameters can be represented by a set of equations from which numerical methods such as dynamic programming allow biologists to derive algorithms that an optimal forager would implement in order to maximize the caloric intake. These algorithms are then used to predict an animal's behavior. Mathematically speaking, OFT is the translation of decision theory axioms—together with many auxiliary hypotheses—into tractable calorie-maximization algorithms.

Economic models of animal behavior have succeeded in explanation and prediction. For example, such models can predict how birds split their time between defending a territory and foraging (Kacelnik, Houston, and Krebs 1981), or between singing and foraging (Thomas 1999). In their meta-analysis, Sih and Christensen (2001) reexamined 134 foraging studies in laboratory and natural context—experimental and observational—and concluded that although predictive success is not perfect, the predictivity of the theory is relatively high when prey are motionless (the prey can be a plant, seeds, honey, and so on).

Interactive contexts are aptly modeled by game theory, mainly social foraging, fighting and predatory-prey relations (Dugatkin and Reeve 1998; Hansen 1986; Lima 2002). For example, a model of Vickery et al. (1991) predicted that the co-occurrence of three social foraging strategies—producer (gathering nutrients), scrounger (stealing nutrients), and opportunist (switching between producer and scrounger)—occurs only in the very improbable case in which the losses opportunists would incur while foraging would be exactly equivalent to the profit of stealing. The model, however, predicts certain distributions of pairs of strategies that constitute evolutionary stable strategies (ESS), that is, a strategy that cannot be invaded by any alternative competing strategy. The proportion of food patch shared by scroungers, the size of the group, and the degree of compatibility between the scrounger and producer strategy (i.e., if it is easy for the animal to perform both activities) determine the distribution of the strategies in a population—a conclusion that was also confirmed, inter alia, in birds (*Lonchura punctulata*, cf. Mottley and Giraldeau 2000). As predicted by the model, the producer strategy becomes less common when the cost of individual foraging increases.

Recently, behavioral ecologists found that animals could also be modeled as traders in biological markets. Obviously, biological markets do not have symbolic and conventional currency systems, but many interactions between animals are repeated exchanges that institute currencies and market prices. As soon as agents are able to provide commodities for mutual profit, the competition for obtaining commodities creates a higher bid. Animals seek and select partners according to the principle of supply and demand in interspecific mutualism, mate selection, and intraspecific cooperation. An example of the last type is the cleaning market instituted by *Hipposcarus harid* fishes and cleaner-fishes *Labroides dimidiatus*. The "customers" (*Hipposcarus*) use the services of the cleaner to have its parasites removed, whereas the cleaners occasionally cheat and eat the healthy tissues of its customers. Because the cleaners offer a service that cannot be found elsewhere, they benefit from a certain economic advantage. A customer cannot choose to be exploited or not, whereas the cleaner chooses to cooperate or not (thus the payoffs are asymmetric). The customer—a predator fish that could eat the cleaner—abstains from consuming the cleaner in the majority of the cases, given the reciprocal advantage. Bshary and Schäffer (2002) observed that cleaners spend more time with occasional customers than with regular ones and fight for them, as occasional customers are easier to exploit. All this makes perfect economic sense.

Does this bioeconomic logic apply to human beings as well? In part, yes. Human behavioral ecology models agents as optimal foragers subject to a multitude of constraints. Given available resources in the environment of a community, one can generate a model that predicts the optimal allocation of resources. These models are of course more complex than animal ones, as they integrate social parameters like local habits, technology, or economic structures. For instance, models of human foraging were able to explain differences in foraging style between tribes in the Amazonia, given the distance to be traversed and the technology used (Hames and Vickers 1982. Food sharing, labor division between men and women, agricultural cultures, and even Internet browsing (where the commodity is information) can be modeled by human behavioral ecology (Jochim 1988; Kaplan et al. 1984; Pirolli and Card 1999). However, as mentioned previously, humans often base their decision on theoretical rationality, that is, capacities afforded by linguistic competence and hypersociality, and thus there is a sense in which certain important decisions fall outside the scope of the current analysis, because they rely on mechanisms unavailable to other animals. My focus, therefore, is on the shared mechanisms.

Goals, Behavior, and Dopaminergic Information Processing
Although behavioral ecology supports the idea that animals pursue ultimate goals (survival and reproduction) and instrumental ones (e.g., prey and patch choice), it remains silent on the actual implementation of goal orientation. Research in neuroeconomics—the study of the neural mechanisms of decision making (Glimcher 2003a; McCabe 2005; Zak 2004)—suggests that much of goal-oriented neural computation is realized by midbrain dopaminergic systems activity (Egelman, Person, and Montague 1998; Frank and Claus 2006; McCoy and Platt 2005; Montague and Berns 2002; Montague, Hyman, and Cohen 2004; Niv, Daw, and Dayan 2006). Dopaminergic neurons are subcortical neural

structures that modulate cortical activity by both neurotransmission and neuromodulation (through modulation of neurotransmission). Through four different pathways, dopaminergic neurons interact with emotive and cognitive brain areas. Synthesized in the ventral tegmental area (VTA), dopamine is sent to other areas through VTA axons.

Neuroscience has revealed their role in working memory, motivation, learning, decision making, planning, and motor control, and how their dysfunctions cause many decision-making pathologies (Morris et al. 2006; Redish 2004; Schultz 2001; Williams and Dayan 2005). For instance, hyperdopaminergia is involved in pathologies where the focus of attention is exaggerated: schizophrenia, attention deficit hyperactivity disorder, addiction, and obsessive-compulsive behavior. Hypodopaminergia is involved in pathologies of motor control such as Parkinson and dystonias: the patient is impaired in his or her ability to execute intentional movements. Thus the right amount of dopaminergic activity is required for focusing adequately on goals.

Following the distinctions I outlined earlier between goal seeking, goal achieving, and goal orientation, I will suggest how dopaminergic mechanisms are involved in all three activities.

Motivation and Goal Seeking Motivation mechanisms are goal-seeking mechanisms par excellence: it is not necessary to picture what one's dinner will look like in order to find something to eat. A readiness to invest in effort that could lead to nutrients is sufficient. A clear dissociation has been shown between motivation and hedonic impact, often referred to as "wanting" vs. "liking" (Berridge and Robinson 1998; Pecina et al. 2003; apostrophes indicate that these concepts are not equivalent to folk-psychological constructs). Hyperdoparminergic animals become highly motivated in acquiring certain resources such as food ("wanting"), but they do not display orofacial signs of a higher "liking." Like drug addicts, their dopaminergic neurons over-motivate them to acquire certain rewards, and hence these animals are highly motivated even if their "liking" does not increase (Berridge 2003a). Conversely, the behaviors of hypodopaminergic animals show that they still enjoy sucrose's taste (e.g., they will pass their tongue over their lips), but they are not motivated in acquiring some. Dopaminergic mechanisms attribute "incentive salience" to certain cues that, in return, trigger approach behaviors. Besides food or sex, dopaminergic mechanisms also drive individuals to acquire more abstract stimuli like art, money, trust, or revenge (King-Casas et al. 2005; Lohrenz et al. 2007; Montague, King-Casas, and Cohen 2006, 420; Singer et al. 2006). All sources of reward rely on dopaminergic mechanisms.

Prediction Errors and Goal Achieving   Goal-achieving control is realized through mechanisms that detect whether a goal is reached, that is, if things are better, worse, or just as planned. This is exactly what dopaminergic neurons do: they broadcast, in different brain areas, a reward-prediction error signal whenever an unexpected reward or the absence of an expected reward is detected. Moreover, they learn from their mistake: from a series of prediction errors, they learn to predict future rewards.

Computational neuroscience identified a class of reinforcement-learning algorithms that mirror the activity of dopaminergic neurons: temporal-difference (TD) learning algorithms (Niv, Duff, and Dayan 2005; Suri and Schultz 2001; Sutton and Barto 1987, 1998). In its simplest form, a TD model uses sensory inputs to predict a discounted sum of all future rewards. The difference between successive reward predictions is computed and constitutes an error signal. A learning rule then updates a *value function*—a function that maps state-action pairs to numerical values—according to the prediction error signals. In Actor-Critic architectures (a variety of TD method), the error signal also updates a *behavioral policy*. The value function assesses actions; the policy recommends actions based on actual states. Thus, if an action produces a high reward, it will be highly valuated: the TD model will predict that the action will lead to such-and-such reward, and the policy will favor the selection of this action when the appropriate situation will present itself. Whenever an unpredicted reward or absence of predicted reward is detected, the TD algorithm will revise its policy and value function (see Suri 2002 for details about the neural substrates of the Actor and Critic).

TD-learning algorithms are thus neural mechanisms of decision making implemented in dopaminergic systems. They are not the only ones (many competing processes may interact in decision making), but numerous studies confirm their importance for decision making (Montague 2006). They are specifically solicited in situations where simple cues won't be sufficient, and when values are needed.

Cognitive Control and Goal Orientation   In certain situations, learned routines or rule-governed representations won't be enough. When situations are too different, dangerous, and uncertain, or when they require the overcoming of a habitual response, decisions must be explicitly guided by representations. The behavior is then controlled "top-down," not "bottom-up." Acting upon an internal representation—instead of routines—is referred to, in cognitive science, as cognitive control or executive function (Norman and Shallice 1980; Shallice 1988). The agent is led by a representation of a goal and will robustly readjust its behavior in order to maintain the pursuit of a goal. In the Stroop task, for instance, subjects must identify the color of written words such as "red," "blue," or "yellow" printed in different colors (the word and the ink color do not match). The written word, however, primes the subject to focus on the meaning of the word instead of focusing on the ink's color. If, for instance, the word "red" is written in yellow ink, subjects will utter "red" more readily than they say "yellow." There is a cognitive conflict between the semantic priming induced by the word and the imperative to focus on the ink's color. In this task, cognitive control mechanisms ought to give priority to goals in working memory (naming ink color) over external affordances (semantic priming). An extreme lack of cognitive control is exemplified in subjects who suffer from "environmental dependency syndrome" (Lhermitte 1986): they will spontaneously do what their environment indicates or affords them. For instance, they will sit on a chair whenever they see one, or undress and get into a bed whenever they are in the presence of a bed (even if it's not in a bedroom).

Cognitive control is thought to happen mostly in the prefrontal cortex (PFC), an area strongly innervated by midbrain dopaminergic fibers (Duncan 1986; Koechlin, Ody, and Kouneiher 2003; Miller and Cohen 2001; O'Reilly 2006). Prefrontal areas activity is associated with maintenance and updating of cognitive representations of goals. Moreover, impairment of these areas results in executive control deficits (such as the environmental dependency syndrome). Because working memory is limited, however, agents cannot hold everything in their prefrontal areas. Thus the brain faces a trade-off between attending to environmental stimuli (e.g., that may reveal rewards or danger) and maintaining representation of goals, namely, the trade-off between *rapid updating* and *active maintenance* (O'Reilly 2006). Efficiency requires brains to focus on *relevant* information, and again, dopaminergic systems are involved in this process. A good deal of research suggests that dopaminergic activity implements a "gating" mechanism, by which the PFC alternates between rapid updating and active maintenance (Montague, Hyman, and Cohen 2004; O'Donnell 2003; O'Reilly 2006). A higher level of dopamine in prefrontal areas signals the need to rapidly update goals in working memory ("opening the gate"); a lower level induces resistance to afferent signals and thus a focus on represented goals ("shutting the gate"). Hence dopaminergic neurons select which information (goal representation or external environment) is worth paying attention to.

Dopaminergic systems are one of the most common mechanisms of biological decision making in the economy of nature. For instance, TD models closely mimic human, ape, and honeybee choice behavior under uncertainty (Egelman, Person, and Montague 1998; Glimcher, Dorris, and Bayer 2005; Montague et al. 1995); even fruit flies rely on dopaminergic systems to make decisions (Zhang et al. 2007). These systems achieve *goal seeking* (motivation, incentive salience), *goal achieving* (the reward-prediction error signal), and *goal orientation* (gating).

Value and Valuation Mechanisms

Valuation is the process by which a system maps an object, property, or event X onto a value space, and a *valuation mechanism* as the device implementing the matching between X and the value space. I do not mean that values need to be explicitly represented as a space: by valuation *space*, I mean an artifact that accounts for the similarity between values by plotting each of them as a point in a multidimensional coordinate system. Color spaces, for instance, are not conscious representations of colors, but spatial depictions of color similarity along several dimensions such as hue, saturation, and brightness (Gegenfurtner and Kiper 2003). And I do not mean that valuation first represents objects, then adds a value: as argued earlier, evaluative information is prior to semantic information. Valuation and evaluative information are basic modes of mind-world interaction that representations may augment and refine.

The simplest and most common value space has two dimensions: *valence* (positive or negative) and *magnitude*. Valence distinguishes between things that are liked and things that are not. As the preceding section reported, "wanting" and "liking" are dissociated: they tend to coincide, but need not. Thus if X has a negative valence, it does not imply that X will be avoided, but only that it is disliked. Magnitude encodes the level of liking vs. disliking. Other dimensions might be added—temporality (whether X is located in the present, past or future), other- vs. self-regarding, seeking vs. avoiding, basic vs. complex, for instance—but the core of any value system is valence and magnitude, because these two parameters are required to establish rankings. To prefer Heaven to Hell, Democrats to Republicans, salad to meat, or sweet to bitter involves valence and magnitude.

Nature endowed many animals (mostly vertebrates) with rapid and intuitive valuation mechanisms: emotions (Bechara and Damasio 2005; Bechara et al. 1997; Damasio 1994, 2003; LeDoux 1996; Naqvi, Shiv, and Bechara 2006; Panksepp 1998). Although it is a truism in psychology and philosophy of mind and that there is no crisp definition of what emotions are (Faucher and Tappolet 2002; Griffiths 2004; Russell 2003), I will consider here that an emotion is any kind of neural process whose function is to attribute a valence and a magnitude to something else and whose operative mode are *somatic markers*. Somatic markers are bodily states that "mark" options as advantageous/disadvantageous, such as skin conductance, cardiac rhythm, and so on (Bechara and Damasio 2005; Damasio 1994; Damasio et al. 1996). Through learning, bodily states become linked to neural representations of the stimuli that brought about these states. These neural structures may later reactivate the bodily states or a simulation of these states and thereby indicate the valence and magnitude of stimuli. These states may or may not account for many legitimate uses of the word "emotions," but they constitute meaningful categories that could identify natural kinds (Griffiths 1997).

More than irrational passions, affective states are phylogenetically ancient and adaptively significant valuation mechanisms. Since Darwin (1896), many biologists, philosophers, and psychologists have argued that they have adaptive functions such as focusing attention and facilitating communication (Cosmides and Tooby 2000; Ekman 1972; Griffiths 1997). As Antonio Damasio and his colleagues discovered, subjects impaired in affective processing are unable to cope with everyday tasks, such as planning meetings (1994). They lose money, family, and social status. However, they were completely functional in reasoning or problem-solving tasks. Moreover, they did not feel sad for their situation, even if they perfectly understood what "sad" means, and seemed unable to learn from bad experiences. They were unable to use affect to aid in decision making, a hypothesis that entails that in normal subjects, affect *does* aid in decision making. These findings suggest that decision making needs affect, not as a set of convenient heuristics, but as central information-processing mechanisms. Without affect, it is possible to *think* efficiently, but not to *decide* efficiently; affective areas, however, are solicited in subjects who learn to recognize logical errors (Houde and Tzourio-Mazoyer 2003).

Affects—especially the so-called basic or core ones such as anger, disgust, liking, and fear (Berridge 2003b; Ekman 1999; Griffiths 1997; Russell 2003; Zajonc 1980)—are prominent explanatory concepts in neuroeconomics. The study of valuation mechanisms reveals how the brain values certain objects (e.g., money), situations (e.g., investment, bargaining), or parameters (risk, ambiguity) of an economic nature. Three kinds of mechanisms are typically involved in neuroeconomic explanations:

1. *Core affect mechanisms*, such as fear (amygdala), disgust (anterior insula), and pleasure (nucleus accumbens), encode the magnitude and valence of stimuli.
2. *Monitoring and integration mechanisms* (ventromedial/mesial prefrontal, orbitofrontal cortex, anterior cingulate cortex) combine different values and memories of values together.
3. *Modulation and control mechanisms* (prefrontal areas, especially the dorsolateral prefrontal cortex), modulate or even override other affect mechanisms.

Of course, there is no simple mapping between psychological functions and neural structures, but cognitive neuroscience assumes a dominance and a certain regularity in functions. Disgust does not *reduce* to insular activation, but anterior insula is significantly involved in the physiological, cognitive, and behavioral expressions of disgust. There is a bit of simplification here—due to the actual state of science—but enough to do justice to our best theories of brain functioning. To put it succinctly, I assume an "evolutionary mechanistic functionalism," according to which there is a many-to-many mapping between cognitive functions (e.g., perception, valuation, emotions) and neural structures, that these functions are hierarchy of mechanisms, and that basic function have been preserved by evolution and reemployed in novel domain. Thus a cognitive function may recruit many brain areas, and a brain area (more precisely, the set of syntactic computations it implements) may be involved in many cognitive (intentional) functions. For instance, social cognition recruits the anterior cingulate cortex, the amygdala, the prefrontal cortex, the medial temporal lobe, and so on (Lieberman 2007), and the dorsolateral prefrontal cortex is involved in motor planning, organization, working memory, attention, and many executive functions. As Michael Anderson showed (2007a, 2007b), if this perspective is not strictly localizationist (one-to-one mapping between function and structure), it is not strictly holistic (functional equipotentiality in brain structures). If we adopt an evolutionary conception of the brain (as a set of homologies shaped by selection pressures whose function can be reascribed), then "older" brain areas—that is, those shared by common ancestors to all brainy animals—will likely be redeployed in many cognitive function, and "recent" cognitive functions will likely be more distributed. As Anderson forcefully argued, both predictions are largely supported by neurocience. Language (in humans) recruits many areas (compared to smell), and the amygdala is involved in a wide range of cognitive functions (compared to dorsolateral prefrontal cortex).

I will here review two cases of individual and strategic decision making and will show how affective mechanisms are involved in valuation. (The material for this part is partly drawn from Hardy-Vallée 2007.)

In a study by Knutson et al. (2007), subjects had to choose whether they would purchase a product (visually presented), and then whether they would buy it at a certain price. Though desirable products caused activation in the nucleus accumbens, activity is detected in the insula when the price is seen as exaggerated. If the price is perceived as acceptable, a lower insular activation is detected, but mesial prefrontal structures are more strongly elicited. The activation in these areas was a reliable predictor of whether subjects would buy the product: prefrontal activation predicted purchasing, and insular activation predicted the decision of not purchasing. Thus purchasing decision involves a trade-off, mediated by prefrontal areas, between the pleasure of acquiring (elicited in the nucleus accumbens) and the pain of purchasing (elicited in the insula). A box of chocolates—a stimulus presented to the subjects—is located in the high-magnitude, positive-valence regions of the value space, and the same chocolate box priced at $80 is located in the high-magnitude, negative-valence regions of the space.

In the ultimatum game, a "proposer" makes an offer to a "responder" who can either accept or refuse the offer. The offer is a split of an amount of money. If the responder accepts, he or she keeps the offered amount and the proposer keeps the difference. If the responder rejects it, however, both players get nothing. Orthodox game theory recommends that proposers offer the smallest possible amount, and that responders should accept every proposition, but all studies confirm that subjects make fair offers (about 40 percent of the amount) and reject unfair ones (less than 20 percent; Oosterbeek, Sloof, and Van de Kuilen 2004). Brain scans of people playing the ultimatum game indicate that unfair offers trigger, in the responders' brains, a "moral disgust": the anterior insula is more active when unfair offers are proposed, and insular activation is proportional to the degree of unfairness and correlated with the decision to reject unfair offers (Sanfey et al. 2003). Moreover, unfair offers are associated with greater skin conductance (van 't Wout et al. 2006). Visceral and insular responses occur only when the proposer is a human: a computer does not elicit such reactions. In addition to the anterior insula, the dorsolateral prefrontal cortex (DLPFC) is also recruited in the ultimatum). When there is more activity in the anterior insula than in the DLPFC, unfair offers tend to be rejected, but they tend to be accepted when DLPFC activation is greater than that of the anterior insula.

These two experiments illustrate how neuroeconomics might begin to decipher the value spaces and how affective mechanisms contribute to valuation. (It should be noted that nonaffective mechanisms also contribute to valuation and that assessing whether these values are appropriate and justified may requires a theoretical form or rationality.) Although human valuation, shaped in large part by the "game of giving and asking for reasons," is more complex than the simple valence-magnitude space, this "neuro-utilitarian" framework is useful for interpreting imaging and behavioral data: more specifically, we need an explanation for insular activation in purchasing and ultimatum decisions, and the most simple and informative, as of today, is that it triggers an aversive feeling. More generally, it also reveals that the human value space is profoundly social: humans value fairness and reciprocity. Cooperation and altruistic punishment (punishing cheaters at a personal cost when the probability of future interactions is null), for instance, activate the nucleus accumbens and other pleasure-related areas (Rilling et al. 2002; de Quervain et al. 2004). People like to cooperate and make fair offers. Neuroeconomic experiments also indicate how value spaces can be similar across species. It is known, for instance, that in humans, losses elicit activity in fear-related areas such as the amygdala (Naqvi, Shiv, and Bechara 2006). Because capuchin monkeys' behavior also exhibits loss aversion (i.e., a greater sensitivity to losses than to equivalent gains), behavioral evidence and neural data suggests that the neural implementation of loss aversion in primates shares common valuation mechanisms and processing (Chen, Lakshminarayanan, and Santos 2006). The primate—and maybe the mammal or even the vertebrate—value space locates resource losses in a particular region.

Conclusion: On Natural Rationality

In this chapter, I tried to capture the unity of biological decision making and of evaluative information processing. I argued that it must be understood as entrenched in a richer theoretical framework, that of Darwin's economy-of-nature. According to this perspective, animals can be modeled as economic agents and their control systems can be modeled as economic devices. All living beings are thus deciders, strategists, or traders in the economy of reproduction and survival.

When he suggested that nature is an economy, Darwin paved the way for a stronger interaction between biology and economics. One of the consequences of a bioeconomic approach is that decision making becomes an increasingly important topic. The usual, commonsense construal of decision making suggests that it is inherently tied to human characteristics—to language, in particular. If that is the case, then talk of animal decisions is merely metaphorical. However, behavioral ecology showed that animal and human behavior is constrained by economic parameters and coherent with the economy-of-nature principle. Neuroeconomics suggests that neural processing follows the same logic. Dopaminergic systems drive animals to achieve certain goals; affective mechanisms place goals and actions in value spaces. Although these systems have been extensively studied in humans, they are not peculiar to them. Humans display a unique complexity of goals and values, but this complexity relies partly on neural systems shared with many other animals: for instance, the nucleus accumbens and the amygdala are common to mammals. Brainy animals evolved an economic decision-making organ that allows them to cope with complex situations. As Gintis remarks, the complexity and the metabolic cost of central nervous systems coevolved in vertebrates, which suggests that despite their cost, brains are designed to make adaptive decisions (2007, 3).

Theoretically, this chapter suggests that the concept of decision making should be analyzed in a manner similar to the analysis given of the concept of *cooperation*. Nowadays, the evolutionary foundations, neural substrates, psychological mechanisms, formal modeling, and philosophical analyses of cooperation constitute a coherent—although not unified—field of inquiry.[2] The nature of prosocial behavior, from kin selection to animal cooperation to human morality, is best understood by adopting a naturalistic stance that highlights both the continuity of the phenomenon and the human specificity. Biological decision making deserves the same eclecticism.

Talking about *biological* decision making comes at a certain conceptual price. As many philosophers have pointed out, whenever one is describing actions and decisions, one is also presupposing the rationality of the agent (Davidson 1980; Dennett 1987; Popper 1994). When we say that agent A *chose* X, we suppose that A had reasons, preferences, and so on. The default assumption is that preferences and actions are coherent: the first caused the second, and the second is justified by the first. The rationality philosophers are referring to, however, is a complex cognitive faculty that requires language and propositional attitudes such as beliefs and desires. When animals forage their environment or select preys, patches, or mates, no one presupposes that they entertain beliefs or desires. There is nonetheless a presupposition that "much of the structure of the internal mental operations that inform decisions can be viewed as the product of evolution and natural selection" (Real 1994, 4). Thus, to a certain degree, the neuronal processes concerned with evaluative information are effective and efficient; otherwise, natural selection would have discarded them. I label these presuppositions, and the mechanisms they might reveal, "natural rationality." Natural rationality is a possibility condition for the concept of biological decision making and the economy-of-nature principle. One needs to presuppose that there is a natural excellence in the biosphere *before* studying decisions and constraints, and that evaluative information processing percolates in all the tree of life.

More than a logical prerequisite, natural rationality concerns the descriptive and normative properties of the mechanisms by which humans and other animals make decisions. Most concepts of rationality examine only the descriptive or the normative side, and hence tend to describe cognitive/neuronal processes without concern for their optimality, or state ideal conditions for rational behavior. For instance, though classical economics considers rational-choice theory either as a normative theory or a useful fiction, proponents of bounded rationality or ecological rationality refuse to characterize decision making as optimization (Chase, Hertwig, and Gigerenzer 1998; Gigerenzer 2004; Selten 2001). Others advocate a strong division of labor between normative and descriptive projects. For example, Tversky and Kahneman concluded from their studies of human bounded rationality that the normative and descriptive accounts of decision making are two separate projects that "cannot be reconciled" (Tversky and Kahneman 1986, S272). The perspective I suggest here is that we should expect an overlap between normative and descriptive theories and that the existence of this overlap is warranted by natural selection. On the normative side, we should ask what procedures and mechanisms biological agents should follow in order to make effective and efficient decisions given the constraints of the economy of nature. On the descriptive side, we must assess whether a procedure succeeds in achieving goals or, conversely, what goals could a procedure aim to achieve. If there is no overlap between norms and facts, then either norms should be reconceptualized or facts should be scrutinized: it might be the case that the norms are unrealistic or that we did not identify the right goal or value.

This accounts contrasts with those of philosophers who, like Dennett (1987) or Davidson (1980), construe rationality as an idealization and also researchers, like Cosmides and Tooby, who preach the elimination of this concept because of its idealized status (1994). According to this account, rationality should be conceived not as an a priori postulate in economy and philosophy, but as an empirical and multidisciplinary research program. Quine once said that "creatures inveterately wrong in their inductions have a pathetic but praiseworthy tendency to die out before reproducing their kind" (1969, 126). Whether it is true for *inductions* is still open to debate, but I suggest that it clearly applies to *decisions*.

Acknowledgments

I am grateful to George Terzis and Robert Arp, for helpful comments on an earlier draft and to various readers of my Natural Rationality blog (NaturalRationality.blogspot.com) for their ideas, criticisms, questions, and suggestions. Funding for research was provided by the Social Science and Humanities Research Council of Canada.

Notes

1. I am grateful to the editors for the idea of a value-based account of information.

2. See, for instance, how neuroscience, game theory, economic, philosophy, psychology, and evolutionary theory interact (Fehr and Fischbacher 2002, 2003; Hauser 2006; Penner et al. 2005).

References

Anderson, Michael L. 2007a. Evolution of cognitive function via redeployment of brain areas. *Neuroscientist* 13 (1): 13–21.

Anderson, Michael L. 2007b. The massive redeployment hypothesis and the functional topography of the brain. *Philosophical Psychology* 20 (2): 143–174.

Audi, Robert. 2001. *The architecture of reason: The structure and substance of rationality.* Oxford: Oxford University Press.

Baron, Jonathan. 2000. *Thinking and deciding.* Cambridge: Cambridge University Press.

Bechara, Antoine, and Antonio R. Damasio. 2005. The somatic marker hypothesis: A neural theory of economic decision. *Games and Economic Behavior* 52 (2): 336–372.

Bechara, Antoine, Hanna Damasio, Daniel Tranel, and Antonio R. Damasio. 1997. Deciding advantageously before knowing the advantageous strategy. *Science* 275 (5304): 1293–1295.

Berridge, Kent C. 2003a. Irrational pursuits: Hyper-incentives from a visceral brain. In *The psychology of economic decisions*, vol. 1, ed. Isabelle Brocas and Juan D. Carrillo, 17–40. Oxford: Oxford University Press.

Berridge, Kent C. 2003b. Pleasures of the brain. *Brain and Cognition* 52 (1): 106–128.

Berridge, Kent C., and Terry E. Robinson. 1998. What is the role of dopamine in reward: hedonic impact, reward learning, or incentive salience? *Brain Research: Brain Research Reviews* 28 (3): 309–369.

Bohan Broderick, Paul. 2004. On communication and computation. *Minds and Machines* 14 (1): 1–19.

Bowler, Peter J. 1976. Malthus, Darwin, and the concept of struggle. *Journal of the History of Ideas* 37 (4): 631–650.

Brandom, Robert B. 2000. *Articulating reasons: An introduction to inferentialism.* Cambridge, Mass.: Harvard University Press.

Bshary, Redouan, and Daniel Schäffer. 2002. Choosy reef fish select cleaner fish that provide high-quality service. *Animal Behaviour* 63 (3): 557–564.

Burnet, Thomas. ca. 1692/1965. *The sacred theory of the Earth.* Carbondale, Ill.: Southern Illinois University Press.

Camerer, Colin. 2000. Prospect theory in the wild. In *Choice, values, and frames*, ed. Daniel Kahneman and Amos Tversky, 288–300. New York: Cambridge University Press.

Cartwright, Nancy. 1989. *Nature's capacities and their measurement.* Oxford: Oxford University Press.

Chase, Valerie M., Ralph Hertwig, and Gerd Gigerenzer. 1998. Visions of rationality. *Trends in Cognitive Sciences* 2 (6): 206–214.

Chen, M. Keith, Venkat Lakshminarayanan, and Laurie R. Santos. 2006. How basic are behavioral biases? Evidence from capuchin monkey trading behavior. *Journal of Political Economy* 114 (3): 517–537.

Cosmides, Leda, and John Tooby. 1994. Better than rational: Evolutionary psychology and the invisible hand. *American Economic Review* 84 (2): 327–332.

Cosmides, Leda, and John Tooby. 2000. Evolutionary psychology and the emotions. In *Handbook of Emotions*, ed. Michael Lewis, Jeannette M. Haviland-Jones, and Lisa Feldman Barrett, 91–115. New York: Guilford.

Damasio, Antonio R. 1994. *Descartes' error: Emotion, reason, and the human brain*. New York: Putnam.

Damasio, Antonio R. 2003. *Looking for Spinoza: Joy, sorrow, and the feeling brain*. London: Harcourt.

Damasio, Antonio R., Hanna Damasio, and Yves Christen. 1996. *Neurobiology of decision-making*. New York: Springer.

Darwin, Charles. 1859/2003. *On the origin of species by means of natural selection*. New York: Fine Creative Media.

Darwin, Charles. 1887. Autobiography. In *The life and letters of Charles Darwin, including an autobiographical chapter*, vol. 1, ed. Francis Darwin, 26–106. London: John Murray.

Darwin, Charles. 1896. *The expression of the emotions in man and animals*. New York: D. Appleton.

Davidson, Donald. 1980. *Essays on actions and events*. Oxford: Oxford University Press.

Dawkins, Richard. 1976. *The selfish gene*. New York: Oxford University Press.

de Quervain, Dominique J.-f, Urs Fischbacher, Valerie Treyer, Melanie Schellhammer, Ulrich Schnyder, Alfred Buck, and Ernst Fehr. 2004. The neural basis of altruistic punishment. *Science* 305 (5688): 1254–1258.

Dennett, Daniel C. 1987. *The intentional stance*. Cambridge, Mass.: MIT Press.

Dennett, Daniel C. 1995. *Darwin's dangerous idea: Evolution and the meanings of life*. New York: Simon & Schuster.

Dretske, Fred I. 1981. *Knowledge and the flow of information*. Cambridge, Mass.: MIT Press.

Dugatkin, Lee Alan, and Hudson Kern Reeve. 1998. *Game theory and animal behavior*. New York: Oxford University Press.

Dukas, Reuven. 2004. Evolutionary biology of animal cognition. *Annual Review of Ecology Evolution and Systematics* 35: 347–374.

Duncan, John. 1986. Disorganisation of behaviour after frontal lobe damage. *Cognitive Neuropsychology* 3 (3): 271–290.

Egelman, David M., Christophe Person, and P. Read Montague. 1998. A computational role for dopamine delivery in human decision-making. *Journal of Cognitive Neuroscience* 10 (5): 623–630.

Ekman, Paul. 1972. *Emotion in the human face: Guide-lines for research and an integration of findings*. New York: Pergamon Press.

Ekman, Paul. 1999. Basic emotions. In *Handbook of cognition and emotion*, ed. Tim Dalgleish and Mick Power, 45–60. Sussex, U.K.: Wiley & Sons.

Eliasmith, Chris, and Charles H. Anderson. 2003. *Neural engineering: Computation, representation, and dynamics in neurobiological systems*. Cambridge, Mass.: MIT Press.

Faucher, Luc, and Christine Tappolet. 2002. Fear and the focus of attention. *Consciousness and Emotion* 3 (2): 105–144.

Fehr, Ernst, and Urs Fischbacher. 2002. Why social preferences matter: The impact of non-selfish motives on competition, cooperation and incentives. *Economic Journal* 112 (478): C1–C33.

Fehr, Ernst, and Urs Fischbacher. 2003. The nature of human altruism. *Nature* 425 (6960): 785–791.

Floridi, Luciano. 2004a. Information. In *The Blackwell guide to the philosophy of computing and information*, ed. Luciano Floridi, 40–61. Oxford: Blackwell.

Floridi, Lucanio. 2004b. On the logical unsolvability of the Gettier problem. *Synthese* 142 (1): 61–79.

Frank, Michael J., and Eric D. Claus. 2006. Anatomy of a decision: striato-orbitofrontal interactions in reinforcement learning, decision making, and reversal. *Psychological Review* 113 (2): 300–326.

Gayon, Jean. 2003. From Darwin to today in evolutionary biology. In *The Cambridge Companion to Darwin*, ed. Jonathan Hodge and Gregory Radick, 240–264. Cambridge: Cambridge University Press.

Gegenfurtner, Karl R., and Daniel C. Kiper. 2003. Color visions. *Annual Review of Neuroscience* 26: 181–206.

Ghiselin, Michael T. 1978. The economy of the body. *American Economic Review* 68 (2): 233–237.

Ghiselin, Michael T. 1995. Perspective: Darwin, progress, and economic principle. *Evolution; International Journal of Organic Evolution* 49 (6): 1029–1037.

Ghiselin, Michael T. 1999. Darwinian monism: The economy of nature. In *Sociobiology and bioeconomics: The theory of evolution in biological and economic theory*, ed. Peter Koslowski, 7–24. Berlin: Springer.

Giere, Ronald N. 2004. How models are used to represent reality. *Philosophy of Science* 71 (5): 742–752.

Gigerenzer, Gerd. 2004. Fast and frugal heuristics: The tools of bounded rationality. In *Blackwell handbook of judgment and decision making*, ed. Derek J. Koehler and Nigel Harvey, 62–88. Oxford: Blackwell.

Gintis, Herbert. 2007. A framework for the integration of the behavioral sciences. *Behavioral and Brain Sciences* 30 (1): 1–16.

Glimcher, Paul W. 2003a. *Decisions, uncertainty, and the brain: The science of neuroeconomics*. Cambridge, Mass.: MIT Press.

Glimcher, Paul W. 2003b. The neurobiology of visual-saccadic decision making. *Annual Review of Neuroscience* 26: 133–179.

Glimcher, Paul W., and Aldo Rustichini. 2004. Neuroeconomics: The consilience of brain and decision. *Science* 306 (5695): 447–452.

Glimcher, Paul W., Michael C. Dorris, and Hannah M. Bayer. 2005. Physiological utility theory and the neuroeconomics of choice. *Games and Economic Behavior* 52 (2): 213.

Griffiths, Paul E. 1997. *What emotions really are: The problem of psychological categories.* Chicago: University of Chicago Press.

Griffiths, Paul E. 2004. Emotions as natural and normative kinds. *Philosophy of Science* 71 (5): 901–911.

Hames, Raymond B., and William T. Vickers. 1982. Optimal diet breadth theory as a model to explain variability in Amazonian hunting. *American Ethnologist* 9 (2): 358–378.

Hammerstein, Peter, and Edward H. Hagen. 2005. The second wave of evolutionary economics in biology. *Trends in Ecology & Evolution* 20 (11): 604–609.

Hansen, Andrew J. 1986. Fighting behavior in bald eagles: A test of game theory. *Ecology* 67 (3): 787–797.

Hardy-Vallée, Benoit. 2007. Decision-making: A neuroeconomic perspective. *Philosophy Compass* 2 (6): 939–953.

Hardy-Vallée, Benoit. Forthcoming. Decision-making in robotics and psychology: A distributed account. *New Ideas in Psychology.*

Hauser, Marc D. 2006. *Moral minds: How nature designed our universal sense of right and wrong.* New York: Ecco.

Hauser, Marc D., Noam Chomsky, and W. Tecumseh Fitch. 2002. The faculty of language: What is it, who has it, and how did it evolve? *Science* 298 (5598): 1569–1579.

Hestmark, Geir. 2000. Oeconomia Naturae L. *Nature* 405 (6782): 19.

Hirshleifer, Jack. 1978. Natural economy versus political economy. *Journal of Social and Biological Structures* 1 (4): 319–337.

Hodgson, Geoffrey M. 2001. Bioeconomics. In *Encyclopedia of Political Economy,* ed. Phillip Anthony O'Hara, 37–41. London: Routledge/Taylor & Francis.

Houde, Olivier, and Nathalie Tzourio-Mazoyer. 2003. Neural foundations of logical and mathematical cognition. *Nature Reviews: Neuroscience* 4 (6): 507–514.

Jochim, Michael A. 1988. Optimal foraging and the division of labor. *American Anthropologist* 90 (1): 130–136.

Johnson-Laird, Philip N., and Eldar Shafir. 1993. The interaction between reasoning and decision making: an introduction. *Cognition* 49 (1–2): 1–9.

Kacelnik, Alejandro, Alasdair I. Houston, and John R. Krebs. 1981. Optimal foraging and territorial defence in the Great Tit (*Parus major*). *Behavioral Ecology and Sociobiology* 8 (1): 35.

Kahneman, Daniel, and Amos Tversky. 1979. Prospect theory: An analysis of decision under risk. *Econometrica* 47 (2): 263–291.

Kahneman, Daniel, and Amos Tversky. 1991. Loss aversion in riskless choice: A reference-dependent model. *Quarterly Journal of Economics* 106 (4): 1039–1061.

Kahneman, Daniel, and Amos Tversky, eds. 2000. *Choices, values, and frames.* Cambridge: Cambridge University Press.

Kaplan, Hillard, Kim Hill, Kristen Hawkes, and Ana Hurtado. 1984. Food sharing among Ache hunter-gatherers of eastern Paraguay. *Current Anthropology* 25 (1): 113–115.

King-Casas, Brooks, Damon Tomlin, Cedric Anen, Colin F. Camerer, Steven R. Quartz, and P. Read Montague. 2005. Getting to know you: Reputation and trust in a two-person economic exchange. *Science* 308 (5718): 78–83.

Knutson, Brian, Scott Rick, G. Elliott Wimmer, Drazen Prelec, and George Loewenstein. 2007. Neural predictors of purchases. *Neuron* 53 (1): 147–156.

Koch, Christof, and Idan Segev. 2000. The role of single neurons in information processing. *Nature Neuroscience* 3 (November): 1171–1177.

Koechlin, Etienne, Chrystèle Ody, and Frédérique Kouneiher. 2003. The architecture of cognitive control in the human prefrontal cortex. *Science* 302 (5648): 1181–1185.

Krebs, John R., and Nicholas B. Davies. 1997. *Behavioural ecology: An evolutionary approach.* Oxford: Blackwell Science.

LeDoux, Joseph E. 1996. *The emotional brain: The mysterious underpinnings of emotional life.* New York: Simon & Schuster.

Lhermitte, F. 1986. Human autonomy and the frontal lobes. Part II: Patient behavior in complex and social situations: The "environmental dependency syndrome." *Annals of Neurology* 19 (4): 335–343.

Lieberman, Philip. 2007. The evolution of human speech: Its anatomical and neural bases. *Current Anthropology* 48 (1): 39–66.

Lima, Steven L. 2002. Putting predators back into behavioral predator-prey interactions. *Trends in Ecology & Evolution* 17 (2): 70–75.

Lin, Longnian, Guifen Chen, Hui Kuang, Dong Wang, and Joe Z. Tsien. 2007. Neural encoding of the concept of nest in the mouse brain. *Proceedings of the National Academy of Sciences of the United States of America* 104 (14): 6066–6071.

Linnaeus, Carl. 1751. Specimen academicum de oeconomia naturae. *Amoenitas Academicae* 2: 1–58.

Lohrenz, Terry, Kevin McCabe, Colin F. Camerer, and P. Read Montague. 2007. Neural signature of fictive learning signals in a sequential investment task. *Proceedings of the National Academy of Sciences of the United States of America* 104 (22): 9493–9498.

Lyell, Charles. 1853. *Principles of geology; Or, the modern changes of the earth and its inhabitants considered as illustrative of geology.* London: J. Murray.

MacArthur, Robert H., and Eric R. Pianka. 1966. On optimal use of a patchy environment. *American Naturalist* 100 (916): 603–609.

Mameli, Matteo. 2005. The inheritance of features. *Biology and Philosophy* 20 (2–3): 365–399.

Maynard Smith, John. 2000. The concept of information in biology. *Philosophy of Science* 67 (2): 177–194.

McCabe, Kevin. 2005. Neuroeconomics. In *Encyclopedia of cognitive science*, ed. Lynn Nadel, 294–298. New York: Wiley InterScience.

McCoy, Allison N., and Michael L. Platt. 2005. Expectations and outcomes: decision-making in the primate brain. *Journal of Comparative Physiology A. Neuroethology, Sensory, Neural, and Behavioral Physiology* 191 (3): 201–211.

McFarland, David, and Thomas Bösser. 1993. *Intelligent behavior in animals and robots*. Cambridge, Mass.: MIT Press.

Miller, Earl K., and Jonathan D. Cohen. 2001. An integrative theory of prefrontal cortex function. *Annual Review of Neuroscience* 24: 167–202.

Millikan, Ruth Garrett. 1984. *Language, thought, and other biological categories: New foundations for realism*. Cambridge, Mass.: MIT Press.

Millikan, Ruth Garrett. 1996. Pushmi-Pullyu representations. In *Mind and morals: Essays on cognitive science and ethics*, ed. Larry May, Marilyn Friedman, and Andy Clark, 145–161. Cambridge, Mass.: MIT Press.

Montague, P. Read. 2006. *Why choose this book? How we make decisions*. New York: Penguin.

Montague, P. Read, and Gregory S. Berns. 2002. Neural economics and the biological substrates of valuation. *Neuron* 36 (2): 265–284.

Montague, P. Read, and Steven R. Quartz. 1999. Computational approaches to neural reward and development. *Mental Retardation and Developmental Disabilities Research Reviews* 5 (1): 86–99.

Montague, P. Read, Peter Dayan, Christophe Person, and Terrence J. Sejnowski. 1995. Bee foraging in uncertain environments using predictive hebbian learning. *Nature* 377 (6551): 725–728.

Montague, P. Read, Steven E. Hyman, and Jonathan D. Cohen. 2004. Computational roles for dopamine in behavioural control. *Nature* 431 (7010): 760–767.

Montague, P. Read, Brooks King-Casas, and Jonathan D. Cohen. 2006. Imaging valuation models in human choice. *Annual Review of Neuroscience* 29: 417–448.

Morgan, Mary S., and Margaret Morrison. 1999. *Models as mediators: Perspectives on natural and social sciences*. Cambridge: Cambridge University Press.

Morris, Genela, Alon Nevet, David Arkadir, Eilon Vaadia, and Hagai Bergman. 2006. Midbrain dopamine neurons encode decisions for future action. *Nature Neuroscience* 9 (July): 1057–1063.

Mottley, Kieron, and Luc-Alain Giraldeau. 2000. Experimental evidence that group foragers can converge on predicted producer-scrounger equilibria. *Animal Behaviour* 60 (3): 341–350.

Naqvi, Nasir, Baba Shiv, and Antoine Bechara. 2006. The role of emotion in decision making: A cognitive neuroscience perspective. *Current Directions in Psychological Science* 15 (5): 260–264.

Niv, Yael, Nathaniel D. Daw, and Peter Dayan. 2006. Choice values. *Nature Neuroscience* 9 (August): 987–988.

Niv, Yael, Michael O. Duff, and Peter Dayan. 2005. Dopamine, uncertainty and TD learning. *Behavioral and Brain Functions* 1:6. http://www.behavioralandbrainfunctions.com/content/1/1/6.

Norman, Donald A., and Tim Shallice. 1980. *Attention to action: Willed and automatic control of behavior.* San Diego: Center for Human Information Processing, University of California.

O'Reilly, Randall C. 2006. Biologically based computational models of high-level cognition. *Science* 314 (5796): 91–94.

O'Donnell, Patricio. 2003. Dopamine gating of forebrain neural ensembles. *European Journal of Neuroscience* 17 (3): 429–435.

Oosterbeek, Hessel, Randolph Sloof, and Gijs Van de Kuilen. 2004. Differences in Ultimatum Game experiments: Evidence from a meta-analysis. *Experimental Economics* 7 (2): 171–188.

Panksepp, Jaak. 1998. *Affective neuroscience: The foundations of human and animal emotions.* New York: Oxford University Press.

Pecina, Susana, Barbara Cagniard, Kent C. Berridge, J. Wayne Aldridge, and Xiaoxi Zhuang. 2003. Hyperdopaminergic mutant mice have higher "wanting" but not "liking" for sweet rewards. *Journal of Neuroscience* 23 (28): 9395–9402.

Penner, Louis A., John F. Dovidio, Jane A. Piliavin, and David A. Schroeder. 2005. Prosocial behavior: Multilevel perspectives. *Annual Review of Psychology* 56: 365–392.

Pianka, Eric R. 2000. *Evolutionary ecology.* San Francisco: Benjamin Cummings.

Pirolli, Peter L., and Stuart K. Card. 1999. Information foraging. *Psychological Review* 106 (4): 643–675.

Popper, Karl R. 1994. Models, instruments, and truth: The status of the rationality principle in the social sciences. In *The myth of the framework: In defence of science and rationality*, ed. M. A. Notturno, 154–184. London: Routledge.

Quine, W. V. O. 1969. *Ontological relativity and other essays.* New York: Columbia University Press.

Real, Leslie A., ed. 1994. *Behavioral mechanisms in evolutionary ecology.* Chicago: University of Chicago Press.

Redish, A. David. 2004. Addiction as a computational process gone awry. *Science* 306 (5703): 1944–1947.

Rilling, James K., David A. Gutman, Thorsten R. Zeh, Giuseppe Pagnoni, Gregory S. Berns, and Clinton D. Kilts. 2002. A neural basis for social cooperation. *Neuron* 35 (2): 395–405.

Russell, James A. 2003. Core affect and the psychological construction of emotion. *Psychological Review* 110 (1): 145–172.

Sanfey, Alan G., James K. Rilling, Jessica A. Aronson, Leigh E. Nystrom, and Jonathan D. Cohen. 2003. The neural basis of economic decision-making in the ultimatum game. *Science* 300 (5626): 1755–1758.

Schabas, Margaret. 2005. *The natural origins of economics*. Chicago: University of Chicago Press.

Schultz, Wolfram. 2001. Reward signaling by dopamine neurons. *Neuroscientist* 7 (4): 293–302.

Searle, John R. 2001. *Rationality in action*. Cambridge, Mass.: MIT Press.

Selten, Reinhard. 2001. What is bounded rationality? In *Bounded rationality: The adaptive toolbox*, ed. Gerd Gigerenzer and Reinhard Selten, 13–36. Cambridge, Mass.: MIT Press.

Shafir, Eldar, Itamar Simonson, and Amos Tversky. 1993. Reason-based choice. *Cognition* 49 (1–2): 11–36.

Shannon, Claude E., and Warren Weaver. 1949. *The mathematical theory of communication*. Urbana: University of Illinois Press. Reprinted and repaginated 1963.

Shallice, Tim. 1988. *From neuropsychology to mental structure*. Cambridge: Cambridge University Press.

Sih, Andrew, and Bent Christensen. 2001. Optimal diet theory: When does it work, and when and why does it fail? *Animal Behaviour* 61 (2): 379–390.

Singer, Tania, Ben Seymour, John P. O'Doherty, Klaas E. Stephan, Raymond J. Dolan, and Chris D. Frith. 2006. Empathic neural responses are modulated by the perceived fairness of others. *Nature* 439 (7075): 466–469.

Smith, Adam. 1759/2002. *The theory of moral sentiments*. Cambridge: Cambridge University Press.

Suri, Ronald E. 2002. TD models of reward predictive responses in dopamine neurons. *Neural Networks* 15 (4): 523–533.

Suri, Ronald E., and Wolfram Schultz. 2001. Temporal difference model reproduces anticipatory neural activity. *Neural Computation* 13 (4): 841–862.

Sutton, Richard S., and Andrew G. Barto. 1998. *Reinforcement learning: An introduction*. Cambridge, Mass.: MIT Press.

Sutton, Richard S., and Andrew G. Barto. 1987. A temporal-difference model of classical conditioning. In *Proceedings of the Ninth Conference of the Cognitive Science Society*, 355–378. Hillsdale, N.J.: Lawrence Erlbaum Associates.

Thagard, Paul. 1992. *Conceptual revolutions*. Princeton: Princeton University Press.

Thagard, Paul. 2007. Abductive inference: From philosophical analysis to neural mechanisms. In *Inductive reasoning: Experimental, developmental, and computational approaches*, ed. Aidan Feeney and Evan Heit, 226–247. Cambridge: Cambridge University Press.

Thaler, Richard H. 1980. Toward a positive theory of consumer choice. *Journal of Economic Behavior & Organization* 1 (1): 39–60.

Thomas, Robert J. 1999. Two tests of a stochastic dynamic programming model of daily singing routines in birds. *Animal Behaviour* 57 (2): 277–284.

Tooby, John, and Leda Cosmides. 2005. Conceptual foundations of evolutionary psychology. In *The handbook of evolutionary psychology*, ed. David M. Buss, 5–67. Hoboken, N.J.: Wiley.

Tversky, A., and D. Kahneman. 1986. Rational choice and the framing of decisions. *Journal of Business* 59 (4): S251–S278.

van 't Wout, Mascha, René S. Kahn, Alan G. Sanfey, and André Aleman. 2006. Affective state and decision-making in the ultimatum game. *Experimental Brain Research* 169 (4): 564–568.

Vickery, William L., Luc-Alain Giraldeau, Jennifer J. Templeton, Donald L. Kramer, and Colin A. Chapman. 1991. Producers, scroungers, and group foraging. *American Naturalist* 137 (6): 847–863.

White, Donald W., Lawrence M. Dill, and Charles B. Crawford. 2007. A common, conceptual framework for behavioral ecology and evolutionary psychology. *Evolutionary Psychology* 5 (2): 275–288.

Williams, Jonathan, and Peter Dayan. 2005. Dopamine, learning, and impulsivity: A biological account of attention-deficit/hyperactivity disorder. *Journal of Child and Adolescent Psychopharmacology* 15 (2): 160–179.

Zajonc, Robert B. 1980. Feeling and thinking: Preferences need no inferences. *American Psychologist* 35 (2): 151–175.

Zak, Paul J. 2004. Neuroeconomics. *Proceedings of the Royal Society B: Biological Sciences* 359 (1451): 1737–1748.

Zhang, Ke, Jian Zeng Guo, Yueqing Peng, Wang Xi, and Aike Guo. 2007. Dopamine-mushroom body circuit regulates saliency-based decision-making in Drosophila. *Science* 316 (5833): 1901–1904.

10

Information Theory and Perception: The Role of Constraints, and What Do We Maximize Information About?

Roland Baddeley, Benjamin Vincent, and David Attewell

Because human and nonhuman primates are highly visual animals, an important step in understanding how their brains operate is to understand how they see: that is, how they transform the structured, colored, and dynamic pattern of light surrounding them into knowledge about the world they live in. Over the last twenty years, a number of problems associated with understanding vision have used information-theoretic techniques to help us gain insight into perception (Shannon 1948). This chapter will describe, in a biased way, one particular strand of this work: how information theory has been used to shed light on the processes and representations present at the earliest stages of vision.

There are three reasons why early vision is a particularly good system to which to apply information theoretic concepts. First, we know a lot about what it does, even if we don't always know precisely why it does it. The pioneering physiological work of Hubel and Wiesel (1962) gave us a good grasp of the basic properties of the representations of early vision, and since then, the neurons in the earliest and largest cortical area involved in vision, V1 (sometimes also known as the striate cortex or area 17), have probably been more studied than those in any other part of the brain. These cells are sensitive to certain parameters of the local pattern of light arriving at the areas of the retina they are interested in, such as color, orientation, and motion. The parameter values to which a given cell is sensitive is known as its *receptive field* (Lennie 2003a). Using increasingly sophisticated methods, such cell properties have been not only qualitatively but also quantitatively described. In addition to our good physiological understanding of early vision, the fact that displaying simple but very well-controlled visual stimuli is relatively easy means that early vision has been studied using the methods of psychophysics more than any other sense. As a result of all this intensive work, we have a lot of data, both physiological and psychophysical, with which we can compare the predictions of any information theory–inspired model of early vision.

The second reason why early vision is amenable to study using information theory is that it is possible to generate estimates of its "natural" input. Information theory calculations often require estimates of the statistics of the signal to be transmitted (Simoncelli and Olshausen 2001) and, for vision, these estimates can be made from natural images and videos. This can be done in more or less sophisticated ways, from simply approximating input in terms of some "representative natural images" taken from the web, to approaches that take into account eye movements (Rucci, Edelman, and Wray 2000; Rucci and Casile 2005), the animal's environmental niche (Warrant 1999), and possibly the effects of the observer's motion through the environment. Importantly, the logistic and theoretical problems associated with making approximations to the input probability distribution for vision are not insurmountable. In contrast, applying information theory approaches to problems in the cognitive domain, for example, may be more challenging due to the difficulty of defining precisely the input to this system.

The last reason why an information theoretical approach may be appropriate is that at least at the earliest stages of vision (e.g., the transmission of information from the eye to the cortex), there exist rather severe bottlenecks for information transmission (discussed shortly). A system that is under severe constraints about how much information can be transmitted is more likely to be, in some sense, efficient. Although this is not so for all stages of early vision, at least some parts of the system have, on the face of it, more information to transmit than would be trivially possible without some sort of efficient (in terms of information theory) recoding. Comparing representations of the world observed in early vision with various optimal information theory-based representations has, therefore, at least some chance of success.

The rest of this chapter will therefore begin with a brief description of some pertinent properties of early vision. We will then describe the basic approach that has been employed when using information theory–based ideas to gain a better understanding of early visual representations, and illustrate this approach with three basic models. These models all share a number of assumptions about the nature of early visual representations and all use approximations to the visual input based on sampling from natural images. They differ, however, in that they propose that different constraints are important in understanding early perception. Last, we will give a very brief overview of some problems with the approach, together with some potentially interesting new directions to pursue.

The Nature of the Early Visual System

The main modeling effort that we will describe is concentrated on understanding the earliest stages of visual information processing. These are possibly the most intensively studied of all brain functions, and any short summary is bound to smooth over many potentially important details. However, some background is required to understand the subsequent modeling.

Early visual processing can be thought of as having three stages. In the first stage, the light in the retina is detected (via rod and cone cells), and the resulting signal is normalized. In the second stage, the signal is recoded, and transmitted from the retina to the cortex. This information is transmitted first using cells known as *retinal ganglion cells*, and then via a subcortical area called the *lateral geniculate nucleus* (LGN), whose exact function, despite a large number of theories, is not well understood (Derrington 2000). We will therefore ignore it. In the third and final stage, the information arrives in the first cortical area (V1), and is represented by a vastly greater number of cells than previously, which make a number of properties (or "features") of the local visual input explicit rather than implicit. In most layers of V1, the visual input is represented in terms of the presence of orientated edges, with cells firing at a high rate when presented with an edge or line within its local "receptive field," and not when no such edge or line is present.

On average, we point our eyes to a new location three times a second, and the information inherent in the light coming from the new direction has to be communicated to higher levels of the cortex. The basic problem the retina faces during the first stage of early vision is that it has to deal (given time to adapt) with light intensities that vary over nine orders of magnitude. This is beyond the dynamic range of all the subsequent processing stages. So, after detecting the light, the light level at every location is transformed with a transform that is not badly approximated over the most important range by a logarithm (Valeton and van Norren 1983). This is then followed by luminance adaptation; the detected light level is normalized by subtracting the average (log) light level in the local region over the recent past from the signal. This processing means that even with very large changes in the illuminant (e.g., the sun going behind a cloud), we can still see. This normalized version of the input needs then to be transmitted to later stages for further processing.

Though the eyes are (obviously) at the front of the head, the first cortical area involved in vision (V1) is at the back of the head. This means that the information has to be transmitted over a comparatively long distance (for the cortex). Given that there are very large numbers of rods and cones, it is unfeasible to send this information as a set of raw image measurements. Instead, the image is compressed by squeezing it through a narrow bottleneck of only ~1 million retinal ganglion cells per eye, which form the optic nerves. Each of these neurons is in charge of representing only a small portion of the visual field. Each neuron represents the difference between the sum of two local luminance averages of different spatial extents. This kind of sensitivity profile is known as a *center-surround* receptive field organization, and the transformed representation it produces is transmitted, via retinal ganglion cells and the LGN, to the early visual cortex (V1). This center-surround representation is fairly fundamental in vision and is one of the features that we would like to explain using information theory arguments.

The earliest representations of the image are not orientation-sensitive. Only when the visual information reaches V1 can the representation used by the majority of cells be thought of as coding for the orientation of local edge or line segments. The neurons in most layers of V1 are also sensitive to many other aspects of the visual input (motion, color, stereo, etc.), but it is this basic local edge-detection property that most modeling has attempted to explain.

This, then, gives us the basic phenomena to be explained: (1) a local center-surround representation of the input coming from the retina to the cortex, and (2) a representation in terms of local edges in the earliest area of the cortex. How can information theory help us here?

The Basic Logic of the Approach

The three models we next present all have the same basic structure. They all propose that these early visual representations maximize the amount of information transmitted about the world, subject to some set of constraints. They all agree that it is important to have a realistic model of the input distribution (the images we normally see), rather than taking as our signal random variations of light. They also all agree that representations should be thought of in terms of linear filters: modeling receptive fields as weighted sums of the pixels of the represented images they are to process. The one way that they differ is in the proposed nature of the constraints that prevent the trivial solution of maximizing information by simply having an output representation identical to the input.

Minimizing the Number of Units or the Amount of Time to Code: Principal Components Analysis

This model is based on the following assumptions. The first is that the input to the system—the probability distribution of natural images that are usually viewed by the visual system—can be well approximated in terms of a multivariate Gaussian (normal) distribution. This is equivalent to saying that if we know the means of the image intensities and the correlation between image measurements made at different locations by the photoreceptors, averaged across a large set of representative images, then we know everything that is to be known about the distribution—there is no additional non-Gaussian structure. This assumption is not in fact true (images do have such non-Gaussian structure; it's what stops them all from looking a bit like clouds), but it does have the virtue of simplicity.

The second assumption is that the noise on each of the receptors in the retina is Gaussian and independent between receptors.

The third and last assumption, and one that requires more explanation, is that the main constraint on information transmission is the limited number of linear filters available to represent the images. This is counterintuitive, as there are in fact far more neurons in V1 than cells in the retina. There are three ways to justify this assumption.

The first justification, which we will call the basic Gaussian justification, is based on its simplicity: given these assumptions, the optimal way of representing the input is in terms of the principal components of the input measurements. This is a well-understood statistical technique that provides us with the opportunity of an analytical understanding of the optimal solution, rather than simply a numerically optimized representation. Even if certain assumptions are not met (such as not having an obvious constraint on the number of output units), knowing the basic Gaussian solution will potentially be informative.

The second justification, which we call the Hebbian learning justification, comes about because of a connection between the optimal solution (the principal components solution), and the result of a number of models of low-level visual development. The well-known Hebb rule (Hebb 1949) specifies how the synapses (weight parameters) in a neural network should be changed depending on experience. That is, it states that neurons that fire together should wire together, and is usually interpreted in terms of a weight update rule (the rate of change of a weight) based on the correlation between two units.

Provided that a neural network is linear, and has some means to make sure that all units represent unrelated features, if it is trained using a number of variants of Hebb's rule, then it will converge when the weights span the same space as the principal components. Because Hebb's rule is a very popular model of visual development, knowing the optimum for this scenario is of interest.

The last reason is more subtle, and we call it the Time Limited Performance Justification. This is based on work by Korutcheva, Parga and Nadal (1997). They analyzed a binary system in terms of maximizing information transmission when presented with Gaussian input. A remarkable property of such a binary system is that as the ratio between the number of output and input units increases, binary processing becomes equivalent to linear processing; it is equivalent to principal components analysis. There are far more cells in V1 (output units) than there are cells in the retina (input units). Neurons communicate with spikes, so, when observed over a very short time window, neurons can either spike or not—they are effectively binary. This means that even if neurons in V1 are not well approximated by linear filters, as long as there are a lot of them (and there are), and we consider their behavior only over very short time scales (so they can either fire or not), if they are maximizing information with their input, then this will be equivalent to the whole system operating as if it were performing principal components analysis. Note that this statement does not make any predictions about the receptive fields observed physiologically or about psychophysical performance when subjects have a large amount of time, but does make strong predictions about the system's performance when operating over very short timescales (less than 30 ms).

This model leads us to a hypothesis: the representation of an image in early vision is in terms of its principal components. How do we go about testing this idea? One possible method is as follows: (1) Collect a set of images representative of the kind of world that vision is used to process. (2) Randomly sample local patches (say windowed 16 × 16-pixel patches) from these images as if eye movements were being made to them. (3) Calculate the principal components of these patches. (4) Compare these with what is known about the physiology or psychophysics of early vision. This is the procedure that Baddeley and Hancock

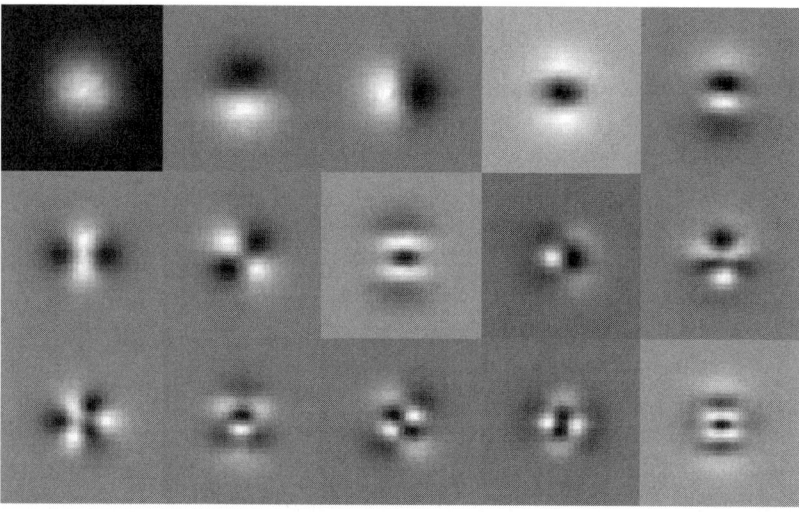

Figure 10.1
The first fifteen local principal components of a large collection of natural scenes based on work reported in Baddeley and Hancock 1991. The components are presented in terms of the amount of image variance they account for, with the component at the top left accounting for the most and components on the top row explaining more than lower rows. As can be seen, the early components (e.g., components 2–7) could be thought of as edge or line detectors of various types (responding optimally when stimulated with a line). Despite this, they do not resemble in detail receptive fields found in V1 (or retinal ganglion cell receptive fields).

(1991) followed, and the resulting two-dimensional filters are shown in figure 10.1.

How do these filters compare to what is known about physiology? Though there is some resemblance (see components 2–7), many of the components are a poor match. More importantly, although components 2–7 *qualitatively* could be described as edge detectors, *quantitatively* they are far from a good match to what is known about receptive fields. This limitation is important, because developmental models such as that of Linsker (1986) converge to the principle components of the input. The approximate "edge detector–like" properties of these models have been argued to be evidence, but the actual properties of even the components that do resemble receptive fields are quantitatively rather different (see later discussion for models that do make more accurate predictions).

This leaves the last suggestion: that the entire system, when given only very short periods of time, appears to resemble the principal components. To understand how to test this idea, a few words on what determines the nature of the components are in order. It turns out that there are three important factors that determine the nature of the components of natural images.

First, the power spectra of the input images are important. Principle components attempt to capture as much of the variance as possible, and the power spectra of the images tells us where this variance is. It is a relatively robust characteristic of natural images, from a wide range of sources, that the power at a given frequency is inversely proportional to its frequency (Field 1987): the vast majority of the power in natural images is at low frequencies. This means that the components also have to be tuned to low spatial frequencies in order to capture input variance.

The second determinant of the components is the nature of the window used when sampling local image patches. Though this determinant has nothing to do with image statistics, it can have a large effect on the derived components. Simply sampling square patches from images always means that any structure will be aligned to the sampling window. Therefore, to avoid artificially imposing structure on components, we windowed the samples with a Gaussian approximation before subsequent processing. This leaves the last characteristic: how anisotropic the image statistics of the world are. In particular, how rapidly does the correlation between image intensity measurements decay as a function of angle?

In artificial images, the correlation decays can be equal for all directions, but in naturalistic images, particularly those of wide open landscapes, the correlation decays considerably quicker in the vertical direction compared to the horizontal (Baddeley 1996). This difference results in the structure of vertical and horizontal components being different. In particular, two components can be thought of as line detectors: in figure 10.1, components 4 and 6 respond optimally when a light or dark line at the appropriate orientation is placed in the center. These two components are not simply rotations of each other, but have slightly different structure. If the components provide a good description of the whole system operating at very short intervals, a match to this difference in structure should be measurable psychophysically. In fact, this proved to be true. In some extremely labor-intensive experiments, in which subjects had to detect a single differently orientated line in the presence of a large number of other lines, Foster and Ward (1991) found that at very short presentation times (40 ms) subjects' performance was well summarized by two line-detection mechanisms operating. Figure 10.2

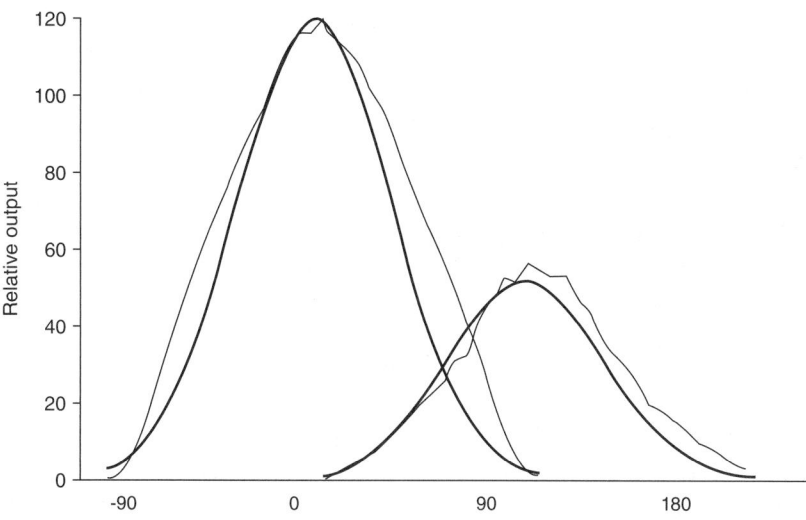

Figure 10.2
Does the orientation tuning of components 4 and 6 in any way match that estimated with short time presentation psychophysics? Shown is the response of these two components to an optimal bar rotated through 180 degrees (thin line), and for comparison the orientation tuning inferred experimentally by psychophysical methods for very short time presentations (thick line), based on the work of Foster and Ward (1991). As can be seen, though the components do not match individual receptive fields, the match of the system's orientation sensitivity is nearly perfect.

shows the response of these two mechanisms as a function of orientation, together with the orientation tuning of the two line-detecting principle components. As can be seen, the match is essentially perfect.

To conclude this section on principal components-based models: models that propose that information about images is maximized, and that the main constraint on communication is the number of filters to represent the image, result in a model that essentially extracts the principal components of images. The early components vaguely quantitatively resemble what is known about early receptive fields, but qualitatively they are not a good match. When people have only extremely short amounts of time to view things, then aspects of the system as a whole do appear to be consistent with information maximization. However, we very rarely only have 40 ms to view something. Given longer viewing times, human performance changes (improves), and the use of a binary code is no longer justified. Therefore, the principal components model is not, in general, a satisfactory model of human performance.

Beginning to Make the Model More Realistic: Metabolic Constraints on Firing Rate (or Sparsity)

That there are more neurons in V1 than receptors in the retina forms a specific coding challenge (Lewicki and Sejnowski 1998), and this abundance of V1 neurons is problematic for the approach described previously. Thus, to respond to this challenge, we are led to a second model in which instead of constraining the *number* of neurons representing the image, we place a constraint on their *level of activity*.

There are a number of justifications for this. By far the simplest is a metabolic argument: given that there are a vast number of neurons in V1 (the largest for any cortical area), they will have a significant cost in terms of energy consumption. As well as minimizing energy consumption, simply to minimize the amount of food we need to find and eat, highly metabolically active neural tissue also presents both resource transport and heat dissipation problems (Falk 1992; Corrard 1999). The brain is a highly metabolically active organ (Rolfe and Brown 1997), and any representation that required less energy would be favored by evolution, as long as it did not hurt performance too much.

There are a number of factors that determine the energy consumption of a piece of the brain (Attewell and Laughlin 2001); indeed, simply maintaining neurons and using synapses uses resources (discussed shortly). Here, however, we reduce energy consumption by minimizing the activity of the neurons; that is, we minimize their average firing rate. The idea is simple: rather than simply looking for the representation that captures as much information about the image as possible, we simultaneously try to minimize the amount of energy used to represent images, where the main form of energy use we concentrate on is the energy used to fire a neuron. If we again use a Gaussian approximation to the information, this optimization is very simple to implement in terms of a neural network learning rule. This was done first in Fyfe and Baddeley (1995), where we used a neural network to extract the filters with very peaked output distributions. In these simulations, we optimized a statistical quantity called *kurtosis* rather than minimizing the absolute output, but the effect is very similar. We, however, calculated only the first three optimal filters, and the nature of the full solution was, therefore, not completely clear. Our work was followed by that of two groups (Harpur and Prager 1996; Olshausen and Field 1996), again both using neural networks, and although the algorithm proposed by Harpur and Prager has computational advantages over both ours and the Olshausen and

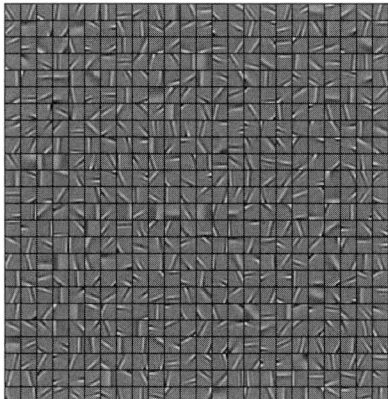

Figure 10.3
"Edge detectors" as derived by simultaneously optimizing the information transmitted about images, while minimizing the average firing rate of the cells. As can be seen, the result of such an optimization is a large number of local, oriented edge detectors that provide a good qualitative match to the receptive fields found in V1. If motion is also taken into account, as shown by van Hateren (1983), the match is quantitative as well.

Field algorithm (one of their methods does not require a prewhitening filter), the results from the Olshausen and Field model were presented particularly nicely (and appeared in *Nature*), so the result is usually attributed to them. The basic result, the optimal receptive fields, when optimized to maximize information transmission while minimizing energy consumption, is shown in figure 10.3.

This, although not obvious to someone not working in the field, was a major breakthrough. Unlike the principal components, which only very approximately resembled the receptive fields found in V1, when the fact that real-world images move was taken into account, the modeled receptive fields provided both a good qualitative and quantitative account of those found in V1 (van Hateren and van der Schaaf 1998).

Given this match, are there any other characteristics of V1 that lend weight to this interpretation? One prediction concerns the probability distribution of firing rates of cells when stimulated with natural scenes (rather than the artificial stimuli more commonly used). It can be shown that, given a constraint on the average firing rate, the neuron firing rate distribution that maximizes the output entropy (and hence potential information transmission), is an exponential distribution (Lennie 2003b). This prediction is easy to test, and in Baddeley et al. (1997) we carried out the simplest experiment: we recorded the distribution of firing rates

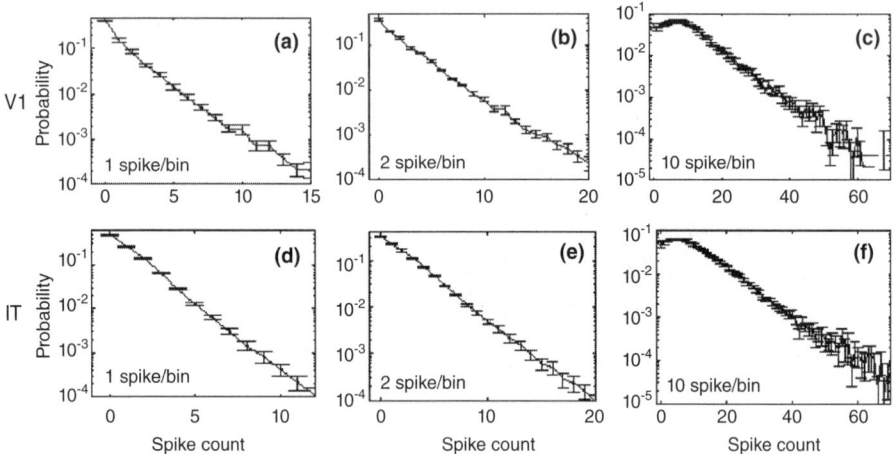

Figure 10.4
The firing rate distributions of cells in two cortical areas (V1, and IT, an area involved in object recognition), when these cells are processing nature videos. In the noiseless limit, given a constraint of the average firing rate, the cells will be transmitting the maximum amount of information when these distributions are exponential (or, because the y-axis is logarithmic, these plots form straight lines). As can be seen, over an order of magnitude range of time, the spike count distributions are indeed well approximated by an exponential distribution lending support to the idea that cells in V1 (and other areas) are maximizing their information transmission while minimizing their average firing rate. Based on Baddeley et al. 1997.

in cells in V1 while playing those videos judged to be representative of the animals' natural environment. The results are shown in figure 10.4. As can be seen, the distributions are very well fit by exponential distributions, and though there are systematic differences, when the fact that the system is not completely noise-free is taken into account, the fit is essentially perfect.

Before we move onto the final model, one point is worth mentioning. Though we have here interpreted the derived receptive fields in terms of minimizing energy consumption, this is far from the most popular interpretation. More usually, they are interpreted in terms of maximizing *sparsity*. Despite one common measure of sparsity being equivalent to minimizing energy consumption, we are not fans of this interpretation. Sparsity traditionally makes sense when you have a system that has only a very few active elements (in this case, neurons), while the majority are silent. At first, this appears an appropriate definition: neurons are in fact either far more on, or far more off, than predicted if their outputs were Gaussian-distributed. In fact, though, when averaged over even short periods, neurons are hardly ever silent, and compared to distributions with the same firing rate, their distribution is about as varied as it could possibly be, showing very little bimodality as would be predicted by a silent versus nonsilent interpretation. In a way, both interpretations state that neurons should minimize their average activity. But although minimizing metabolic activity unambiguously specifies how to quantify this, sparsity has a rather large number of definitions (which, though they give very similar results, is less satisfying), and does not predict that the optimal distribution will be exponentially distributed, as observed, but simply very long-tailed.

Minimizing Synaptic Rather than Firing Rate Energy Consumption

This, then, leaves us with a reasonably satisfying explanation of the receptive field in V1, but we appear to have missed a stage. The retinal ganglion cells do constitute an information bottleneck, and it would seem plausible that their receptive fields could be explained in a similar framework. Their center-surround receptive fields, however, do not resemble those found when minimizing average firing rate. Based on previous research (Vincent and Baddeley 2003; Vincent et al. 2005), our working hypothesis is to assume that because the information transmission bottleneck operating is so extreme, no meaningful energy saving can be made by constraining the firing rate. In contrast, as shown in figure 10.5, rather large energy consumption savings can be made by minimizing the energy used by synapses (which is quantified by the sum of the filter coefficients).

Constraints on synaptic energy consumption, which uses up a significant proportion of the entire brain's energy budget, were explored by Vincent and Baddeley (2003). When a cost function that takes into account synaptic (rather than firing rate) energy use is maximized, the optimal representations are very different, and are shown in figure 10.6. Quantitatively (and qualitatively), these representations provide a very good characterization of the observed receptive fields of retinal ganglion cells.

Furthermore, the explicitly energy-saving approach was applied to simultaneously calculate the optimal receptive fields for a simplified two-layer retinocortical visual system (Vincent et al. 2005). Although it is commonly assumed that the oriented V1 receptive fields derive appropriate input from multiple center-surround neurons, as far as we are aware, this was the first study to confirm that such an arrangement is in fact an optimal solution to something. But in order to do this, it was insufficient

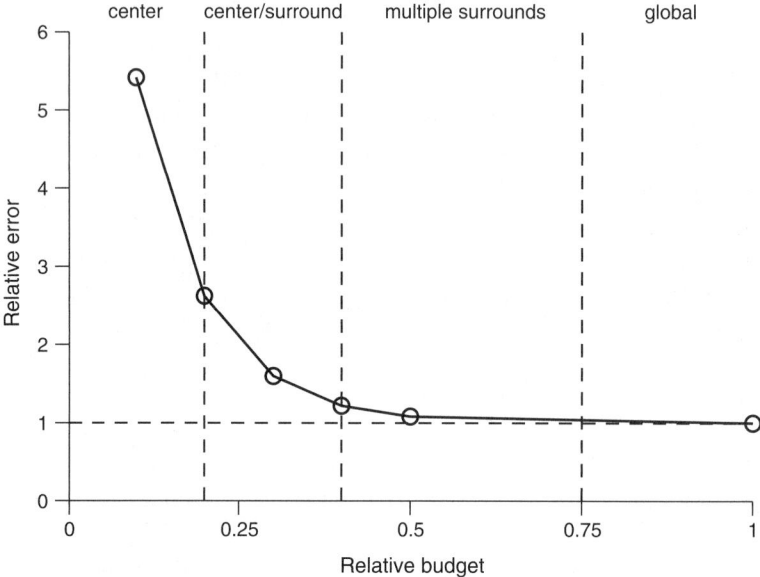

Figure 10.5
Large saving of the synaptic budget can be made at relatively little cost in terms of the amount of information transmitted. Shown are the results of six filter systems, each optimized with a different "budget" to be spent on synapses, where a budget of one corresponds to the optimal unconstrained solution. As can be seen, a 70 percent saving of metabolic cost associated with synapses can be made without severely affecting information transmission. When there is no constraint, the network simply spans the space of the principal components (and does not match observed physiology). In contrast, when only 20 to 40 percent of the energy is spent on synapses, the receptive fields in contrast form localized receptive fields. Examples of these are shown in figure 10.6. Based on Vincent and Baddeley 2003.

to simply maximize information; metabolic costs also had to be considered.

In summary, maximizing information about the natural input has provided a reasonable way of approaching the nature of the early cortical representations. If we assume that the main constraint is simply the number of units (or that the system is operating over a very short time, so neuronal outputs can be treated as binary), the optimal solution is the principal components of the input. This does not account for the form of any observed early receptive fields, but does explain short presentation time psychophysical performance. If we allow a larger number of neurons to represent an image, but minimize their average firing rate, then the optimal receptive fields provide a very good account of those measured

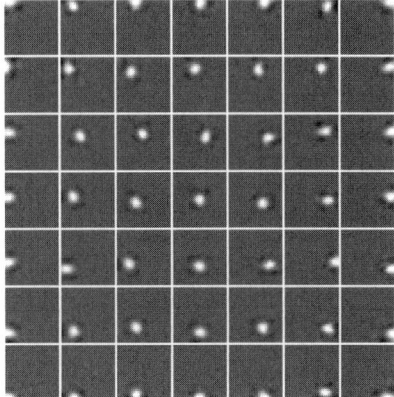

Figure 10.6
The optimal receptive fields of a system simultaneously maximizing information transmission while minimizing the amount of energy used on synaptic transmission. As can be seen, the receptive fields have a local center-surround organization, and the parameters of these receptive fields provide a good match to the properties of retinal ganglion cells. Based on Vincent and Baddeley 2003.

in V1. If, however, the constraint is placed on the metabolic cost of the synapses, then the optimal representation provides a good match to the receptive fields of the retinal ganglion cells.

Problems and Where Next?

The results presented in this discussion provide at least a potentially interesting framework for understanding early perception, but there is one very important theoretical worry that needs to be more directly addressed if we are to make progress. The worry is that we have treated the input as the pattern of light that arrives at the eye. It is easy to forget when studying vision; animals fundamentally do not care about light, but about the world that generated that light. This means that maximizing the information about the light is a sensible strategy only if all aspects of the variation of light are equally informative about behaviorally important aspects of the world. This is very unlikely to be true.

Unfortunately, approaches based on maximizing the information about the behaviorally important aspects of the world require estimates not of the statistics of natural images (which is easy), but the statistics of the behaviorally relevant aspects of the world that the animal needs to know about. It is the information about the world that should be

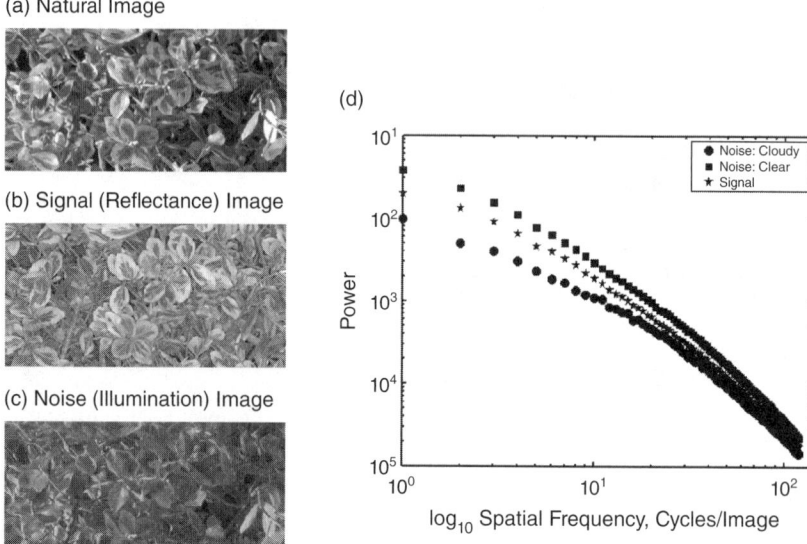

Figure 10.7
Information about the world is more important than information about light. Although at an early stage of research, we have been attempting to decompose images into variation that is relevant to properties of the world, and variation that is not informative. This figure shows part of the process, where, using parallel illumination of a scene, we can decompose it into the contribution due to reflectance, and the contribution due to illumination. The three panels to the left show: a natural image (a), which can be regarded as the product of the reflectance signal (image b), and illumination noise (image c). Panel (d) shows the power spectra of natural signal and noise images, averaged over horizontal and vertical orientations. As can be seen, the majority of variation in natural images is due to variations in illumination and does not tell us about the properties of the world. The spatial frequency spectra of illumination variation and reflectance variation are also very similar, meaning that it cannot be removed by simple linear filtering. This result is robust across a number of illumination environments. Representations that maximize the information about the indirectly observed reflectance will have to perform processing on the images (we are using Bayesian techniques) rather than simply recode them in order to maximize information.

maximized, not the information about the light generated by that world. This presents two difficulties: specifying what aspects of the world are behaviorally relevant, and quantifying them. Despite these technical difficulties and a number of false starts, we believe we are making progress on one particularly simple aspect of the world: the surface reflectance properties of objects in the world.

This chapter is not the place to describe our latest information-theoretical measurements, but figure 10.7 shows the most successful method we have developed to measure, in naturalistic settings, not only the characteristics of images, but also of the illumination and objects' surface reflectance properties that are associated with those images. This is just an example, but we believe that making models that maximize the information presented not about images, but about the behaviorally important aspects of those images, will be required if we are to make further progress in understanding how and why we have the early representations of the world that we do.

References

Attewell, David, and Simon B. Laughlin. 2001. An energy budget for signaling in the grey matter of the brain. *Journal of Cerebral Blood Flow and Metabolism* 21 (10): 1133–1145.

Baddeley, Roland J., and Peter J. B. Hancock. 1991. A statistical analysis of natural images matches psychophysically derived orientation tuning curves. *Proceedings of the Royal Society B: Biological Sciences* 246 (1317): 219–223.

Baddeley, Roland J. 1996. The correlational structure of natural images and the calibration of spatial representations. *Cognitive Science* 21 (3): 351–372.

Baddeley, Roland, J. L. F. Abbott, Michael C. A. Booth, Frank Sengpiel, Tobe Freeman, Edward A. Wakeman, and Edmund T. Rolls. 1997. Responses of neurons in primary and inferior temporal visual cortices to natural scenes. *Proceedings of the Royal Society B: Biological Sciences* 264 (1389): 1775–1783.

Corrard, François. 1999. Selective brain cooling. *Archives de Pediatrie* 6 (1): 87–92.

Derrington, Andrew M. 2000. Vision: Can colour contribute to motion? *Current Biology* 10 (7): R268–R270.

Falk, Dean. 1992. *Braindance: New discoveries about human origins and brain evolution.* New York: Henry Holt.

Field, David J. 1987. Relations between the statistics of natural images and the response properties of cortical cells. *Journal of the Optical Society of America. A, Optics and Image Science* 4 (12): 2379–2394.

Foster, David H., and Patrick A. Ward. 1991. Horizontal-vertical filters in early vision: Predict anomalous line-orientation identification frequencies. *Proceedings of the Royal Society B: Biological Sciences* 243 (1306): 83–86.

Fyfe, Colin, and Roland J. Baddeley. 1995. Finding compact and sparse- distributed representations of visual images. *Network* 6 (3): 333–344.

Harpur, George, and Richard Prager. 1996. Development of low entropy coding in a recurrent network. *Network* 7 (2): 277–284.

Hebb, D. O. 1949. *Organization of behavior: A neuropsychological theory*. New York: Wiley.

Hubel, David H., and Torsten N. Wiesel. 1962. Receptive fields, binocular interaction and functional architecture in the cat's visual cortex. *Journal of Physiology* 160 (1): 106–154.

Korutcheva, Elka, Nestor Parga, and Jean-Pierre Nadal. 1997. Information processing by a noisy binary channel. *Network* 8 (4): 405–424.

Lennie, Peter. 2003a. Receptive fields. *Current Biology* 13 (6): 216–219.

Lennie, Peter. 2003b. The cost of cortical computation. *Current Biology* 13 (6): 493–497.

Lewicki, Michael S., and Terrence J. Sejnowski. 1998. Learning overcomplete representations. *Neural Computation* 12 (2): 337–365.

Linsker, Ralph. 1986. From basic network principles to neural architecture: Emergence of spatial opponent cells. *Proceedings of the National Academy of Sciences of the United States of America* 83 (19): 7508–7512.

Olshausen, Bruno A., and David J. Field. 1996. Emergence of simple-cell receptive field properties by learning a sparse code for natural images. *Nature* 381 (6583): 607–609.

Rolfe, David F. S., and Guy C. Brown. 1997. Cellular energy utilization and molecular origin of standard metabolic rate in mammals. *Physiological Reviews* 77 (3): 731–758.

Rucci, Michele, and Antonino Casile. 2005. Fixational instability and natural image statistics: Implications for early visual representations. *Network (Bristol, England)* 16 (2–3): 121–138.

Rucci, Michele, Gerald M. Edelman, and Jonathan Wray. 2000. Modeling LGN responses during free-viewing: A possible role of microscopic eye movements in the refinement of cortical orientation selectivity. *Journal of Neuroscience* 20 (12): 4708–4720.

Shannon, Claude E. 1948. A mathematical theory of communication. *Bell System Technical Journal* 27 (3, 4): 379–423, 623–656.

Simoncelli, Eero P., and Bruno A. Olshausen. 2001. Natural image statistics and neural representation. *Annual Review of Neuroscience* 24 (March): 1193–1216.

Valeton, J. Matthee, and Dirk van Norren. 1983. Light adaptation of primate cones: An analysis based on extracellular data. *Vision Research* 23 (12): 1539–1547.

van Hateren, J. H. 1983. Spatiotemporal contrast sensitivity of early vision. *Vision Research* 33 (2): 257–267.

van Hateren, J. Hans, and Arjen van der Schaaf. 1998. Independent component filters of natural images compared with simple cells in primary visual cortex. *Proceedings of the Royal Society B: Biological Sciences* 265 (1394): 359–366.

Vincent, Benjamin T., and Roland Baddeley. 2003. Synaptic energy efficiency in retinal processing. *Vision Research* 43 (11): 1283–1290.

Vincent, Benjamin T., Roland Baddeley, Tom Troscianko, and Iain Gilchrist. 2005. Is the early visual system optimised to be energy efficient? *Network* 16 (2–3): 175–190.

Warrant, Eric J. 1999. Seeing better at night: Life style, eye design and the optimum strategy of spatial and temporal summation. *Vision Research* 39 (9): 1611–1630.

11

Attention, Information, and Epistemic Perception

Nicolas J. Bullot

Attention became a topic studied in experimental psychology by the end of the nineteenth century. With the subsequent development of psychology, interdisciplinary research on attention became an integral part of the cognitive and medical sciences (Posner and Raichle 1994; Parasuraman 1998; Wright 1998; Braun, Koch, and Davis 2001; Handy, Hopfinger, and Mangun 2001). Meanwhile, attention continues to raise a wide range of philosophical questions concerning, for example, sensory-motor control, perceptual reference, language understanding, social intentionality, and the neural correlates of consciousness. This chapter focuses on a question that is fundamental to bridging the gap between epistemology and biology: what is the role of attention in the acquisition of knowledge?

To address this problem, I will outline a theory grounded in what I call the *attentional constitution principle* (ACP). This principle asserts that attention is constitutive of humans' perceptual knowledge about individuals (i.e., objects and persons). The ACP expands research on perception and demonstrative identification, which originated in the writings of thinkers such as Peirce (1932–1935), Russell (1910), Sellars (1944, 1959), Dretske (1969, 1981, 2000), Evans (1982), Peacocke (1983, 1991, 1992), and Campbell (2002, 2004). Its method is grounded in the thought that the epistemology of empirical beliefs should mesh with the psychobiology of attention in order to explain how human agents navigate and analyze their environment. In contrast to the nonbiological epistemology of knowledge or the nonepistemological psychobiology of attention, the ACP holds that the function of human attention is mainly to serve perceptual knowledge through the extraction of causal information.

Section 11.1 formulates the ACP. Section 11.2 introduces a concept of information that is useful and relevant to the theory. Specifically, I distinguish causal information from semantic information and information processing. Section 11.3 introduces the argument from cognitive access to lend support to the ACP. This argument relies on premises (justified in sections 11.4 and 11.5) stating that overt and covert forms of attention are necessary for establishing direct cognitive access to target individuals and for extracting causal information relative to such individuals while they are perceived. From this analysis, it follows that the use of attention is necessary for assessing the truth value of empirical beliefs and linguistic information reports about perceived individuals. This argument raises the challenge of discovering a theory whereby the *epistemic* use of attention is explained. Sections 11.5 and 11.6 suggest that the procedural theory of attention can explain the epistemic and pragmatic roles of attentional systems in the extraction of causal information. The procedural theory characterizes attention as a multicomponent system that controls sensory-motor routines for solving action and epistemic requests, and thus for seeking, extracting, and using causal information available in the organism's environment.

11.1 The Attentional Constitution Principle of Singular Perceptual Knowledge

This chapter studies singular perception and singular action. I employ the term *singular* to refer to acts that are directed at individuals. Here, the term *individual* is used to denote a particular material thing that persists, changes, or grows and is located in the spatiotemporal world. There are two classes of individual: inanimate *objects* (e.g., artifacts) and intentional *agents* (e.g., human persons). Such objects or agents follow continuous paths in space and time, have cohesive parts, and have the power to affect other individuals. We can not only perceive them at different locations, or moving to new locations, but we can also identify persisting individuals across changes in their appearance or location.

Individuals present a unique set of properties determining their fundamental ground of difference (i.e., the material ground determining their uniqueness, identity over time, and singular causation). This ground of difference is that which is to be known in singular knowledge and that which is causally relevant for guiding the performance of singular actions. The notion of a fundamental ground of difference of an individual has philosophical roots that go back at least to Spinoza or Leibniz.

In recent philosophy of mind and language, it is analyzed by Evans (1982) and Campbell (1993, 2002). We can link the idea that individuals have a fundamental ground of difference with the thought that material individuals possess singular causal powers, or singular causation (see, e.g., Ellis 2000; Shoemaker 1984). In section 11.2, I link this singular causation to causal information and suggest, in section 11.6, that the function of epistemic attention is to extract causal information relative to the singular causation of individuals.

I borrow this use of the term *singular* from the philosophy of language and the semantic theory of singular terms, in which the fact that the human mind is directed at particular things (individuals) has been recognized as a key factor to address in order to explain the intentionality of the human mind (see Bullot and Rysiew 2007; Frege 1892; Strawson 1956; Devitt 1974; Kripke 1980; Evans 1982).

In the study of perceptual knowledge, the term *singular* is useful to stress that perceptions and actions are usually directed at individual things, in the sense that they use mechanisms that point toward, track over time, or come into contact with particular persons or objects. This idea can be expressed in this principle:

P1 dependence-on-tracking of true empirical beliefs and perceptual knowledge The acquisition of perceptual knowledge of (of nonaccidentally true empirical beliefs about) individuals depends on the tracking and perceptual-demonstrative identification of individuals.

P1 expresses a condition admitted in various versions by the realist accounts of perceptual knowledge (Sellars 1944, 1959; Strawson 1959; Quinton 1973, 1979; Dretske 1967, 1969, 1995b; Evans 1982; Campbell 1993, 2002; Peacocke 2001, 2003). In philosophy, the notion of *perceptual-demonstrative identification* is traditionally understood as a mental act (a form of intentionality) in which an individual is identified on the basis of its current perception (McDowell 1984; Pettit and McDowell 1986; Woodfield 1982). For instance, this kind of identification happens when, on the basis of your perceptual experience, you identify that "this is your mother." In the phrase "perceptual demonstrative identification," the term *demonstrative* is used to indicate that this kind of identification frequently occurs with a thought that contains a demonstrative term, which is labeled *demonstrative thought* in philosophy.

Perceptual-demonstrative thoughts have a form such as "This is F," in which "this" is a demonstrative term and "is F" is a predicate that refers to a concept or an attribute ascribed to the referent of the demonstrative term. Arguably, such thoughts are associated with the evaluation of propositions expressed by observations and information reports about what is perceived. Demonstrative thoughts can identify, locate, or describe the referents (or targets) of our perceptual experiences. For instance, if, at a crowded conference, your colleague points to a person and utters "This is Donald Broadbent," your grasp of the meaning of the demonstrative proposition depends on your ability to visually pick out and identify the referent of the demonstrative.

The condition of *tracking* is mentioned in P1 because perceptual tracking seems to be a necessary condition of perceptual-demonstrative identification. This notion of tracking refers to the ability to identify and reidentify an individual (as remaining the same individual) over a certain period of time, in spite of changes in its intrinsic or relational properties (e.g., aging or changes in appearances and location). For instance, the demonstrative identification of Donald Broadbent (in the situation described previously) requires you to succeed in visually tracking Donald Broadbent over a certain time and series of limited changes. In section 11.5, I will distinguish perceptual tracking from more epistemic forms of tracking. The theory that I propose posits that attentional systems integrate different forms of tracking through the extraction of and storage of causal information about the identity and location of target individuals. This form of tracking can be termed "integrated tracking" (for more on this, see Bullot 2006; Bullot and Rysiew 2007; Bullot and Droulez 2008).

I accept P1 and believe it corresponds to a relatively widespread view in the philosophical epistemology of perceptual knowledge. Here, I will focus on a more specific claim, which can be expressed in this principle:

Principle of the attentional constitution of singular perceptual knowledge **Attentional systems are constitutive of human agents' perceptual knowledge of individuals in their ecological environment.**

Because there are close ties between perceptual knowledge of individuals and action planning, the claim can be associated with this more comprehensive thesis:

Generalized *attentional constitution principle* **Attentional systems are constitutive of the links between singular perceptions (i.e., perceptions directed at individuals) and singular actions (i.e., actions directed at individuals).**

How are we to understand this notion of *constitutive* relation? The strongest form of constitutive relation would be identity. In such a case, the thesis would mean that some attentional processes in perception are identical to some knowledge-acquisition processes, or that the performance of some attention system is a necessary and sufficient condition of some perceptual knowledge. A weaker constitutive relation is a part-whole relationship. On this interpretation, the thesis would mean that some attentional systems are necessary parts of systems for the acquisition of perceptual knowledge. According to the ACP, in either its weak or strong forms, the study of attention is required to explain the genesis of agents' perceptual knowledge and its relations to their actions.

Now, the ACP can be paired with the thought that the attentional capacities in our hominin ancestors evolved to have the function to efficiently keep track of, and act on, individuals present in their environment such as prey or predator animals, or even tools (Arp 2006, 2008), and that this evolution was an adaptative response to environmental pressures. Before presenting an argument (sections 11.4 to 11.6) and a procedural theory (sections 11.6 and 11.7) to lend support to the ACP, I will clarify the information-theoretic approach I propose to use (section 11.2), along with a few basic problems relative to the concept of attention (section 11.3).

11.2 Causal, Semantic, and Mathematical Notions of Information

Phenomenological descriptions of mental acts (e.g., de Biran 1804/1988; Hatfield 1998; Husserl 1995; Merleau-Ponty 1945) may support some version of ACP. However, the justification of this principle is probably more forcefully achieved if we combine phenomenology with an information-theoretic approach to attention. When the term "attention" is used in connection with perception in information-theoretic psychology, it frequently denotes the selection of information for further analysis. In neurobiology, this can be formulated in terms of the selection of information for global availability across neural networks in which modulation correlates with conscious attentive perception (Dehaene and Naccache 2001; Dehaene et al. 2006; Kastner and Ungerleider 2000; Posner 1994; Somers et al. 1999; O'Craven et al. 1997; Handy, Hopfinger, and Mangun 2001).

This concept of attention qua mental selection has been central in the descriptions of attention since at least the work of William James (1890) and other psychologists of the late nineteenth century (Helmholtz 1867; Ribot 1908; James 1890; Titchener 1908; Sully 1898; Hatfield 1998). However, a major step in the theoretical study of attention occurred with the description of mental selection in information-theoretic terms. This description originates in the research inspired by the mathematical theory of communication and carried out in England and the United States during and after the Second World War (Miller 1981, 2003). Information-theoretic approaches conceive of human persons, brains, and minds as information-seeking or information-processing systems. This focus on information has introduced a number of innovations in the understanding of perception. For instance, the information-theoretic approach provided researchers with a framework departing from behaviorist theories (e.g., Neisser 1967; Posner 1994) and the atomistic theories of sense-data or sensations (e.g., Gibson 1966; Miller and Johnson-Laird 1976). This has proven to be a useful move to study the cognitive and active dimensions of human perception.

Information-theoretic approaches in biological psychology, philosophy, and other fields of cognitive science have become remarkably diverse (Miller 1956; Broadbent 1958; Dretske 1981, 1994; Adams 2003). As a result, the variety of information concepts can generate methodological intricacies. For instance, talk about information is ambiguous when it does not specify whether the concept of information in use refers to an objective (or mind-independent) property of physical facts or a subjective (or mind-dependent) construction of the mind. As an attempt to prevent conceptual slips, I will distinguish three classes of information-theoretic concepts.

1. *Causal information* The first class is relative to causal, environmental (Gibson 1966) or material information (Bogdan 1988), or natural meaning (Dretske 1988; Grice 1957; Millikan 1984). I will use the concept of causal information to refer to an objective property of certain facts or structures of the material world, which is to have constant (or invariant) connections with other facts or structures of the world. If a particular component A is constantly connected to component B (or has the propensity to lead to B), A can be viewed as carrying causal information relative to B by virtue of its constant connection to B. One can, therefore, apprehend A as a *carrier* (or vehicle) of causal information relative to B, of which specific characteristics may vary according to ontological kinds (Bogdan 1988). Causal information refers to an objective connection between A and B in which existence is independent of, and prior to, the knowledge that one may obtain about A or B.

Depending on specific ontological levels and terminologies, such connected components A and B may be apprehended as events, individuals, facts, or situations. Their connections may be regarded as constant conjunctions, causal links, or laws of nature. Specifically, the concept of causal information can be accommodated to a variety of ontological accounts of causation—for example, causation qua singular causation, laws of nature, counterfactual dependencies, or statistical regularities (Ellis 2000; Israel and Perry 1990). Although this concept of information is in the spirit of important points introduced by Fred Dretske (1981, 1994), the notion of causal information is distinct from Dretske's notion of information, because the latter is primarily conceived of from the standpoint of Shannon's mathematical theory of communication (1948) and thus cannot be reduced to causation.

I will focus on the case in which the basic carriers of causal information are material individuals such as material objects, biological organisms, and human persons. Here, I will assume that the fundamental ground of difference of an individual depends on the singular causal powers of that individual, and that such powers carry causal information about numerous other facts. Consider the case of human persons. Objective facts involving human individuals carry causal information about other facts, because the former are constantly connected to the latter. For example, the fact that you are a living adult human person carries causal information about numerous facts relative to your biological organism, such as the causal facts that you were born, that your body is made of cells, or that your cells contain DNA. You remain a carrier of this causal information regardless of whether you cognitively access the specifics of that causal information, which is carried by your organism and your DNA. Such causal information is an objective property of the fundamental material ground of difference of your own particular biological organism.

One can conceive of human folk knowledge and scientific knowledge as the extraction and analysis of distinct sorts of information (e.g., Bogdan 1988; Dretske 1981; Israel 1988; Israel and Perry 1990). An argument that supports this information-theoretic approach is that human subjects continuously communicate their knowledge through what one can term, after Israel and Perry (1990), *information reports*.

Information reports are sentences about what certain causal facts indicate about other facts. Information reports are omnipresent in the communication about forensic evidence, archeological or historical archives, or clinical medical knowledge among many other forms of empirical inquiries grounded in the scrutiny of material individuals. For instance, given the regular connection between the fact that a human agent manipulates an object and the fact that fingerprints are left on the surfaces of the object, a human fingerprint on a knife carries causal information relative to the fact that the knife has been manipulated by a particular human individual. This can be expressed in these information reports: (1) this fingerprint indicates that somebody has manipulated the knife, and (2) the fact that this fingerprint has these specific patterns indicates that Jack has manipulated the knife.

Such reports have a specific structure (Israel 1988; Israel and Perry 1990). The referent of the noun phrase in (1) refers to the carrier of causal information (or a part thereof). The noun phrase used to describe the carrier of causal information can be the referent of a demonstrative phrase. The proposition introduced by the *that*-clause, which refers to the fact that is indicated by the primary carrier of causal information, can be thought of as the *informational content* of the linguistic report (Israel and Perry 1990).

2. *Semantic information* Information reports are paradigm cases of the building of singular knowledge through the conversion of causal information into semantic contents, or semantic information. One can use the concept of *semantic information* (or intentional information), which is distinct from causal information, to refer to the property of that which has the function to carry intentional content or meaning. Semantic information can thus be understood in teleological terms—that is, through an analysis of the functions of the carrier of semantic information—and comes in different varieties of natural or conventional carriers of semantic information (e.g., Bogdan 1988; Dretske 1988, 1995b; Millikan 1984). Numerous theories in cognitive science view phenomena such as experiences, emotions, thoughts, information reports, or cultural contents either as possessing or processing semantic information. This use raises the problem of specifying the ways minds convert causal information into semantic information, or extract semantic information from causal information.

3. *Formal-mathematical information* A third class of information-theoretic notions includes the formal concepts of information, which have been initially introduced as mathematical tools for measuring the performance of communicating devices. The classical notion, in this category, was introduced by the mathematical theory of communication of Shannon (1948) and Shannon and Weaver (1949). In the latter,

information is a measure of one's freedom of choice when one selects a message (the logarithm of the number of available choices or of probabilities).

The conceptual relations of the different classes of information-theoretic concepts are notoriously knotty. A number of thinkers have expressed concerns about the risk of conflating the colloquial notion of semantic information with the formal concepts of the mathematical theory of communication (see, e.g., Bar-Hillel 1955; Partridge 1981; Wicken 1987). In addition, the project of grounding the theory of meaning and intentionality in the mathematical theory of communication—which has tempted Dretske (1981), among others (see, e.g., Adams 2003)—remains contentious (Dretske 1994). Similar concerns have been expressed about the project of using the concept of information to describe genetic coding (Godfrey-Smith 1999, 2000a, 2000b; Griffiths 2001; Maynard Smith 2000). Moreover, methodological debates on the concept of information are also found in psychology. For instance, there is a striking contrast between the information-theoretic frameworks of Donald Broadbent (1958, 1971, 1982) and James Gibson (1966, 1979).

In his seminal book *Perception and Communication*, Broadbent (1958) borrows the term *information*, with a constellation of other notions, from the mathematical theory of communication (e.g., information source, channel, signal, noise, and capacity). However, it seems fair to view Broadbent's approach (1958, 1971, 1982) as a global strategy to analyze psychological activities relative to semantic information (or content) empirically rather than an attempt to develop Shannon's mathematical theory of information or the theory of causal information. The key concepts originating from Broadbent's school in psychology are the notions of information processing and of processing levels (e.g., Kosslyn 1994; Newell 1990), which are theoretical concepts used to analyze the semantic processing of information. Through the concept of information processing, attention can be defined as analysis for further detailed processing (Kosslyn 1994; Treisman 1969, 1988). The notion of information processing performed by psychological faculties is primarily used to model the functional architecture of the mind/brain activities that underpin the possibility of semantic information. Its use is aimed at naturalizing semantic information.

In contrast to Broadbent (1958), Gibson (1966, 1979) develops an approach to perception that explicitly departs from the mathematical theory of communication and focuses on what Gibson terms *environmental information*, which is a concept roughly equivalent to the concept of causal information. He holds that "the information for perception is not transmitted" and "does not consist of signals, and does not entail a sender and a receiver" (1966, 63). Gibson's concept of environmental information refers to invariant regularities, structures, or specificities, which are present in the organism's environment and which can be extracted in perception and action. These invariant regularities are found in a variety of "ambient arrays" of energy, such as the ambient optical array (Gibson 1979, 65–91) or the acoustic array, which are fully objective and described by physical laws (also see Stoffregen and Bardy 2001). He maintains that "when we say that information is conveyed by light, or by sound, odor, or mechanical energy, we do not mean that the source is literally conveyed as a copy or replica" because "the sound of a bell is not the bell and the odor of cheese is not cheese" (Gibson 1966, 187) and the perspective projection of the faces of an individual is not the individual itself. However, in all these cases "a property of the stimulus is univocally related to a property of the object by virtue of physical laws" (Gibson 1966, 187), and this is what Gibson labels environmental information.

Gibson's environmental information, therefore, is a form of causal information. It describes the invariant or law-like structure of the physical world. For instance, in light structured by the environment, "the information lies in the *structure* of ambient light, that is, in its having an *arrangement* or being an *array*" (Gibson 1966, 208; also Gibson 1979, 47–64). The ecological psychology of perception develops the idea that, in perceptual exploration, the organism "picks up" causal information (Gibson 1966, 250–265) in the sense that it detects and explores the invariant structure of its environment.

There is a discrepancy between Broadbent's and Gibson's approaches to information: Broadbent's information-processing view primarily accounts for the mental operations performed on semantic information (e.g., storage in memory systems), and Gibson's ecological approach holds that the function of perception is to extract causal information. In spite of this apparent dilemma, I will suggest that the information-theoretic insight of each approach can be resolved by considering that attention is a key component of the translation of causal information into semantic information via the control of information-processing routines.

11.3 Common Assumptions about Attentional Systems and Information

In this chapter, the main idea I wish to convey is that attention is fundamental in the acquisition of perceptual knowledge because of its role in the conversion of causal information into semantic information. Support for this thought is found mainly in the theories that consider attention to be a faculty of selection for further information-processing. The early research on such an approach to attention has been primarily carried out in Broadbent's school, which hypothesized that the need for selective attention arises from certain basic limited processing capacity in the brain. This conception was developed by Broadbent (1958, 1971, 1981, 1982) and other pioneers of cognitive psychology such as George A. Miller (1956), Ulric Neisser (1967), Anne Treisman (1969), Neville Moray (1969), and Michael Posner (1978; also Kahneman 1973; Parasuraman and Davies 1984; Cowan 1995; Parasuraman 1998; Pashler 1998; Wright 1998; Braun, Koch, and Davis 2001).

In this view, the selective character of attentional operations is a consequence of global information-processing limitations. This approach is usually combined with closely related assumptions, which have been critically pinpointed by cognitive scientists, such as Allport (1993) and others (e.g., Desimone and Duncan 1995; Gibson 1966, 1979; Neisser and Becklen 1975). The defining assumptions of the early models of information processing (I follow Allport's analysis with a few changes) are, primarily, these statements about computational resources and control:

A1 The concept of *attention* refers to a processing resource, which is limited in quantity and must be allocated selectively. This conception originates in Broadbent's notion of attention as a "selective filter" that feeds a "limited capacity channel" (Broadbent 1958; Treisman 1969; Pashler 1998; Desimone and Duncan 1995; Duncan 1984, 1996; Humphreys, Duncan, and Treisman 1999).

A2 Attention is a necessary condition for certain kinds of processes, which are *controlled* processes; attention is not necessary for other kinds of processes, which are *automatic* processes (Kahneman 1973; Posner 1978, 1982, 1994; Shiffrin 1997; Shiffrin and Schneider 1977).

In addition, these early models usually endorse a set of assumptions about the problem of the unity and variety of attention, which can be expressed as follows:

A3   As a limited processing resource, attention is unitary.

If one rejects the unity of attention, the attempt to characterize cognitive processes in terms of those that (discretely) do, or do not, require attention becomes an ambiguous enterprise. In spite of giving some credence to A3, the early theories usually recognize that the division or scission of attention was possible, and thus tend to accept this statement:

A4   Attention is a unitary resource that can, in some circumstances, be "divided," but such a division demands a specific effort (it costs more resources; see Pashler 1998).

It cannot be taken for granted that the propositions A1 to A4 are compatible, as the ostensible tension between A3 and A4 may illustrate. Although such propositions pertain to controversial debates, they have oriented psychological research toward a series of traditional questions (Allport 1993; Findlay and Gilchrist 2003). A first question is the problem of the spatial or temporal "location" of selection: what is the *locus* (or place, stage) of attentional selection? Does the intervention of attention take place in or at an *early* or *late* stage in the (temporal or sequential) ordering of information processing? The debate opposing psychological theories of the early and late selection presupposes the idea that there is a specific location in which, or where, the "unitary" attention intervenes. Another problem, raised by A2, is this: what are the processes which do, or do not, require attention?

There are reasons to approach propositions A1 to A4 with caution. Consider A3 as an example. There might be good reason to describe the phenomenology of attentive perceptual experience as being unitary. However, A3 is neither a phenomenological claim nor a reductive claim about phenomenology. It is a psychological claim about the functional architecture of attention and the human mind/brain. It maintains that there is a single attention mechanism in the brain that is independent of other cognitive systems, such as sensory-motor control and memory. As such, A3 is debatable. Alternative approaches hold that *the* faculty of attention depends on *multiple selection* systems, and show that attention is reducible to the performance of a variety of sensory-motor and cognitive systems that can carry out a variety of mental procedures, acts, or routines.

The aim of the basic research strategy in biological psychology is to analyze the faculty of attention in terms of functional units and, thus, in terms of *multiple* mechanisms or systems (e.g., Parasuraman 1998; Parasuraman and Davies 1984; Posner 1994; Kahneman 1973; Posner

1978, 1982, 1994; Shiffrin 1997; Shiffrin and Schneider 1977). For instance, in a statement that reflects a common approach in psychology and neurobiology, Parasuraman affirms that "attention is not a single entity but the name given to a finite set of brain processes that can interact, mutually, and with other brain processes, in the performance of different perceptual, cognitive, and motor tasks" (1998, 3). Although there is no completely established taxonomy of attention, Parasuraman proposes the relative independence of three components of the attention faculty, which are selection, vigilance, and control. Given its focus on naturalistic approaches to attention, I will use phrases such as "attentional systems" or "systems of attention" to convey the idea that, at least in naturalistic accounts, the theoretical understanding of the faculty of attention requires the examination of a varied hierarchy of functional units or mechanisms.

This admission of the plurality of attentional systems opens the path to a wide range of questions. These questions originate in the attempt to understand attention systems as a *hierarchy* of selection and control procedures that shape tracking and action. Such control procedures may include the agent's endogenous selection of intentions and goals, of individuals to be tracked or of features to be analyzed. They may also include the selection operated by mechanisms that can prioritize the perception of salient unexpected events, which are usually termed *exogenous attention* to indicate that they are not deliberately or endogenously controlled by the attentive agent. The distinction between endogenous and exogenous factors in the control of attention is another fundamental assumption, which can be expressed as follows:

A5 The faculty of attention can be controlled by endogenous or exogenous mechanisms.

This assumption distinguishes between endogenous and exogenous shifts of attention (A5 should not be conflated with A2). It has long been described by phenomenological analyses that attention can undergo involuntary shifts (see, e.g., Hatfield 1998). This has led to the distinction between *automatic*, or *reflex*, and *voluntary*, or *willed*, attention, within various lexical idioms (James 1890; Wundt 1897; Titchener 1899; Folk and Gibson 2001; Driver and Spence 1998, 2004; Spence 2001). The distinction supports the naturalist stance about the plurality of attentional systems because endogenous and exogenous controls may be distinct with respect to their phenomenology, psychological mechanisms, and neural correlates.

Another assumption, which can serve as a guide for the study of attention systems, can be expressed as follows:

A6 The faculty of attention encompasses *overt* and *covert* forms of selection.

In psychology, the concept of *overt attention* refers to gestures and actions associated with observable activities of attentive tracking such as listening, touching, smelling, tasting, or looking (see, e.g., Posner 1980; Spence 2001; Findlay and Gilchrist 2003). Paradigmatically, overt attention coincides with the displacement or adjustment of a sensory organ to explore target individuals in the organism's environment. As pointed out by Gibson (1966), each perceptual system (i.e., the basic orienting system, the auditory system, the haptic system, the taste-smell system, and the visual system) uses specific modes of overt attention. With respect to the visual system, overt attention is the activity of looking at individuals, which is performed by observable eye movements (patterns of saccades and fixations). Eye fixations are usually tightly bound to cognitive operations under progress (Yarbus 1967; Kowler 1995; O'Regan 1992; Ballard et al. 1997; Triesch et al. 2003; Findlay and Gilchrist 2003). In contrast to overt attention, the concept of *covert attention* refers to the internal and cognitive consequences of the selection that are not so readily observable.

11.4 An Argument from Cognitive Access in Support of the ACP

Although the information-processing approach has provided a fundamental impetus to a functional description of the faculty of attention, a theory grounded in the assumptions A1 to A4 does not explain how humans acquire knowledge of individuals from the extraction of causal information. To provide such an explanation alongside a foundation for the ACP, I will propose an argument from cognitive access and outline a more appropriate theory of attention.

The argument from cognitive access in support of the principle runs as follows: humans' empirical beliefs and perceptual knowledge about target individuals depend on having direct cognitive access to such individuals in order to track and identify them through direct perceptual acquaintance (see P1 in section 11.1). The act of directing attention at a target individual is a necessary condition of having direct cognitive access to this individual (for the orienting of attention at a target provides access to causal information relative to such target). Therefore, attention is a necessary condition of the perceptual knowledge of individuals.

The crux insight of the argument is that selection by perceptual attention institutes direct cognitive access to targets of de re epistemic attitudes, such as perceptual identification, demonstrative thoughts, and empirical beliefs. Such a cognitive access is made possible by specific information-processing procedures performed by attentional systems. Only such attentional procedures can retrieve causal information available in the organism's environment.

To be consistent with naturalist constraints, this argument must be grounded in a biologically plausible account of attention and cognitive access. For this, I propose to ground the argument in the distinction between overt and covert attention (see A6, section 11.3). On that basis, the argument from cognitive access can be expressed as follows:

P2 Perceptual tracking and perceptual-demonstrative identification of an individual i necessarily require a direct cognitive access to i's properties (i.e., some of i's intrinsic or relational properties that carry causal information).

P3 To obtain direct cognitive access to i's properties, an intentional agent must perform search actions and acts of *overt attention* (or overt attentive tracking) in order to introduce and maintain i into at least one of his or her sensory fields and track i.

P4 To obtain direct cognitive access to i's properties, an intentional agent must track i and select i by *covert attention* to analyze some of i's properties and assess propositions (e.g., expressed by information reports) about i.

From the fact that demonstrative identification depends on direct cognitive access (premise P2), and that direct cognitive access depends on acts of overt (premises P3) and covert attention (premise P4), we can conclude that

P5 Acts of overt and covert attention are necessary conditions of the perceptual tracking and demonstrative identification of an individual i.

Arguably, given P5 and P1 (see section 11.1), which states that human perceptual knowledge depends on perceptual tracking and demonstrative identification (the form of which is "This i is F"), it is possible to conclude that ACP is true. Attentional systems are constitutive of agents' perceptual knowledge of individuals. This reasoning concludes that attention is a necessary condition of perceptual knowledge, because of its necessary contribution to the assessment of propositions (expressed by information reports or beliefs) grounded in the perceptual-demonstrative identification "This i is F."

Proposition P2 expresses a commonly received epistemological thought. Perceptual-demonstrative identification of an object is usually defined—in a sense related to Russell's notion of knowledge by acquaintance (1910)—as an identification whose success depends on the actual perception of the individual to be identified. Such perceptual identification may be performed, by excellence, through its localization and analysis in the visual field, or in some other sensory field (Dretske 1969; Evans 1982; Clark 2000, 2004a, 2004b; Rollins 2003; Kaplan 1989a, 1989b; McGinn 1981; Wettstein 1984; Reimer 1992; Siegel 2002). According this tradition, perceptual-demonstrative identification cannot occur in the absence of veridical perception of the target individual, and to perceive in a veridical manner, an individual requires having direct perceptual access to some of its properties. This direct access possesses a cognitive value, because it determines the cognitive significance of the representation of the target individual (e.g., Campbell 2002). In perceptual-demonstrative identification, perceptual access serves the epistemic goals of the agent—such as to verify an empirical belief expressed in information reports such as (1) and (2) in section 11.2.

Premises P3 and P4 and my comments on P2 refer to the establishment of a proper direct cognitive access to a target individual. How are we to understand this notion? I will follow the common understanding of cognitive access in terms of global availability for mental acts such as recognition, reasoning, and the rational guidance of action and speech (Evans 1982; Baars 1988; Block 1995, 2001). I am using the term "direct" to restrict the discussion to perception and the perceptual retrieval of causal information. Hence, in this use, we can declare that a human agent a has direct cognitive access to an individual i when, in virtue of a's current perception of i, some properties of i and their related causal information are available to a's mind/brain for use in identification, localization, reasoning, and the rational guidance of a's action and speech.

Two kinds of analyses may appear as conflicting accounts of this perceptual access. As a first kind, the conceptualist and intentionalist accounts stress the roles of conceptual capacities for individuating the target of perception (Wiggins 1997, 2001; McDowell 1990, 1996; Kaplan 1989a; Reimer 1992; Siegel 2002). For instance, a few of them emphasize the role of sortal concepts in the spatiotemporal delineation of the demonstrative's referent (Wiggins 1997, 2001). In another kind of analysis, the explanation of cognitive access is conducted in an analysis of the nonconceptual mechanisms or contents that allow the perceiver to be "anchored" onto the target via sensory-motor skills (Cussins 2003a, 2003b; Clark 2000, 2004a, 2004b; Gunther 2003; Pylyshyn 2003).

If an account of cognitive access for identification is restricted to only one of the two types of explanation, it may be at risk of circularity with regard to the analysis of access, because access might then be conceived as a purely conceptual/descriptive process without grounding in external individual, or as a purely sensory-motor anchoring without grounding in conceptual thought (see Strawson 1959; Peacocke 1992; Pylyshyn 2003; Bullot, Casati, and Dokic 2005). I suggest that the interest in studying overt tracking and covert attention in this context is that the faculty of attention is likely to explain the missing link between the two explanations. It should help us have a better understanding of how conceptual and nonconceptual abilities interact to determine direct cognitive access. Thus, in the argument from cognitive access to preclude any circularity in the analysis, attention is viewed as a mediating faculty of control that articulates the conceptual and nonconceptual conditions of the cognitive access to individuals.

11.5 Cognitive Access and the Tracker's Goal-Directed Movements (Justification of P3)

Let us focus on proposition P3, which asserts that the perceptual and cognitive access to an individual is dependent on *overt* attention. The statement can be justified on the ground that (1) the preparation and initiation of cognitive access is dependent on a wide spectrum of spatial actions and motor behavior needed for tracking a target (for reaching the situation where searched causal information can be made available), and that (2) such actions can be viewed as sequences of information-seeking acts of overt attentive tracking.

Consider the case of a person who is looking for another individual i (e.g., a partner, a lost artifact, a building), to act upon i or verify an information report about i. Call the former agent the *tracker* and the latter the *target*. By the former definition of *direct cognitive access*, if, as a tracker, one intends to obtain direct cognitive access to i, one must put oneself in a situation of directly perceiving i and extract causal information relative to i. Assuming the existence of the target, a tracker can be presented with approximately two cases.

The first case can be termed *sustained perceptual absence*: the target i is apart or very distant from the tracker's perceptual field and cannot be perceived at the moment. The second case can be termed *perceptual proximity*: the target *i* is either present within the tracker's perceptual fields or present in the region surrounding the tracker's body, that is, its peripersonal space (Maravita, Spence, and Driver 2003).

In a situation of sustained perceptual absence of the target, in which the tracker maintains the intention to obtain direct cognitive access to *i*, the tracker must move its body (and, thus, the sensors thereof) in order to track down *i*'s present location until *i* is reached, found or caught. This spatial search may require small-scale spatial actions (e.g., performing a saccade) or large-scale spatial actions (e.g., displacements in case of migrations or pilgrimages). These spatial actions can be described in different frames of reference.

For example, in the case of a pilgrimage, the pilgrim, as tracker, may have to move across lengthy territories before finally reaching a particular holy target. In this example, a useful description of the tracker's move to *i*'s location may be made according to allocentric reference frames, for example, by means of a map. The initial phase of the tracker's search corresponds to bodily movements initiated toward a still imperceptible target. Prior to a successful end of the search, the tracker's behavior is not grounded in the direct perception of *i*, although it may use, of course, the perceptual tracking of clues relative *i*'s location (e.g., the perception of maps or of signs that carry causal or semantic information relative to the target location). Thus, this kind of search can be termed *epistemic tracking* (Bullot 2006; Bullot and Rysiew 2007) because it aims at reinstating the perception of a target via epistemic means, which may use memories, reasoning, and communication about the target's identity and location.

Notice that epistemic tracking is organized to prepare cognitive access associated with the direct perceptual-attentive tracking of *i*, although in epistemic tracking, the tracker's target may not at all be available in the tracker's peripersonal space. Still, there may be concrete, perceptual clues or signs within that space (e.g., Bullot and Droulez 2008; Sutton 2006) that might eventually lead the tracker to a location where the target *i* can be directly perceived. In this case, some of the evidence that leads the tracker to its target is still perceptual (and therefore not just epistemic in a nonperceptual sense) although the target *i* itself is not directly perceived or within the tracker's peripersonal space.

Consider, now, the case of clear perceptual proximity with the target. To prepare cognitive access to *i*, when the tracker is sufficiently close to *i*'s location and *i* is available within its perceptual fields, the tracker must perform another class of bodily movements, which are the typical tracking actions described by the concept of overt attention (see A6, section 11.3). It includes movements such as the displacement and orientation of the sensory organs (e.g., eyes, hands, ears) in order to focus on the properties of the target. In addition to these movements, the preparation and optimization of direct cognitive access imply attention:attention and inhibition of the movements the inhibition or modification of competing movements. For example, when the tracker arrives in the target's proximity, the tracker may suspend its locomotion or change its way of breathing. The description of these preparatory movements has been developed in the theories of sensory-motor consequences of attention since the end of the nineteenth century.

In addition to James (1890, 434–438), this reference to the motor and overt consequences of attentional selection is found in a number of other authors. Sully (1898, 82) describes attention as an active mode of consciousness that affect certain motor process. Ribot (1908, 3) writes that the mechanism of attention "is primarily motor, i.e. [attention] always acts on muscles, mainly in the form of a stop." Similar assumptions are held in contemporary theories of vision such as the motor (or premotor) theories of visual attention, which describe spatial covert attention as a preparation of saccadic eye movements (e.g., Rizzolatti, Riggio, and Sheliga 1994). A comprehensive classification of the possible acts of overt attention for each perceptual system can be found in Gibson (1966).

William James, for instance, analyzes the overt and organic phenomena that accompany the procedures that seek for causal information through the attentional tracking of a target. He noticed that when we look or listen we accommodate our eyes and ears involuntarily, and we turn our head and body as well. Similarly, when we taste or smell we adjust the tongue, lips and respiration to the target. James concludes that in all these acts of overt attention "besides making involuntary muscular contractions of a positive sort, we inhibit others which might interfere with the result—we close the eyes in tasting, suspend the respiration in listening, etc." (1890, 435).

We can conclude from all these examples that various classes of tracking actions and attentional movements prepare perceptual access to a target (and to related causal information) and, consequently, prepare the target's perceptual tracking and demonstrative identification. This appears to be a sufficient reason to admit P3.

11.6 Cognitive Access, Epistemic Attention, and the Procedural Theory (Justification of P4)

This section is a justification of proposition P4, which states that an agent must track and select an individual i by covert attention in order to obtain direct cognitive access to i's properties. Here, the rationale is that the description of overt attentive tracking is insufficient to explain direct cognitive access to an individual, because the description of overt attention does not fully account for the epistemic uses of attentional acts.

In spite of their complementary status, overt and covert attention must be kept distinct (see Posner 1980; Findlay and Gilchrist 2003). For instance, the description of overt and goal-directed attentional behavior in vision is mainly the description of eye movements. There is an emerging consensus to acknowledge that, in an unconstrained context, the description of eye movements cannot unambiguously reveal the covert cognitive tasks performed by the tracker (Ballard et al. 1997; Findlay and Gilchrist 2003). For instance, the fact that a tracker is looking in the direction of an individual i does not necessarily imply that the tracker is currently paying attention to i to identify i. Empirical evidence supports this point, and, subsequently, the distinction between overt and covert attention.

The seminal psychological argument for the distinction originates from a series of experiments conducted by Michael Posner and collaborators (Posner 1978, 1980; Posner, Nissen, and Ogden 1978; Posner, Snyder, and Davidson 1980). They have found convincing evidence that human trackers can shift covert visual attention independently of their eye movements (i.e., overt visual attention). Specifically, they demonstrated that reaction times to visual targets selected by covert attention were faster for spatial locations that had been previously cued in a context of unchanged fixation. These experiments are usually interpreted as evidence that covert and overt attention can be uncoupled in vision.

In addition, other experimental findings indicate that directing the eyes at an individual does not imply the identification or memorization of this individual. As described in perceptual phenomena such as *attentional blink* (Shapiro 2001; Shapiro and Terry 1998), *change blindness* (O'Regan, Rensink, and Clark 1999; Simons and Rensink 2005a, 2005b), or *inattentional blindness* (Mack and Rock 1998), a sensory organ can be directed precisely toward a target without this target being consciously noticed, identified, or recalled. For instance, the concept of inattentional blindness (Mack and Rock 1998) refers to the failure to detect the presence of an entirely visible stimulus (such as a red square or a mobile bar) presented in the region of fixation. To study this phenomenon, Mack and Rock used a paradigm based on the principle of situating the subjects in a position where they would neither pay attention to nor expect to see an individual thing—termed *critical stimulus*—but nonetheless would look at the region within which this thing is presented. Inattentional blindness illustrates a case in which directing the gaze toward a target does not imply the target's identification or, more weakly, does not imply the capacity to submit a verbal report on its identity.

Taken together, these findings suggest that overt attentional behavior and covert attentional information-processing are two distinct, but (jointly) necessary, conditions of direct cognitive access. Overt attentional selection, such as the action of looking at i, does not necessarily imply an epistemic analysis of i's properties aimed at identifying or locating i. Therefore, the nature of covert mental procedures must be elucidated in order to account for the epistemic dimension of the access to i's properties. However, what is the nature of covert acts of attentional selection, and how do they relate to overt attentional tracking? To answer these questions, I will introduce the concept of *epistemic attention*.

In the remainder of this chapter, the phrase *(perceptual) epistemic attention* will refer to the capacities to identify or locate individuals that are currently perceived, and to disclose facts about them through the extraction of causal information from their perceived properties. In this usage of the term "epistemic," I conform to a tradition in the epistemology of perception represented mainly by Dretske's concepts of *epistemic seeing* (1969) and *meaningful perception* (1995a). In the following analysis, an example of the use of epistemic attention is the primary epistemic seeing in Dretske's (1969, 72–93) sense, which refers to knowledge that "(this) i is F" based on direct perception by the perceiver of the fact that i satisfies a perceptual predicate F (and the perceiver is justified in thinking that i is F because he or she perceives the fact that i is F). The perceptual reidentification of an object is another example (Strawson 1959; Treisman 1992).

In contrast to the epistemic tracking carried out when the target is not perceived, by definition, an act of perceptual epistemic attention requires the analysis of some properties of a perceived individual. Thus, perceptual epistemic attention can be performed only while the target individual is being tracked within a sensory field. How shall we analyze the acts of epistemic perceptual attention?

Given its definition, the central requirement for epistemic attention is this: to qualify as "epistemic," an attentional system must be necessary to the acquisition of knowledge. Thus, in perception, epistemic attention must contribute to the ability of the tracker to form nonaccidental true beliefs (e.g., Nozick 1981) and information reports (Israel 1988; Israel and Perry 1990) about perceived individuals. Consequently, epistemic attention must bestow on the tracker an ability (1) to flexibly assess the truth value of perceptual and demonstrative beliefs and (2) to revise, or update, such beliefs as a function of information made available through perceptual analysis of the target's properties.

When assessed with regard to this epistemological constraint, the early models of selective attention—those that accept A1, A2, and A4—appear limited. Their explanatory scope is too narrow, because they do not account for anything like an ability to assess whether a perceptual-demonstrative proposition is true or false. Models associated with this tradition—such as "attention-as-a-filter" (Broadbent 1982; Eriksen and St. James 1986; Brefczynski and DeYoe 1999; Driver and Baylis 1989; Valdes-Sosa et al. 1998), "visual-attention-as-a-spotlight" (Broadbent 1982; Eriksen and St. James 1986; Brefczynski and DeYoe 1999; Driver and Baylis 1989; Valdes-Sosa et al. 1998), or "attention-as-a-spatial-window" (Treisman 1988; Kosslyn 1994)—restrict their descriptions to early stages of perceptual selection, which would not qualify as vehicles of truth-assessable mental states or truth-assessing mental processes. Furthermore, they may not view attention as a system that functions to extract causal information.

There is also another substantive reason why the early models fail. To explain perceptual verifications or falsifications, you need to analyze the control of sensory-motor procedures used to perform perceptual verifications, such as directing your eyes toward i to verify whether i is F. Arguably, to account for procedures of perceptual verifications, one must have an account of the *executive control* of sensory-motor systems as performance systems for verification procedures. However, as a matter of historical fact, early cognitive models of selective attention were developed in relative independence from the theories of motor control. The executive control for epistemic/cognitive purposes is not an important theme of the early theories of the "locus" of attention selection. For instance, Broadbent's (1958, 297–301) early-selection theory identifies attention to a selective filter that feeds a single limited-capacity information channel and would not directly control the system's effectors—see the information-flow diagram in Broadbent (1958, 299). As a result, this kind of model does not offer predictions about the control of sensory-motor systems for epistemic purposes.

Can one provide a biologically plausible conception of attention that accounts for the perceptual verification or falsification of propositions expressed by empirical beliefs and information reports? My proposal is that a positive answer is possible within the framework of the "procedural/executive" theory of attention (e.g., Ballard et al. 1997; Campbell 2002; Cavanagh 2004; Gray 2000; Logan 1985; Miller and Johnson-Laird 1976; Posner 1994; Tomasello, Carpenter, and Liszkowski 2007), which has different foundational principles from that of early models of attention.

The *procedural*[1] or *executive* theory I propose hypothesizes that attention-driven perceptual processing corresponds to strategies[2] built by each tracker for satisfying requests about individuals. Specifically, perceptual epistemic attention uses perceptual analyzers to generate semantic information about perceived individuals from the cognitive processing of causal information. The semantic information-processing guides the acquisition of singular perceptual representations and the performance of singular actions. (Singular representations are sometimes termed "object files" or singular event "files."[3]) Consistent with the attentional constitution principle (section 11.1), attention is identified with procedures constitutive of singular perceptions and singular actions. Although this account incorporates the semantic concept of information processing, it does not restrict attention to the mere process of filtering out information as suggested by the attention-as-a-filter model (Broadbent 1958).

The core principles of this procedural theory (PT) of attention can be formulated as follows:

PT1 The faculty of *attention* encompasses a set of executive and cognitive systems whose function is to perform singular perceptions (e.g., tracking and identifying a currently perceived individual through the use of a mental file) and singular actions (e.g., acting on a presently available individual).

As a component of the faculty of attention, one can isolate the faculty of perceptual epistemic attention as follows:

PT1′ The faculty of *perceptual epistemic attention* is a system of executive and cognitive procedures that can implement exploratory strategies in order to (1) extract semantic information from causal information; (2) track and identify individuals in perception; and (3) assess the truth value of information reports and empirical beliefs about such individuals and their relations.

In PT1′, the reference to the ability to assess the truth value of propositions expressed by reports and empirical beliefs is what justifies the use of the adjective "epistemic." Perceptual epistemic attention is at the root of epistemic perception understood as the ability to perceive facts, form perceptual-demonstrative beliefs, and express linguistic information reports. The nature of attentional control can be specified as follows:

PT2 Perceptual epistemic attention is a control system that builds *attentional exploratory strategies*, which are hierarchical procedures of information-processing that combine (1) instructions that can be termed "epistemic requests" and "action requests"; and (2) specialized information-seeking operations termed "perceptual routines," "motor routines," or "sensory-motor routines" that allow the tracker to solve or satisfy the epistemic and action requests (as a function of specific context- or task-dependent combinations).

An *epistemic request* is a control procedure—that is, an instruction or command—that instructs a sensory-motor system or perceptual analyzer to extract semantic information about a perceptually tracked individual(s) from available causal information. As a component of a way to come to believe a particular perceptual-demonstrative proposition, the aim of an epistemic request is to solve a specific problem about perceived individuals. A paradigmatic example of the resolving of an epistemic request is the evaluation of perceptual predicates embedded in a demonstrative proposition.

Think about this task: imagine that a human person must act as a tracker and place a dozen eggs that are spread out on a table crowded with other kinds of objects inside a box. To perform the task, the tracker must iterate and solve an epistemic request that may be expressed in public language by the question "Is this an egg?" The procedural theory proposes that human trackers solves this request by assessing whether a perceptual predicate—which one may term $\text{E{\small GG}}(i)$—is satisfied by the objects they are looking at in a serial fashion. Thus, the ability to verify perceptual-demonstrative propositions depends on an ability to assess the truth-value of mental structures built from perceptual predicates ascribed to individuals. An examples of such predicates is Ovoid(i), which is satisfied when the target individual i is egg-shaped. Ovoid(i) is procedurally assessed as satisfied by the current target of attention if the tracker's perceptual epistemic attention can extract from the analysis from i's faces the fact that i is egg-shaped. Perceptual predicates can relate to any other perceivable aspect of the target, such as its spatial relations with neighboring individuals—for instance, Above(i,k) is satisfied when i is above k; Collinear(i,k) is satisfied when i and k are collinear, and so forth.[4]

Similarly, in the action domain, to account for the fact that the trackers' actions are singular (i.e., are directed at individuals), an action request can be represented under the form of an action-predicate assigned to an individual. An *action request* is a procedure that controls motor routines through the use of a repertoire of action predicates, which may be initiated in the context of the performance of a hierarchy of actions. An action predicate can be represented as Grasp-A(i), in which the structure Grasp-A( ) refers to a sensory-motor mechanism that can control the grasping of the individual i in argument position (see Rizzolatti and Arbib 1998, 192).

Requests presuppose the activity of a control system that can assess the semantic information obtained from the performance of a command. Their study relates to the cognitive aspects of sensory-motor control.[5]

An additional key concept in PT2 is the notion of routines. Epistemic and action requests can be solved through the uses of a variety of sensory-motor and cognitive routines. The concept of *routines* refers to hierarchical information-processing systems of elementary mental analyzers or sensory-motor operations that must be carried out to resolve epistemic or action requests—the notion of "perceptual analyzers" has been introduced by Treisman (1969). Routines are structured and hierarchical abilities, in which the iterated retrieval of information in familiar tasks does not impose extreme demands on the tracker's capacities. The routines constitute the basic elements of the repertory of the practical aptitudes of a tracker, and these elements are regarded as stable once acquired after training. The development of some routines is likely to be driven by innate mechanisms. The routines are selected during the action according to the demands of the ongoing task and are controlled by hierarchical structures of goals.

This concept of routine is to be understood with regard to the theories of the sensory-motor capacities.[6] It presents a kinship relation to other notions such as sensory-motor primitives (Ballard et al. 1997), functional routines (Ballard et al. 1997, 735–737; Kosslyn 1994), sensory-motor contingencies (O'Regan and Noë 2001), scripts and microscripts (Schank 1996, 1999), or procedures of haptic exploration (Klatzky and Lederman 1999, 171–172, 174–177)—the aim of which aim is to understand the formation and structure of the skills that allow trackers to carry out epistemic perceptions and actions.

The procedural theory can be summarized by this hypothesis of tracking by epistemic attention:

H A human tracker must carry out exploratory strategies of epistemic attention (i.e., assemble epistemic and action requests with relevant routines) directed at an individual i in order to seek for task-relevant information, build a singular representation of i, and verify or falsify empirical beliefs and linguistic information reports about i.

If H is correct, it provides a relatively new way to justify P3 and P4, and, subsequently, the attentional constitution principle. Direct cognitive access to an individual (see P3 and P4) depends on overt (P3) and covert (P4) attention. On the procedural theory, the cognitive and epistemic processes of covert attention are analyzed as procedures performed by epistemic perceptual attention: cycles of hierarchically organized requests and routines, which are constitutive of the tracker's perceptual verifications. Specific perceptual and conceptual routines explain the extraction of task-relevant causal information related to targets' properties such as shape, color, or acoustic activities. The procedural theory predicts that human agents, as epistemic and perceptual trackers, have routine recourse to probing behaviors and perceptual verification about causal information and individuals in their environment.

In contrast to the nonbiological epistemology of knowledge or the nonepistemological psychobiology of attention, the procedural theory can address epistemological problems in a conceptual framework that is consistent with a naturalist philosophy and a biological investigation. One further advantage of this theory is that it can accommodate the seminal insights of both Broadbent's and Gibson's schools (see section 11.2). Broadbent's school insists that one must account for semantic information in terms of activities of functional information-processing units of the perceiver's brain. The procedural theory addresses this point in its account of information-processing routines. Gibson's school states that perceptual knowledge derives from exploratory actions performed by perceptual systems that seek for and pick out (causal) information in the environment. The procedural theory addresses this point in its account of the epistemic exploratory procedures. The procedural theory can thus reconcile the two approaches in its suggestion that the attentional control of sensory-motor systems and perceptual analyzers provide human trackers with direct cognitive access to individuals qua carriers of causal information, and with the means for extracting semantic information from causal information.

11.7 Biology and the Procedural/Executive Theory of Epistemic Attention

One may object that the procedural theory of epistemic attention cannot be grounded in biology. This concern can be set aside for at least two reasons. First, the theory raises specific biological questions. Second, it leads to hypotheses that can be (and have been) assessed via methods of experimental psychobiology.

To establish the first point, it should suffice to mention a few prevalent questions. A popular issue belongs to neurophysiology: what are the neural bases of executive attention? The question is now an integral part of the biological sciences of attentional control, which use of a variety of methods for specifying the neural correlates of attentive experience and the role of the prefrontal cortex in executive control (e.g., Duncan 2001). For instance, Posner and his collaborators (Bush, Luu, and Posner 2000; Posner 1994; Posner and DiGirolamo 1998; Posner and Raichle 1994) suggest that the neural bases of executive attention involve frontal structures, including the anterior cingulated, that act on different brain areas and account for attention as a control system.

Furthermore, the procedural theory also raises the biological question of determining the evolution of human attention systems in phylogeny and ontogeny. The ontogenetic development of attention systems is studied through specific experimental methods in developmental psychology and neuroscience. Some studies have investigated the relation of attentional systems to heritable traits (e.g., Fan et al. 2003; Rueda et al. 2005).

With respect to phylogeny, the problem of understanding the effects of evolutionary pressure on the evolution of attentional systems remains, to the best of my knowledge, to be further investigated. The attentional systems specific to some of our human ancestors may have been primarily selected to keep track of individual agents and objects. Specifically, a reliable attentional ability for tracking animate individuals is advantageous in terms of ecological fitness, because it provides the tracker with efficient ways to hunt prey and to detect predators (e.g., New, Cosmides, and Tooby 2007). The attentional tracking of animate individuals is also crucial for understanding the biological bases of social cognition, because this kind of tracking is a requirement of learning social hierarchy, participating in collective actions, and learning language by means of joint attention (e.g., Bruner 1983; Tomasello 1995; Tomasello et al. 2005).

Second, the procedural theory can be, and has already been, developed and assessed through the methods of experimental psychobiology and neuroscience. This point can also be illustrated with behavioral research on the deployment of executive attention in the interactive tasks of daily human life. A good example of this kind is the experimental research associated with the "deictic theory of vision" proposed by D. Ballard and collaborators (Ballard 1997; Ballard et al. 1992, 1997). The core thesis of the deictic theory is that the eyes are used deictically. Their use of the term "deictic" refers to the ability of certain sensory-motor mechanisms and actions to serve as a means of direct cognitive access to (causal) information available in the organism's environment.

Ballard et al. (1997, 726–730) use the term "pointers" to refer to this mechanism of direct cognitive access. They argue that the use of pointers is essential to the performance of cognitive and epistemic procedures. Eye fixation is conceived as eye pointing directed at a referent (a target for the cognitive access to causal information in the tracker's environment), and illustrates the use of pointers in the sensory-motor domain. In addition, selection by covert attention is a neural pointer that interacts with eye fixations (Ballard et al. 1997, 725–726). Eye fixations are known to be particularly important when vision interfaces with cognitively, or epistemically, controlled action. This deictic theory is a procedural theory, in the sense defined by principles PT1 and PT2 (section 11.6), of singular perceptions and actions. With its enhanced visual resolution that occurs through foveal vision, the fixation on an external individual serves singular perception. Moreover, a fixation presents this advantage to allow "the brain's internal representations to be implicitly referred to an external point" (Ballard et al. 1997, 724), which can serve in the control of singular actions.

The deictic theory rests on the concept of deictic strategies. Fixations are parts of more overarching hierarchical structures termed "deictic strategies" or "do-it-where-I'm-looking" strategies (Ballard et al. 1997, 725). In a way consistent with PT2, the deictic theory stipulates that a *deictic strategy* is a sequential combination of routines, which use discreet deictic pointers to solve epistemic requests and activate action requests. The paradigm example of a pointer in the sensory-motor domain is eye fixations. Fixations serve singular cognition and action, because they provide direct cognitive access to the referent of the pointing and an addressing mechanism to control motor routines directed at the same referent. Furthermore, Ballard et al. (1997, 729–730, 735–737) distinguish two basic routines combined in deictic strategies to serve the tracking of individuals through the performance of singular perceptions and actions: the "identification routine" (e.g., trying to identify the target of an eye pointing) and the "location routine" (e.g., trying to locate in the environment the target of a pointer in memory). Therefore, in a summary consistent with the procedural hypothesis H, a deictic strategy employs eye fixations to select the individuals who must be targets of identification or location routines in order to solve epistemic request and fulfill action requests. In agreement with PT2, the binding of each fixation's referent with properties or motor instructions is usefully represented in a predicative form that ties a target individual with a general category or concept (in the case of singular perceptual knowledge) or a motor instruction (in the case of singular action).

To assess the principles of the deictic theory, Ballard and his colleagues (Ballard et al. 1992, 1997) used an artificial manipulative task carried out by mouse-controlled modifications of a computer screen display. In this "block assembly" task, the subjects acted on elements presented in computer display of colored blocks and had the task of assembling a copy of the Model (a top left area of the screen with a few colored blocks) in the Workspace (bottom left area of the screen). The experimental set up allowed the authors to keep a detailed record of both the manipulative actions and the eye scanning of the subject carrying out the task. The data collected from the block assembly task supported a deictic characterization of the underlying cognitive operations. Blocks were invariably *fixated* before they were operated on. Furthermore, there was clear evidence that the preferred strategy involved making minimal demands on any internalized memory.

In the block assembly task, as has been found in other tasks, many more saccades were made than what would seem necessary. The most common sequence observed in the block assembly task was eye-to-model, eye-to-resource, pick-from-resource, eye-to-model, eye-to-construction, drop-at-construction. It is referred to as a Model-Pickup-Model-Drop or MPMD strategy. The first eye-to-model shift would be to acquire the color information of the next block to be assembled, then a suitable block is found in the resource space. The second look at the model informs, or confirms, the location of this block in the model, which is then added to the construction. This second look could be avoided if the location information is also stored on the first look.[7]

It is possible to describe this MPMD deictic strategy in terms of a sequence of demonstrative thoughts, such as: "What is the color of the next square to be moved?" (epistemic request about an individual's color); "This is a green square." (demonstrative proposition about information obtained by a fixation and a color-recognition routine); "Pick a green square in this area up." (action request); "What is the location of the green square in the model?" (epistemic request about the relative location of the element in the model); "It is located at the bottom right location" (demonstrative proposition supported by a location routine).

Another group of studies that accords with the procedural theory has been published by Michael Land and his collaborators. Land used a head-mounted video-based eye tracking system, which enabled a record to be built up of the fixation positions adopted by an observer during a variety of actions. Tasks studied include: driving (Land and Lee 1994), table tennis (Land and Furneaux 1997), piano playing (Land and Furneaux 1997) and making tea (Land, Mennie, and Rusted 1999). The results, obtained during cognitively controlled actions, demonstrate the strength of the principle that the gaze is directed to the points of the scene where causal information is to be extracted (Land, Mennie, and Rusted 1999, 1328).

The aim of Land's analysis of tea making (Land, Mennie, and Rusted 1999) was to determine the pattern of fixations during the performance of a well-learned task in a natural setting, and to classify the types of monitoring action that are associated with eye movements. They used a head-mounted eye-movement video camera, which provided a continuous view of the scene ahead, with a dot indicating foveal direction with an accuracy of about 1 degree. A second video camera recorded the subject's activities from across the room. The authors analyzed the actions performed during the task as a control hierarchy in which the largest units describe the goals and subgoals of the operation. The hierarchy comprises these levels: (L1) main goal: "make the tea"; (L2) subordinate goals: "put the kettle on," "make the tea," "prepare the cups"; (L3) intermediate actions: "fill the kettle," "warm the pot"; (L4) basic actions, object-related actions: "find the kettle k," "lift the kettle k," "remove the lid l of k," "transport k to sink," and so forth; (L5) eye fixations "fixate k at time t."

To analyze the fourth level, the authors introduced the concept of "object-related action" units, which they regard as the basic elements of an action sequence. These units, with very rare exceptions, are carried out sequentially and involve engagement of all sensory-motor activity on the relevant individual object or set of individuals. The eyes move to the object before the manipulation starts. In general, the eyes anticipate the action by about 0.6 sec. During a single object-related action, saccades move the gaze around the object, but when shifting between one object-related action and another, very large saccades can occur. The eye movements could, with only occasional exceptions, be placed into one of the following categories of procedures: "locate x" (locate an object to be used later in the task), which may be represented with a deictic perceptual and action predicate such as LOCATE(x); "direct x" or DIRECT(x, l) (directing the hand or object in the hand to a new location); "guide x" or GUIDE(k, l) (guiding the approach of one object to another such as lid and kettle); "check x" or CHECK(w, k) (checking the state or property of an object such as water level in a pot). The description on these control functions is consistent with the procedural theory, in which they are described as epistemic requests (about the location or identity of certain individuals) or action requests, the aim of which aim is to control the action performed on a contextually relevant individual.

11.8 Concluding Remarks

This chapter formulated the attentional constitution principle (ACP). It introduced both the argument from cognitive access to support this principle and the procedural theory of epistemic attention. The procedural theory accords special epistemic importance to attention, due to its role in the perceptual tracking and demonstrative identification of individuals. It conceives of attention as a system that controls sensory-motor routines to satisfy action and epistemic requests—and, thus, to seek or extract semantic information from causal information available in the organism's environment. Hence, through the deployment of epistemic attention in singular perception, human attentive trackers can "navigate" the informational structure of the world to follow individuals and discover truths about them. Concurrently, through executive attention, trackers can perform singular actions on individuals as a function of their dynamical knowledge of the informational structure of the world.

Notes

1. According to the criteria mentioned in the text, one may consider as possible antecedent versions of the procedural theory the contributions of Miller, Galanter, and Pribram (1960) on plan; Miller and Johnson-Laird (1976) on the relations between perception and language and the evaluation of the perceptual predicates; Dretske (1969, 78–139) on epistemic primary seeing; Evans (1982) on demonstrative identification; Posner on executive attention (Posner 1978, 1994; Posner and DiGirolamo 1998); Ullman (1984) on visual routines; Ballard, Hayhoe and collaborators (Ballard et al. 1997) on the deictic strategies; Campbell (2002, 80, and chaps. 2, 3, 4, 5) on attention in reference; Pylyshyn (2003) on focal attention and visual reasoning. One may also include some theories of cognitive control (Allport 1993; Allport, Styles, and Hsieh 1994; Gopher 1993; Logan 1985; Shallice 1994) and some theories of joint attention (Tomasello et al. 2005; Tomasello, Carpenter, and Liszkowski 2007). Although they differ in many important aspects, the aforementioned works tend to analyze the contribution of perceptual attention to singular knowledge and singular actions, and to describe strategies or methods necessary for the acquisition of knowledge on individuals. In addition, such works may account for the fact that acts of identification by perceptual attention can be of a greater or lesser sophistication (Campbell 2002; Clark 2000, 135; Millikan 1984, 239–256), and can be revised and built from increments added to the singular knowledge already available to the tracker (Dretske 1969, 78–139; Pylyshyn 2001, 135–139).

2. The notion of strategy and of perceptual strategies is used in the executive/procedural theories of attention and action planning; see, for instance, Miller, Galanter, and Pribram 1960, Logan 1985, Gopher 1993, or Ballard et al. 1997.

3. See Kahneman, Treisman, and Gibbs 1992, Hommel et al. 2001, and Bullot and Rysiew 2007 for reviews of the literature on mental singular (object, agent, event) files.

4. The notion of perceptual predicate is used by Minsky and Papert (1969), Miller and Johnson-Laird (1976), Ullman (1984, 1996), Pylyshyn (1989, 2003), Peacocke (1983, 1992), Clark (2000, 2004a), and Hurford (2003). There are, of course, different ways to view the neural or behavioral implementation of perceptual predicates, and I remain neutral about this question. The point of the discussion is that something like perceptual predicates is needed to account for the epistemic dimension of perception, and it is likely that these predicates are assessed by attentional procedures.

5. Some directing ideas of control theory applied to cognitive science originate from electrical engineering (e.g., Craik 1947; MacKay 1951; Poulton 1952). They have been developed in the theory of eye movements (see, e.g., Kowler 1995) and other domains.

6. The concept of routines has been used in cognitive psychology to analyze the architecture of practical skills (Gopher and Koriat 1999; Gray 2000; Kirsh and Maglio 1995; Klahr and Wallace 1970; Monsell and Driver 2000; Schank 1996).

It refers to primitives used in sensory-motor and interactive abilities and perceptual abilities such as visual recognition and analysis (Ballard et al. 1997; Hayhoe 2000; Kosslyn 1994; Ullman 1984) or haptic/tactile recognition (Klatzky and Lederman 1999; Lederman et al. 1990). It has also been used in the analysis of understanding, reasoning, and memory (Bower, Black, and Turner 1979; Schank 1996).

7. On occasions, the second look was omitted, indicating that such use of memory was an option. However, these sequences were much less common than the sequences in which a return was made to the model.

References

Adams, Frederick. 2003. The informational turn in philosophy. *Minds and Machines* 13 (4): 471–501.

Allport, Allan. 1993. Attention and control: Have we been asking the wrong questions. A critical review of the last twenty-five years. In *Attention and performance XIV: Synergies in experimental psychology, artificial intelligence, and cognitive neuroscience*, ed. David E. Meyer and Sylvan Kornblum, 183–218. Cambridge, Mass.: MIT Press.

Allport, Allan, Elizabeth A. Styles, and Shulan Hsieh. 1994. Shifting intentional set: Exploring the dynamic control of tasks. In *Attention and performance XV: Conscious and nonconscious information processing*, ed. Carlo Umiltà and Morris Moscovitch, 421–452. Cambridge, Mass.: MIT Press.

Arp, Robert. 2006. The environments of our Hominin ancestors, tool-usage, and scenario visualization. *Biology and Philosophy* 21 (1): 95–117.

Arp, Robert. 2008. *Scenario visualization: An evolutionary account of creative problem solving*. Cambridge, Mass.: MIT Press.

Baars, Bernard J. 1988. *A cognitive theory of consciousness*. Cambridge: Cambridge University Press.

Ballard, Dana H. 1997. *An introduction to natural computation*. Cambridge, Mass.: MIT Press.

Ballard, Dana, Mary M. Hayhoe, Feng Li, and Steven D. Whitehead. 1992. Hand-eye coordination during sequential tasks. *Philosophical Transactions of the Royal Society B: Biological Sciences* 337 (1281): 331–338.

Ballard, Dana H., Mary M. Hayhoe, Polly K. Pook, and Rajesh P. N. Rao. 1997. Deictic codes for the embodiment of cognition. *Behavioral and Brain Sciences* 20 (4): 723–767.

Bar-Hillel, Yehoshua. 1955. An examination of information theory. *Philosophy of Science* 22 (2): 86–105.

Block, Ned. 1995. On a confusion about a function of consciousness. *Behavioral and Brain Sciences* 18 (2): 227–247.

Block, Ned. 2001. Paradox and cross purposes in recent work on consciousness. *Cognition* 79 (1–2): 197–219.

Bogdan, Radu J. 1988. Information and semantic cognition: An ontological account. *Mind & Language* 3 (2): 81–122.

Bower, Gordon H., John B. Black, and Terence J. Turner. 1979. Scripts in memory for text. *Cognitive Psychology* 11 (2): 177–220.

Braun, Jochen, Christof Koch, and Joel L. Davis, eds. 2001. *Visual attention and cortical circuits*. Cambridge, Mass.: MIT Press.

Brefczynski, Julie A., and Edgar A. DeYoe. 1999. A physiological correlate of the "spotlight" of visual attention. *Nature Neuroscience* 2 (4): 370–374.

Broadbent, Donald E. 1958. *Perception and communication*. London: Pergamon Press.

Broadbent, Donald E. 1971. *Decision and stress*. London: Academic Press.

Broadbent, Donald E. 1981. Selective and control processes. *Cognition* 10 (1–3): 53–58.

Broadbent, Donald E. 1982. Task combination and selective intake of information. *Acta Psychologica* 50 (3): 253–290.

Bruner, Jerome. 1983. *Child's talk: Learning to use the language*. Oxford: Oxford University Press.

Bullot, Nicolas J. 2006. The principle of ontological commitment in pre- and postmortem multiple agent tracking. *Behavioral and Brain Sciences* 29 (5): 466–468.

Bullot, Nicolas J., Roberto Casati, and Jerome Dokic. 2005. L'identification des objets et celle des lieux sont-elles interdépendantes? In *Agir dans l'espace*, ed. Catherine Thinus-Blanc and Jean Bullier, 13–32. Paris: Editions de la Maison des Sciences de l'Homme.

Bullot, Nicolas J., and Jacques Droulez. 2008. Keeping track of invisible individuals while exploring a spatial layout with partial cues: Location-based and deictic direction-based strategies. *Philosophical Psychology* 21 (1): 15–46.

Bullot, Nicolas J., and Patrick Rysiew. 2007. A study in the cognition of individuals' identity: Solving the problem of singular cognition in object and agent tracking. *Consciousness and Cognition* 16 (2): 276–293.

Bush, George, Phan Luu, and Michael I. Posner. 2000. Cognitive and emotional influences in anterior cingulate cortex. *Trends in Cognitive Sciences* 4 (6): 215–222.

Campbell, John. 1993. The role of physical objects in spatial thinking. In *Spatial representation: Problems in philosophy and psychology*, ed. Naomi Eilan, Rosaleen A. McCarthy, and Bill Brewer, 65–95. Oxford: Basil Blackwell.

Campbell, John. 2002. *Reference and consciousness*. Oxford: Clarendon Press.

Campbell, John. 2004. Reference as attention. *Philosophical Studies* 120 (1–3): 265–276.

Cavanagh, Patrick. 2004. Attention routines and the architecture of selection. In *Cognitive neuroscience of attention*, ed. Michael I. Posner, 13–28. New York: Guilford Press.

Clark, Austen. 2000. *A theory of sentience.* Oxford: Clarendon Press.

Clark, Austen. 2004a. Feature-placing and proto-objects. *Philosophical Psychology* 17 (4): 443–469.

Clark, Austen. 2004b. Sensing, objects, and awareness: Reply to commentators. *Philosophical Psychology* 17 (4): 553–579.

Cowan, Nelson. 1995. *Attention and memory: An integrated framework.* Oxford: Oxford University Press.

Craik, K. J. W. 1947. Theory of the human operator in control systems, I: The operator as an engineering system. *British Journal of Psychology* 38 (2): 56–61.

Cussins, Adrian. 2003a. Content, conceptual content, and nonconceptual content. In *Essays on nonconceptual content*, ed. York H. Gunther, 133–163. Cambridge, Mass.: MIT Press.

Cussins, Adrian. 2003b. Postscript: Experience, thought, and activity. In *Essays on nonconceptual content*, ed. York Gunther, 147–159. Cambridge, Mass.: MIT Press.

de Biran, Maine. 1804/1988. *Mémoire sur la décomposition de la pensée (Œuvres, Tome III).* Paris: J. Vrin.

Dehaene, Stanislas, Jean-Pierre Changeux, Lionel Naccache, Jérôme Sackur, and Claire Sergent. 2006. Conscious, preconscious, and subliminal processing: A testable taxonomy. *Trends in Cognitive Sciences* 10 (5): 204–211.

Dehaene, Stanislas, and Lionel Naccache. 2001. Towards a cognitive neuroscience of consciousness: Basic evidence and a workspace framework. *Cognition* 79 (1–2): 1–37.

Desimone, Robert, and John Duncan. 1995. Neural mechanisms of selective visual attention. *Annual Review of Neuroscience* 18 (March): 193–222.

Devitt, Michael. 1974. Singular terms. *Journal of Philosophy* 71 (7): 183–205.

Dretske, Fred I. 1967. Can events move? *Mind* 76 (304): 479–492.

Dretske, Fred I. 1969. *Seeing and knowing.* Chicago: University of Chicago Press.

Dretske, Fred I. 1981. *Knowledge and the flow of information.* Cambridge, Mass.: MIT Press.

Dretske, Fred I. 1988. *Explaining behavior: Reasons in a world of causes.* Cambridge, Mass.: MIT Press.

Dretske, Fred I. 1994. The explanatory role of information. *Philosophical Transactions of the Royal Society A: Mathematical, Physical, & Engineering Sciences* 349 (1689): 59–69.

Dretske, Fred I. 1995a. Meaningful perception. In *An invitation to cognitive science, vol. 2: Visual cognition*, ed. Stephen M. Kosslyn and Daniel N. Osherson, 331–352. Cambridge, Mass.: MIT Press.

Dretske, Fred I. 1995b. *Naturalizing the mind.* Cambridge, Mass.: MIT Press.

Dretske, Fred I. 2000. *Perception, knowledge, and belief: Selected essays.* Cambridge: Cambridge University Press.

Driver, Jon, and Gordon C. Baylis. 1989. Movement and visual attention: The spotlight metaphor breaks down. *Journal of Experimental Psychology: Human Perception and Performance* 15 (3): 448–456.

Driver, Jon, and Charles Spence. 1998. Attention and the crossmodal construction of space. *Trends in Cognitive Sciences* 2 (7): 254–262.

Driver, Jon, and Charles Spence. 2004. Crossmodal spatial attention: Evidence from human performance. In *Crossmodal space and crossmodal attention*, ed. Charles Spence and Jon Driver, 179–220. Oxford: Oxford University Press.

Duncan, John. 1984. Selective attention and the organization of visual information. *Journal of Experimental Psychology: General* 113 (4): 501–517.

Duncan, John. 1996. Cooperating brain systems in selective perception and action. In *Attention and performance XVI: Information integration in perception and communication*, ed. Toshio Inui and James L. McClelland, 549–578. Cambridge, Mass.: MIT Press.

Duncan, John. 2001. An adaptative coding model of neural function in prefrontal cortex. *Nature Reviews: Neuroscience* 2 (11): 820–829.

Ellis, Brian. 2000. Causal laws and singular causation. *Philosophy and Phenomenological Research* 61 (2): 329–351.

Eriksen, C. W., and J. D. St. James. 1986. Visual attention within and around the field of focal attention: A zoom lens model. *Perception & Psychophysics* 40 (4): 225–240.

Evans, Gareth. 1982. *The varieties of reference*, ed. John McDowell. Oxford: Oxford University Press.

Fan, Jin, John Fossella, Tobias Sommer, Yanghong Wu, and Michael I. Posner. 2003. Mapping the genetic variation of executive attention onto brain activity. *Proceedings of the National Academy of Sciences of the United States of America* 100 (12): 7406–7411.

Findlay, John M., and Iain D. Gilchrist. 2003. *Active vision: The psychology of looking and seeing*. Oxford: Oxford University Press.

Folk, Charles L., and Bradley S. Gibson, eds. 2001. *Attraction, distraction and action: Multiple perspectives on attentional capture*. Amsterdam: Elsevier.

Frege, Gottlob. 1892. Über Sinn und Bedeutung. *Zeitschrift für Philosophie und philosophische Kritik* 100: 25–50.

Gibson, James J. 1966. *The senses considered as perceptual systems*. London: George Allen and Unwin.

Gibson, James J. 1979. *The ecological approach to visual perception*. Hillsdale, N.J.: Lawrence Erlbaum Associates.

Godfrey-Smith, Peter. 1999. Genes and codes: Lessons from the philosophy of mind? In *Where biology meets psychology: Philosophical essays*, ed. Valerie Gray Hardcastle, 305–331. Cambridge, Mass.: MIT Press.

Godfrey-Smith, Peter. 2000a. Information, arbitrariness, and selection: Comments on Maynard Smith. *Philosophy of Science* 67 (2): 202–207.

Godfrey-Smith, Peter. 2000b. On the theoretical role of "genetic coding." *Philosophy of Science* 67 (1): 24–44.

Gopher, Daniel. 1993. The skill of attention control: Acquisition and execution of attention strategies. In *Attention and performance XIV: Synergies in experimental psychology, artificial intelligence, and cognitive neuroscience*, ed. David E. Meyer and Sylan Kornblum, 299–322. Cambridge, Mass.: MIT Press.

Gopher, Daniel, and Asher Koriat, eds. 1999. *Attention and performance XVII: Cognitive regulation of performance: Interaction of theory and application.* Cambridge, Mass.: MIT Press.

Gray, Wayne D. 2000. The nature and processing of errors in interactive behavior. *Cognitive Science* 24 (2): 205–248.

Grice, Paul. 1957. Meaning. *Philosophical Review* 66 (3): 377–388.

Griffiths, Paul E. 2001. Genetic information: A metaphor in search of a theory. *Philosophy of Science* 68 (3): 394–412.

Gunther, York H., ed. 2003. *Essays on nonconceptual content.* Cambridge, MA: MIT Press.

Handy, Todd C., Joseph B. Hopfinger, and George R. Mangun. 2001. Functional neuroimaging of attention. In *Handbook of functional neuroimaging of cognition*, ed. Roberto Cabeza and Alan Kingstone, 75–108. Cambridge, Mass.: MIT Press.

Hatfield, Gary. 1998. Attention in early scientific psychology. In *Visual attention*, ed. Richard D. Wright, 3–25. Oxford: Oxford University Press.

Hayhoe, Mary. 2000. Vision using routines: A functional account of vision. *Visual Cognition* 7 (1/2/3): 43–64.

Helmholtz, Hermann von. 1867. *Handbuch der physiologischen* Optik, *1856–1866.* Hamburg: L. Voss.

Hommel, Bernhard, Jochen Müsseler, Gisa Aschersleben, and Wolfgang Prinz. 2001. The theory of event Coding (TEC): A framework for perception and action planning. *Behavioral and Brain Sciences* 24 (5): 849–937.

Humphreys, Glyn W., John Duncan, and Anne Treisman, eds. 1999. *Attention, space and action: Studies in cognitive neuroscience.* Oxford: Oxford University Press.

Hurford, James. 2003. The neural basis of predicate-argument structure. *Behavioral and Brain Sciences* 26 (3): 261–316.

Husserl, Edmund. 1995. *Leçons sur la théorie de la signification.* Trans. John English. Paris: J. Vrin.

Israel, David J. 1988. Commentary: Bogdan on information. *Mind & Language* 3 (2): 123–140.

Israel, David J., and John Perry. 1990. What is information? In *Information, language and cognition*, ed. Philip P. Hanson, 1–19. Vancouver, B.C.: University of British Columbia Press.

James, William. 1890. *The principles of psychology.* New York: Dover Publications.

Kahneman, Daniel. 1973. *Attention and effort.* Englewood Cliffs, N.J.: Prentice-Hall.

Kahneman, Daniel, Anne Treisman, and Brian Gibbs. 1992. The reviewing of object files: Object-specific integration of information. *Cognitive Psychology* 24 (2): 175–219.

Kaplan, David. 1989a. Afterthoughts. In *Themes from Kaplan*, ed. Joseph Almog, John Perry, and Howard Wettstein, 481–564. Oxford: Oxford University Press.

Kaplan, David. 1989b. Demonstratives. In *Themes from Kaplan*, ed. Joseph Almog, John Perry, and Howard Wettstein, 565–614. Oxford: Oxford University Press.

Kastner, Sabine, and Leslie G. Ungerleider. 2000. Mechanisms of visual attention in the human cortex. *Annual Review of Neuroscience* 23 (March): 315–341.

Kirsh, David, and Paul Maglio. 1995. On distinguishing epistemic from pragmatic action. *Cognitive Science* 18 (4): 513–549.

Klahr, David, and John G. Wallace. 1970. An information processing analysis of some Piagetian experimental tasks. *Cognitive Psychology* 1 (4): 358–387.

Klatzky, Roberta, and Susan Lederman. 1999. The haptic glance: A route to rapid object identification and manipulation. In *Attention and performance XVII: Cognitive regulation of performance: Interaction of theory and application*, ed. Daniel Gopher and Asher Koriat, 165–196. Cambridge, Mass.: MIT Press.

Kosslyn, Stephen M. 1994. *Image and brain: The resolution of the imagery debate.* Cambridge, Mass.: MIT Press.

Kowler, Eileen. 1995. Eye movements. In *An invitation to cognitive science, vol. 2: Visual cognition*, ed. Stephen M. Kosslyn and Daniel N. Osherson, 215–265. Cambridge, Mass.: MIT Press.

Kripke, Saul. 1980. *Naming and necessity.* Cambridge, Mass.: Harvard University Press.

Land, Michael F., and David N. Lee. 1994. Where we look when we steer. *Nature* 369 (6483): 742–744.

Land, Michael F., and Sophie Furneaux. 1997. The knowledge base of the oculomotor system. *Philosophical Transactions of the Royal Society B: Biological Sciences* 352 (1358): 1231–1239.

Land, Michael F., Neil Mennie, and Jennifer Rusted. 1999. The role of vision and eye movements in the control of activities of daily living. *Perception* 28 (11): 1311–1328.

Lederman, Susan J., Roberta L. Klatzky, Cynthia Chataway, and Craig D. Summers. 1990. Visual mediation and the haptic recognition of two dimensional pictures of common objects. *Perception & Psychophysics* 47 (1): 54–64.

Logan, Gordon D. 1985. Executive control of thought and action. *Acta Psychologica* 60 (2–3): 193–210.

Mack, Arien, and Irvin Rock. 1998. Inattentional blindness: Perception without attention. In *Visual attention*, ed. Richard D. Wright, 55–76. Oxford: Oxford University Press.

MacKay, D. M. 1951. Mindlike behaviour in artefacts. *British Journal for the Philosophy of Science* 2 (6): 105–121.

Maravita, Angelo, Charles Spence, and Jon Driver. 2003. Multisensory integration and the body schema: Close to hand and within reach. *Current Biology* 13 (1): R531–R539.

Maynard Smith, John. 2000. The concept of information in biology. *Philosophy of Science* 67 (2): 177–194.

McDowell, John. 1984. De re senses. *Philosophical Quarterly* 34 (136): 283–294.

McDowell, John. 1990. Peacocke and Evans on demonstrative content. *Mind* 99 (394): 255–266.

McDowell, John. 1996. *Mind and world*. Cambridge, Mass.: Harvard University Press.

McGinn, Colin. 1981. The mechanism of reference. *Synthese* 49 (2): 157–186.

Merleau-Ponty, Maurice. 1945. *Phénoménologie de la perception*. Paris: Gallimard.

Miller, George A. 1956. The magical number seven, plus or minus two: Some limits on our capacity for processing information. *Psychological Review* 63 (2): 81–97.

Miller, George A. 1981. Trends and debates in cognitive psychology. *Cognition* 10 (1–3): 215–225.

Miller, George A. 2003. The cognitive revolution: A historical perspective. *Trends in Cognitive Sciences* 7 (3): 141–144.

Miller, George A., Eugene Galanter, and Karl H. Pribram. 1960. *Plans and the structure of behavior*. New York: Henry Holt and Company.

Miller, George A., and Philip Johnson-Laird. 1976. *Language and perception*. Cambridge, Mass.: Harvard University Press.

Millikan, Ruth Garrett. 1984. *Language, thought, and other biological categories: New foundations for realism*. Cambridge, Mass.: MIT Press.

Minsky, Marvin Lee, and Seymour Papert. 1969. *Perceptrons: An introduction to computational geometry*. Cambridge, Mass.: MIT Press.

Monsell, Stephen, and Jon Driver, eds. 2000. *Control of cognitive processes: Attention and performance XVIII*. Cambridge, Mass.: MIT Press.

Moray, Neville. 1969. *Attention: Selective processes in vision and hearing*. London: Hutchinson Educational.

Neisser, Ulric. 1967. *Cognitive psychology*. New York: Appleton-Century-Crofts.

Neisser, Ulric, and Robert Becklen. 1975. Selective looking: Attending to visually specified events. *Cognitive Psychology* 7 (4): 480–494.

New, Joshua, Leda Cosmides, and John Tooby. 2007. Category-specific attention for animals reflects ancestral priorities, not expertise. *Proceedings of the National Academy of Sciences of the United States of America* 104 (42): 16598–16603.

Newell, Allen. 1990. *Unified theories of cognition: The William James Lectures.* Cambridge, Mass.: Harvard University Press.

Nozick, Robert. 1981. *Philosophical explanations.* Cambridge, Mass.: Harvard University Press.

O'Craven, Kathleen M., Bruce R. Rosen, Kenneth K. Kwong, Anne Treisman, and Robert L. Savoy. 1997. Voluntary attention modulates fMRI activity in human MT-MST. *Neuron* 18 (4): 591–598.

O'Regan, J. Kevin. 1992. Solving the "real" mysteries of visual perception: The world as an outside memory. *Canadian Journal of Psychology* 46 (3): 461–488.

O'Regan, J. Kevin, and Alva Noë. 2001. A sensorimotor account of vision and visual consciousness. *Behavioral and Brain Sciences* 24 (5): 939–1031.

O'Regan, J. Kevin, Ronald A. Rensink, and James J. Clark. 1999. Blindness to scene changes caused by "mudsplashes." *Nature* 398 (6722): 34.

Parasuraman, Raja, ed. 1998. *The attentive brain.* Cambridge, Mass.: MIT Press.

Parasuraman, Raja, and David R. Davies, eds. 1984. *Varieties of attention.* Orlando: Academic Press.

Partridge, Derek. 1981. Information theory and redundancy. *Philosophy of Science* 48 (2): 308–316.

Pashler, Harold E. 1998. *The psychology of attention.* Cambridge, Mass.: MIT Press.

Peacocke, Christopher. 1983. *Sense and content: Experience, thought, and their relations.* Oxford: Oxford University Press.

Peacocke, Christopher. 1991. Demonstrative content: A reply to John McDowell. *Mind* 100 (397): 123–133.

Peacocke, Christopher. 1992. *A study of concepts.* Cambridge, Mass.: MIT Press.

Peacocke, Christopher. 2001. Does perception have a nonconceptual content? *Journal of Philosophy* 98 (5): 239–264.

Peacocke, Christopher. 2003. Postscript: The relations between conceptual and nonconceptual content. In *Essays on nonconceptual content*, ed. York H. Gunther, 318–322. Cambridge, Mass.: MIT Press.

Peirce, Charles S. 1932–1935. *Collected papers of Charles Sanders Peirce, 1866–1913*, 6 vols., ed. Charles Hartshorne and Paul Weiss. Cambridge, Mass.: Harvard University Press.

Pettit, Philip, and John McDowell, eds. 1986. *Subject, thought, and context.* Oxford: Clarendon Press.

Posner, Michael I. 1978. *Chronometric exploration of mind.* Hillsdale, N.J.: Lawrence Erlbaum Associates.

Posner, Michael I. 1980. Orienting of attention. *Quarterly Journal of Experimental Psychology* 32 (1): 3–25.

Posner, Michael I. 1982. Cumulative development of attentional theory. *American Psychologist* 37 (2): 168–179.

Posner, Michael I. 1994. Attention: The mechanisms of consciousness. *Proceedings of the National Academy of Sciences of the United States of America* 91 (16): 7398–7403.

Posner, Michael I., and Gregory DiGirolamo. 1998. Executive attention: conflict, target detection, and cognitive control. In *The attentive brain*, ed. Raja Parasuraman, 401–423. Cambridge, Mass.: MIT Press.

Posner, Michael I., Michel Nissen, and Michael Ogden. 1978. Attended and unattended processing modes: the role of set for spatial location. In *Modes of perceiving and processing information*, ed. Herbert L. Pick and Elliot Saltzman, 137–158. Hillsdale, N.J.: Lawrence Erlbaum.

Posner, Michael I., and Marcus E. Raichle. 1994. *Images of mind.* New York: Scientific American Library.

Posner, Michael I., Charles R. R. Snyder, and Brian J. Davidson. 1980. Attention and the detection of signals. *Journal of Experimental Psychology. General* 109 (2): 160–174.

Poulton, E. C. 1952. Perceptual anticipation in tracking with two-pointer and one-pointer displays. *British Journal of Psychology* 43 (3): 222–229.

Pylyshyn, Zenon W. 1989. The role of location indexes in spatial perception: A sketch of the FINST spatial-index model. *Cognition* 32 (1): 65–97.

Pylyshyn, Zenon W. 2001. Visual indexes, preconceptual objects, and situated vision. *Cognition* 80 (1–2): 127–158.

Pylyshyn, Zenon W. 2003. *Seeing and visualizing: It's not what you think.* Cambridge, Mass.: MIT Press.

Quinton, Anthony. 1973. *The nature of things.* London: Routledge.

Quinton, Anthony. 1979. Objects and events. *Mind* 88 (350): 197–214.

Reimer, Marga. 1992. Three views of demonstrative reference. *Synthese* 93 (3): 373–402.

Ribot, Théodule. 1908. *Psychologie de l'attention (Dixième édition).* Paris: Félix Alcan.

Rizzolatti, Giacomo, and Michael Arbib. 1998. Language within our grasp. *Trends in Neurosciences* 21 (5): 188–194.

Rizzolatti, Giacomo, Lucia Riggio, and Benecio Sheliga. 1994. Space and selective attention. In *Attention and performance XV: Conscious and nonconscious information processing*, ed. Carlo Umiltà and Morris Moscovitch, 395–420. Cambridge, Mass.: MIT Press.

Rollins, Mark. 2003. Perceptual strategies and pictorial content. In *Looking into pictures: An interdisciplinary approach to pictorial space*, ed. Heiko Hecht, Robert Schwartz, and Margaret Atherton, 99–122. Cambridge, Mass.: MIT Press.

Rueda, M. Rosario, Mary K. Rothbart, Bruce D. McCandliss, Lisa Saccomanno, and Michael I. Posner. 2005. Training, maturation, and genetic influences on the development of executive attention. *Proceedings of the National Academy of Sciences of the United States of America* 102 (41): 14931–14936.

Russell, Bertrand. 1910. Knowledge by acquaintance and knowledge by description. *Proceedings of the Aristotelian Society* 11: 108–128.

Schank, Roger C. 1996. Goal-based scenarios: Case-based reasoning meets learning by doing. In *Case-based reasoning: Experiences, lessons and future directions*, ed. David B. Leake, 295–347. Cambridge, Mass.: MIT Press.

Schank, Roger C. 1999. *Dynamic memory revisited*. Cambridge: Cambridge University Press.

Sellars, Roy Wood. 1944. Causation and perception. *Philosophical Review* 53 (6): 534–556.

Sellars, Roy Wood. 1959. Sensations as guides to perceiving. *Mind* 68 (269): 2–15.

Shallice, Tim. 1994. Multiple levels of control processes. In *Attention and performance XV: Conscious and nonconscious information processing*, ed. Carlo Umiltà and Morris Moscovitch, 395–420. Cambridge, Mass.: MIT Press.

Shannon, Claude E. 1948. A mathematical theory of communication. *Bell System Technical Journal* 27 (3–4): 379–423, 623–656.

Shannon, Claude E., and Warren Weaver. 1949. *The mathematical theory of communication*. Urbana, Ill.: University of Illinois Press.

Shapiro, Kimron, ed. 2001. *The limits of attention: Temporal constraints in human information processing*. Oxford: Oxford University Press.

Shapiro, Kimron, and Kathleen Terry. 1998. The attentional blink: The eyes have it (but so does the brain). In *Visual attention*, ed. Richard D. Wright, 306–329. Oxford: Oxford University Press.

Shiffrin, Richard M. 1997. Attention, automatism, and consciousness. In *Scientific approaches to consciousness*, ed. Jonathan D. Cohen and Jonathan W. Schooler, 49–64. Hillsdale, N.J.: Erlbaum.

Shiffrin, Richard M., and William Schneider. 1977. Controlled and automatic human information processing: II. Perceptual learning, automatic attending, and a general theory. *Psychological Review* 84 (2): 127–190.

Shoemaker, Sydney. 1984. *Identity, cause, and mind*. Cambridge: Cambridge University Press.

Siegel, Susanna. 2002. The role of perception in demonstrative reference. *Philosophers' Imprint* 2 (1): 1–21.

Simons, Daniel J., and Ronald A. Rensink. 2005a. Change blindness, representations, and consciousness: Reply to Noë. *Trends in Cognitive Sciences* 9 (5): 219.

Simons, Daniel J., and Ronald A. Rensink. 2005b. Change blindness: Past, present, and future. *Trends in Cognitive Sciences* 9 (1): 16–20.

Somers, David C., Anders M. Dale, Adriane E. Seiffert, and Roger B. H. Tootell. 1999. Functional MRI reveals spatially specific attentional modulation in human primary visual cortex. *Proceedings of the National Academy of Sciences of the United States of America* 96 (4): 1663–1668.

Spence, Charles. 2001. Crossmodal attentional capture: A controversy resolved? In *Attraction, distraction and action: Multiple perspectives on attentional capture*, ed. Charles L. Folk and Bradley S. Gibson, 231–262. Amsterdam: Elsevier.

Stoffregen, Thomas A., and Benoît G. Bardy. 2001. On specification and the senses. *Behavioral and Brain Sciences* 24 (2): 195–261.

Strawson, P. F. 1956. Singular terms, ontology and identity. *Mind* 65 (260): 433–454.

Strawson, P. F. 1959. *Individuals: An essay in descriptive metaphysics*. London: Methuen.

Sully, James. 1898. *Outlines of psychology*. London: Longmans, Greens & Co.

Sutton, John. 2006. Distributed cognition: Domains and dimensions. *Pragmatics & Cognition* 14 (2): 235–247.

Titchener, Edward B. 1899. *An outline of psychology*. New York: The Macmillan Company.

Titchener, Edward B. 1908. *Lectures on the elementary psychology of feeling and attention*. New York: The Macmillan Company.

Tomasello, Michael. 1995. Joint attention as social cognition. In *Joint attention: Its origins and role in development*, ed. Chris Moore and Philip J. Dunham, 103–130. Hillsdale, N. J.: Lawrence Erlbaum Associates.

Tomasello, Michael, Malinda Carpenter, Josep Call, Tanya Behne, and Henrike Moll. 2005. Understanding and sharing intentions: The origins of cultural cognition. *Behavioral and Brain Sciences* 28 (5): 675–735.

Tomasello, Michael, Malinda Carpenter, and Ulf Liszkowski. 2007. A new look at infant pointing. *Child Development* 78 (3): 705–722.

Treisman, Anne. 1969. Strategies and models of selective attention. *Psychological Review* 76 (3): 282–299.

Treisman, Anne. 1988. Features and objects: The fourteenth Bartlett memorial lectures. *Quarterly Journal of Experimental Psychology* 40 (2): 201–237.

Treisman, Anne. 1992. Perceiving and re-perceiving objects. *American Psychologist* 47 (7): 862–875.

Triesch, Jochen, Dana H. Ballard, Mary M. Hayhoe, and Brian T. Sullivan. 2003. What you see is what you need. *Journal of Vision* 3 (1): 86–94.

Ullman, Shimon. 1984. Visual routines. *Cognition* 18 (1–3): 97–159.

Ullman, Shimon. 1996. *High-level vision: Object recognition and visual cognition*. Cambridge, Mass.: MIT Press.

Valdes-Sosa, Mitchell, Maria A. Bobes, Valia Rodriguez, and Tupac Pinilla. 1998. Switching attention without shifting the spotlight: Object-based attentional modulation of brain potentials. *Journal of Cognitive Neuroscience* 10 (1): 137–151.

Wettstein, Howard. 1984. How to bridge the gap between meaning and reference. *Synthese* 58 (1): 63–84.

Wicken, Jeffrey S. 1987. Entropy and information: Suggestions for common language. *Philosophy of Science* 54 (2): 176–193.

Wiggins, David. 1997. Sortal concepts: A reply to Xu. *Mind & Language* 12 (3–4): 413–421.

Wiggins, David. 2001. *Sameness and substance renewed*. Cambridge: Cambridge University Press.

Woodfield, Andrew, ed. 1982. *Thought and object: Essays on intentionality*. Oxford: Oxford University Press.

Wright, Richard D., ed. 1998. *Visual attention*. Oxford: Oxford University Press.

Wundt, Wilhelm. 1897. *Outlines of psychology*, trans. C. H. Judd. Leipzig: Wilhelm Engelmann.

Yarbus, A. L. 1967. *Eye movements and vision*, trans. Basil Haigh. New York: Plenum Press.

12
Biolinguistics and Information

Cedric Boeckx and Juan Uriagereka

Modern (theoretical) linguistics was born half a century ago in the midst of what is often called the cognitive revolution. Noam Chomsky (1956, 1957), Morris Halle (1995, 2002), Eric Lenneberg (1967), and others distanced themselves from the then-dominant behaviorist paradigm, and reached back to earlier philosophical concerns, using the faculty of language as "a mirror to the mind." Rather than as a list of behaviors or a communication mechanism, language was seen as an organ of the mind, a key to understanding mental life. If psychology is defined as the science of mental life, then linguistics animated by rationalist concerns is best characterized as a branch of psychology/cognitive science, and, as such, a part of biology. This is what the term *biolinguistics* seeks to emphasize.

Biolinguists seek to uncover the laws that determine the nature of the language faculty, the principles behind its development, and the ways in which it is put to use. Ultimately, biolinguists would like to contribute to our understanding of how such a capacity is realized in neural terms, and even how it evolved in the species.

Given their perspective, biolinguists insist on studying language as "I-language," where *I-* refers to individual, internalist, and intensional (Chomsky 1986). Individual and internalist, because the language faculty obviously resides inside someone's head; a mentalist approach is thus in order. Intensional, because one of the key, and still very surprising, properties of human language is a kind of infinity: our language faculty makes it possible to produce and understand an unbounded range of expressions, in principle. As Wilhelm von Humboldt put it, with language, humans make infinite use of finite means (1836/1988). Such finite means (ultimately properties of the finite brain) must be characterized intensionally, in pretty much the way our knowledge of natural numbers is.

In a sense, by asking questions about the nature, development, use, implementation, and evolution of the language faculty, biolinguists are endorsing a research agenda that is very much on a par with how the life sciences approach behavior in animals (compare Tinbergen's [1964] famous "four questions" for ethology).

The Need for a Language Organ: An Argument from Information Theory

Chomsky (1959) took the central message of the ethologists to be the following: learning different things about the world from different kinds of experience requires mental computations, tailored both to what is to be learned and to the kind of experience from which it is to be learned. Therefore, there must be task-specific learning organs, with structures customized both to what they are supposed to extract from experience and to the kind of experience from which they are to extract it. And it is the task of the cognitive scientist to figure out what the specific structures of these learning organs are.

Chomsky's earliest achievement in linguistics was to demonstrate the limitations of early models of language based on information theory, including their heavy reliance on Markovian processes (see Chomsky 1956, 1957; also Miller 2003). That said, Claude Shannon's (1948) precise characterization of the notion of information can illuminate the nature of learning, and show how crucially needed the notion of an innate (biologically given) language organ is (see Gallistel 2001, 2006, 2007; Gallistel and King 2009).

Shannon's (1948) definition of information is very intuitive. For him, the amount of information conveyed by a signal is measured by the amount by which the signal reduces the receiver's uncertainty about the state of the world. If you already know that the first letter of "John" is "J," our telling you that the first letter of "John" is "J" tells you nothing. It conveys no information. But if you didn't know this before, then, of course, our message would be informative. This simple idea has nonintuitive mathematical consequences, though. Specifically, it implies that signaling (conveying information) presupposes prior knowledge on the part of the receiver. For Shannon, information is measured in terms of the difference between what you already know and what you don't (see also Shannon and Weaver 1949). This means that for you to learn something, you must have some prior knowledge against which to measure the thing learned. That is, if the mind of the newborn baby were genuinely a blank slate, with no expectations about the sort of thing she could encounter, then any signal would convey an infinite amount of information. But, according to Shannon's theory, no signal can convey an infinite amount of information in a finite amount of time, so no information would be conveyed, and no learning could take place.

To put this important point differently, in order for us to acquire information about the world from our experience of it, we must have built into our signal-processing capacity the range of possibilities that could be encountered in the world. You may not know in advance what the first letter of John's wife's name is, but you must know that it must be a letter drawn from a finite alphabet. That is already something (you know it is 1 of 26 options). Likewise, a thermometer will be able to register information about external temperature only if it is built for this specific purpose—if it has some notion of temperature somehow built into it. It is because we have some linguistic structure built into us that we can experience language; likewise, it is because we have no notion of infrared color built into us that this remains inaccessible to us. What we can learn depends on what we are predisposed to learn. Our environment is specified by our biology. It is the organization of our brain that sets the limit on the information that we could in principle get from the signal, that is, what we could possibly learn.

Consider the way children learn their language. A child must pay attention to various factors when her caretaker produces an utterance, but, more importantly in the present context, there are also lots of details that she has to ignore: the specific time of day at which the utterance was produced, the clothes the caretaker may be wearing, and so on. In short, she must pay attention to strictly linguistic factors. As a matter of fact, there are even linguistic factors that she has to ignore, so we have learned from decades of linguistic research. For instance, the fact that, say, the second word in the sentence is very short, or bears no stress, or refers to very abstract concepts (factual though it may well be for concrete utterances), is very likely at right angles to what the linguistic rule the child must internalize, for example, concerning what relevant sentences must start with. So all of that potential information must be ignored, lest the wrong rule be entertained.

Chomsky (1986, 80–81) emphasizes the following passage from C. S. Peirce, where the basic point at issue is made clear:

Besides, you cannot seriously think that every little chicken that is hatched has to rummage through all possible theories until it lights upon the good idea of picking something up and eating it. On the contrary, you think the chicken has an innate idea of doing this; that is to say, that it can think of this, but has no faculty of thinking of anything else. The chicken you say pecks by instinct. But if you are going to think every poor chicken endowed with an innate tendency toward a positive truth, why should you think that to man alone this gift is denied? I am sure that you must be brought to acknowledge that man's mind has a natural adaptation to imagining correct theories of some kinds. (Peirce 1934, 414–415)

In forming the idea of picking up food, and not countless other ideas it could hit upon, Peirce's chicken resembles a child learning a linguistic procedure. Specifically, what they share is that the number of hypotheses they consider is restricted. This is more than just hitting upon the right thing; it is also moving away from the wrong sort of generalization.

When we attempt to simulate the learning of a specific task—say, by building a neural network that would simulate what the brain does—it is important to remember what we built into the network to give it a head start. This is just a crucial portion of what is necessary for learning to be successful. Equally important, in fact, is what is not built into it. What allows neural networks to succeed, when they do, is not so much what they have been built to bring to the task, but rather what they are specifically built *not* to bring to it. If neural networks kept track of every property inherent in the data, they would never be able to make any generalizations, let alone the right generalizations.

The function of the language organ, then, is not really to provide a grammar, but rather to provide a set of constraints on what can and cannot be a possible grammar. It is interesting in this regard that Stuart Kauffman and his colleagues have recently argued that the concept of information, if it is to be useful in biology ("biotic information"), should be equated with "Instruction" or, better yet, "Constraint" (see Kauffman et al. 2008). If we take the brain to be an organ of computation, as modern cognitive science does, and if it computes information it gets from the environment (meaning: it learns), then we are already in nativist territory, for there cannot be information processing without any priors. Put another way, the brain could not run without a nontrivial amount of genetically specified structure about the world.

In short, learning would be a total mystery if the brain were a blank slate; the external world would be lethally chaotic. Once this point is recognized, the essential questions of cognitive science are: What does the brain compute? From what data (external stimuli) does it compute?

Although Markovian processes standardly used in information theory are inadequate to capture core properties of NLs, "chunks" of English (or any other language) can indeed be characterized by means of a finite-state machine. Indeed, as Lasnik (2007) has recently emphasized, Chomsky and his associates explicitly argued that such chunks exist and even have interesting properties (see, e.g., Chomsky and Miller 1963). For instance, iterative patterns (*never, never, never surrender!*)—which are among the most common patterns across the world's languages and among the first to emerge in every child's speech—are of this type (see Uriagereka 2008, chap. 6). The same can be said about processes sensitive to strict string adjacency—what are known as "adjunction" and "head-head" dependencies in syntax; such as Chomsky's (1957) Affix Hopping transformation. Such patterns are best characterized in terms of P&R strictly local (SL) bigrams: the narrowest subclass within the finite-state layer. As far as we can tell, there are in fact no more complex FS objects that natural language exploits. If this observation is true, we would like to know why (a point we return to in the next section).

Equally interesting is the fact that, as soon as NLs exploit patterns that are beyond the computational reach of finite-state automata, once again "context-free space" is not utilized by the system. Here too, only one option, still of the simplest kind, appears to be empirically attested. Since at least Chomsky 1970, it has been known that NLs make use of endocentric phrase-structure rules only: the projected label of a phrase must be of the same type as the head of that phrase (e.g., a verb phrase must contain a verb, or a noun phrase, a noun). We claim that this is no more than a pattern of adjacent identical elements defined, this time, over a "vertical" range of nonterminals. Granted, formal languages are not interested in nonterminals (the domain of strong generative capacity in any system), and concentrate instead on just the possible arrangements among the terminal strings (weak generative capacity). But linguists are interested in both, in fact in the nonterminals more than the terminals, although of course standard linguistic data must make use of terminal information of the sort that signal waves carry.

The key observation here is that just like NLs exploit "horizontal" adjacency for patterns of so-called selection (head-head dependencies and the like) and adjunction, NLs also exploit "vertical" adjacency for patterns of so-called projection (head-label dependencies), albeit at a different level of abstraction. The former is adjacency at the terminal string; the latter is adjacency at a level of generalization over terminal strings, the essential phrasal constituencies that all languages manifest,

How does it compute (what are the specific computations it performs)? And how does anything in the brain get translated into observable behavior? Such questions constitute the daily bread and butter of theoretical linguists and their colleagues in interrelated fields (language acquisition, psycholinguistics). In the next section, we turn to some specific aspects of the language organ.

Computational Resources of the Language Faculty

It is a foundational result within biolinguistics that the computational resources of human language must go beyond finite-state automata (Chomsky 1956). How else would one capture the knowledge underlying the basic modification conditions in simple examples like *anti-missile missile*, *anti-anti-missile-missile missile*, *anti-anti-anti-missile-missile-missile missile*, and so on? (see Lasnik 2000 for discussion). Because words in sentences are generally not like beads on a string, Chomsky (1956, 1957) explored computational capacities higher up on what came to be known as the Chomsky hierarchy of automata to capture what humans tacitly know about their languages.

In the wake of recent experiments testing the computational resources in animals' minds, Pullum and Rogers (2006; hereafter P&R) review the nature of the Chomsky hierarchy (CH) and urge linguists and cognitive scientists to go beyond its gross features and "cuts" (finite state, context-free, context-sensitive, and so on). They stress the need to take seriously the fact that each class of the CH is densely populated by possible languages, a result within traditional formal-language theory. Most relevant for our purposes, P&R show that within the finite state class, one finds subclasses of string sets that have interesting computational properties. Specifically, four subclasses are identified: Strictly Local (SL), Locally Testable (LT), Star Free (SF), and (genuine) Finite State (FS). Similar subclasses exist at higher levels of the CH, even if not all are equally well- understood.

Although it is not trivial to adapt results along these lines to the biolinguistic I-languages, we may grant the computational significance of the subclasses that P&R identifies. Starting from this, we suggest that a close look at the structures and processes that natural languages (NLs) present reveals that NLs make sparse use of the classes of string sets available in the CH. More specifically, and in a sense to be explained shortly, NLs exploit only the simplest subclass within each CH layer.

if somewhat obliquely (although this is certainly less oblique in some languages than in others).

In the same general vein, continuing with our approach of strong generative capacity in NL systems, it has been well-known since Aravind Joshi's early work on these matters (Joshi 1985) that even when NLs patterns enter the context-sensitive range, again, only a small portion of that space is explored. In our view it continues to be, in some fairly clear sense, the very first cut at that level of complexity: what is often called the "mildly" context-sensitive level (admittedly this notion hasn't been successfully characterized in mathematical terms). Here, another notion of adjacency, appropriately relativized to the class of even more abstract patterns that arise at such levels of complexity, appears to be relevant: strict subjacency, a notion introduced in Chomsky (1973), interpreted in more or less the terms we are suggesting, as in Berwick and Weinberg (1984). Because this is a bit harder to visualize, we will give a simple example.

The issue is *when* discontinuous dependencies, of the sort arising when topicalizations are created (*this, I can't see* from a structure like *I can't see this*), can proceed without difficulties for speakers. To understand the basic concept of so-called locality for the relevant long-distance relations, we have to bear in mind that syntactic units cluster in cycles for which computations recur. The most obvious such cycle corresponds to a proposition, the domain where truth-values are ascertained. Surprisingly, discontinuous dependencies can occur across such cycles (*this, I believe [I can't see]* from a structure like *I believe [I can't see this]*), but not always. For example, **this, I wonder [whether I can see]* is judged deviant by most English speakers.

One simple way to describe the pattern above is in terms of thinking of the structures of propositions as possessing the onion-layer structure of Russian matryoshka dolls containing other such dolls. We can think of such objects as being *sub*-jacent (a neologism deriving from *ad*-jacent, emphasizing the idea that the successive domains are contained one within the next). Whenever long-distance dependencies are subjacent, they succeed. In contrast, when some grammatical element prevents a subjacent relation (and instead appropriate dependencies must be stated leaping over intermediate layers of structure), the result is more or less deviant. That is the traditional intuition. In slightly more technical terms, we can imagine long-distant dependencies as relating nonterminals across subjacent projections (sets of nonterminals). One may call this "diagonal" adjacency (crossing cycles within the syntactic computation).

The patterns that emerge in NLs thus appear to be always of the simplest kind, no matter what the level of abstraction they are being examined at: adjacent terminal bigrams, then moving on to nonterminal bigrams of a kind (although this is inexact talk, in that this move already displaces us into the domain of the system's strong generative capacity), and finally to sets of nonterminals, for which a certain "bigramic" condition still obtains, now expressed at the level of computational cycles (so still more removed from directly observable data). In a nutshell: we find head-head (horizontal) adjacency (characterizable as the simplest finite state pattern), head-projected (vertical) label adjacency (characterizable as the simplest context-free pattern involving nonterminals), and nonhead long-distance dependencies across (diagonally) adjacent phrases (characterizable as the simplest context-sensitive pattern involving cycles).

Readers familiar with the linguistic literature focusing on locality conditions in syntax (for review, see Lasnik, Uriagereka, and Boeckx 2005) will recognize that the three layers of adjacency we have discussed constitute the basis for standard principles like the Head Movement Constraint, Relativized Minimality, Successive Cyclicity, and so on. The three kinds of adjacency discussed in this section can be represented by means of the following finite-state representations (always understanding that each representation is more abstract than the previous, and only the first of this yields standard objects of the sort that formal language theory operates with):

(1) ba ba

(2) X° XP

(3) $[\alpha F]X \ [\alpha F]Y$

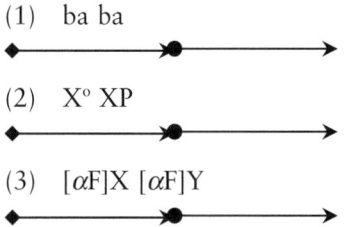

The machine depicted in (1) is the standard finite-state representation. The one in (2) depicts the formation of a domination path for so-called projected structures. The one in (3) is meant to capture the feature-matching relation achieved over so-called c-command paths, of the sort standardly used in models capturing context-sensitive dependencies (various systems use slightly different, though roughly equivalent, notations).

Design Considerations, or How Much Information Must Be Built into UG

It is easy to make sense of the emerging bigram patterns unearthed in the previous section in the context of the most recent approach within biolinguistics: the minimalist program (Chomsky 1995, and much subsequent work; see Boeckx 2006 for an overview). Basically, the claim in the previous section amounts to this: The *simplest* pattern *within each class (CH-cut)* is all that exists. Such a conclusion fits snugly with other minimalist considerations, for reasons that will become obvious very shortly.

The minimalist program refers to a family of approaches exploring a conjecture, first explicitly formulated by Noam Chomsky in the early 1990s, concerning the nature (design features) of the human language faculty. As we noted previously, since Chomsky's early works, the language faculty has been treated as an organ of the mind in order to begin to make sense of one "big fact" about human beings: short of pathology or highly unusual environmental circumstances, they all acquire at least one language by the time they reach puberty, in a way that is remarkably uniform and relatively effortless.

The acquisition of language is all the more remarkable when we take into account the enormous gap between what human adults (tacitly) know about their language and the evidence that is available to them during the acquisition process. It should by now be obvious that the linguistic input a child receives is radically impoverished and extremely fragmentary. It is in order to cope with this "poverty of stimulus" that Chomsky took humans to be biologically endowed with an ability to grow a language, much like ethologists have done to account for the range of behaviors that animals display. The theory about the biological equipment that makes language acquisition possible is called Universal Grammar (UG), taking the term from the rationalist tradition.

Under Chomsky's impetus, recently biolinguists have also explored the possibility that the amount of genetic information needed to capture some central properties of the language faculty may in fact be minimal. In this view, relevant properties would emerge, epigenetically as it were, from what Chomsky (2005) has called "third factor" considerations (essentially, biophysics). This is based on an idea that Lewontin (2000) stresses for the design of all organisms: it is based on (1) their genetic endowment, (2) the contribution of the environment (including culture in the case of humans), and (3) principles of growth and laws of form that transcend the limits of "genomic" nativism.

The central question animating minimalist research is whether the computational system at the core of language is optimally designed to meet the demands on the systems of the mind/brain it interacts with. By "optimal design," biolinguists intend to explore the idea that all properties of the computational system of language can be made to follow from minimal structural specifications, aka "bare output conditions"—the sort of properties that the system would have to have to be usable at all (e.g., all expressions generated by the computational system should be formatted in a way that the external systems can handle). Put yet another way, the computational system of language, minimalistically construed, would consist solely of the most efficient algorithm to interface with the other components of the mind, the simplest procedure to compute (generate) its outputs (expressions) and communicate them to the organs of the mind that will interpret them and allow them to enter into thought and action.

Though those concerns have been articulated only during the last couple of decades, Chomsky has always stressed the possibility that "some aspects of a complex human achievement [may be the result of] principles of neural organization that may be even more deeply grounded in physical law" (1965, 59). The issue here is to what extent the design of FL (and, beyond it, complex biological systems in general), exhibit a law-like character, determined in large part by very general properties of physical law, and mathematical principles in an even broader sense. Chomsky remarks that "throughout the modern history of generative grammar, the problem of determining the character of FL has been approached 'from top down': How much must be attributed to UG to account for language acquisition? The M[inimalist] P[rogram] seeks to approach the problem 'from bottom up': How little can be attributed to UG while still accounting for the variety of I-languages attained" (2007, 12).

Concretely, approaching UG from below means that the inventory of basic operations at the core of FL must be reduced to a minimum, and much of the richness previously attributed to UG has to be reevaluated: it must be shown to be the result of simple interactions, or else either attributed to the external mental systems that the core computational system of language interacts with, or to "third factors" in the earlier sense.

Evolutionary biology has so far underemphasized the roles of physical and mathematical laws in shaping the form and structure of living organisms, but it is well-known that these constitute powerful constraints for both phylogenies and ontogenies (see Gould 2002). To the extent that third factors can account for properties of the language faculty, obviously less genetic information needs to be posited in the case of language. Key properties of language may then emerge epigenetically.

The bigram hypothesis explored in the previous section fits well with minimalist considerations: if NLs are for some reason forced to go beyond the finite state layer to code dominance in addition to precedence (perhaps a case of virtual conceptual necessity to meet interface demands on the sound side and on the meaning side), and beyond the context-free layer to establish local checking relations (perhaps another case of virtual conceptual necessity, an instance of so-called Last Resort considerations to satisfy "legibility" conditions at the interfaces), at each point economy dictates that the simplest subclass within each computational layer be chosen. The need to go progressively more abstract within the CH may be dictated by interface demands, but sticking to the simplest (bigramic) solution within each level would be a third factor consideration.

How Syntax Informs the Rest of the Mind

In this final section, we would like to examine the nature of the information sent by natural language syntax to other cognitive systems. This is the question of interface at the heart of the minimalist program, and also the question of meaning in natural language.

At first blush, the following seems correct. The word *cat* refers to the sort of animal that Fido hates, and the class of things that Felix is a member of. And we understand the sentence "potatoes are in the closet" as true if indeed there are potatoes in the closet, and as false, otherwise. It is, thus, tempting to say that words like *cat* refer to certain entities in the world, and sentences like "potatoes are in the closet" refer to states of affairs that are either true or false. Such statements about meaning are very common indeed. But they confuse meaning and use. As Strawson (1950) said, words don't refer (to things in the world), it's people that refer with words.

It is true that one can use the word *cat* to refer to Felix, and that one can say "potatoes are in the closet" to indicate that this state of affairs indeed obtains, but we shouldn't conclude from this that meaning is reference, and that understanding sentences means knowing the conditions under which they may be true. One shouldn't confuse knowledge *of* language and knowledge *about* language; that is to say, one should distinguish between what speakers know by virtue of having a language faculty, and what they know by virtue of being a language-user in some environment.

Making truth-evaluable assertions, referring to things, and inferring from statements and questions—in short, conveying information about the world out there—are some of the things we can do with words and sentences, but this is highly context-dependent, and thus highly variable. Already here we should be skeptical about taking reference and truth-conditions as the key notions for natural language meanings. Scientists (correctly) tend to shy away from variable facts of this sort, and instead they try to uncover stable phenomena: simple, tractable things they can hope to render intelligible. As Paul Pietroski stresses, whether (an utterance of) a sentence is true or false is what is known as an "interaction effect" (2005). It is the result of a host of factors, a hodgepodge of psychological states and physical circumstances that one may expect will forever resist interesting theorizing, the same way specific trajectories of apples falling from trees or stones thrown at a target will forever resist interesting theorizing. Physicists are right to abstract away from them; semanticists should, in our view, do the same.

If information about the world is not what words and sentences provide, what is (linguistic) meaning? The best hypothesis we currently have about what the meanings of words and sentences are treats meanings as "instructions to build concepts," to activate concepts we have independently of language (and that we may share with other species), or build new ones that would be (literally) unthinkable without language. On meanings as instructions, see Chomsky (1995, 2000), who builds an analogy with the more standard take on phonology providing instructions to the articulators, that is, tongue, lips, and so on (Halle 1995, 2002; also see Pietroski, forthcoming, for a recent exploration of this hypothesis). Words and sentences can thus be thought of as procedures that impose certain "mental traffic patterns" among concepts; that is, they enforce perspectives on the way we think, and on the way we judge things as true or false (without determining truth or falsity), ways that are in large part as unique to us as language itself is.

It is now standard to take the mind to consist of modules (Fodor 1983). Pylyshyn (1984, 1999) argues that one property stands out as being the real signature of a module: the encapsulation of the processes inside the module from both cognitive influence and cognitive access. This is referred to as the "cognitive impenetrability" of a given module. The hypothesis we would like to entertain here is that natural language plays a significant role in actually *breaking* the bounds of modularity.

Following several researchers (among them Spelke 2003; Carruthers 2002, 2006; Pietroski 2005; Mithen 1996; Arp 2008; Boeckx, forthcoming), we take language to combine information from various modules (core knowledge systems) that would otherwise be isolated. Language instructs the mind to construct concepts that would otherwise be impossible, because their parts would be encapsulated. This allows the formation of novel molecules of thought, as it were. Experimental work by Spelke and her colleagues (for overviews, see Spelke 2003 and Carruthers 2006) suggests that human language provides a mold within which mental transactions across modular boundaries are carried out.

For example, whereas animals possess both an approximate number sense and a subitizing system for small numerosities, only humans can achieve the precision of the subitizing system for big numbers (e.g., Baroody and Dowker 2003). Likewise, whereas other animals can orient themselves either by relying on a geocentric position (say, the northern corner of the experimental setting (Gallistel and Cramer 1996), or on a particular landmark (say, right next to a box), only humans seem to be able to flexibly combine both kinds of information (e.g., looking for a particular geocentric position oriented with respect to a particular landmark; say, north of the box).

The discussion in this section leads to the hypothesis that our distinct kind of intelligence ("thought") is due to our capacity to flexibly combine representations that would otherwise remain isolated. Language offers the clearest example of an algebraic system that combines (indefinitely). The mode of combination happens not to depend on its conceptual roots: we combine events, colors, numbers, emotions, and numerous other entities in sentences, and treat them all like x's and y's, thereby transcending the limits of domain-specific systems that can only combine an x with another x.

With language in humans, biology has found a way to develop a system that instructs the mind to fetch a concept here and a concept there, and combine them even if the concepts by themselves wouldn't fit naturally; their "word-clothings" make them fit together. The roots of our knowledge are ancient and, for all anyone can tell, continuous with other species. But our kind of thought, our creative bent, as it were, required the evolution of lexicalization, which applies a uniform format to concepts that would never combine otherwise. Human language, under this hypothesis, takes the form of a central processing unit that creates a lingua franca—a genuine language of thought that is simple natural language—out of the mutually unintelligible dialects of thoughts that are the core (nonlinguistic) knowledge systems.

If these reflections are on the right track, they help understand what makes humans unique (thought-wise). Marc Hauser (2008) characterizes what makes us unique (what he dubs "humaniqueness") thus:

1. The ability to combine and recombine different types of information and knowledge in order to gain new understanding
2. To apply the same rule or solution to one problem to a different and new situation
3. To create and easily understand symbolic representations of computation and sensory input
4. To detach modes of thought from raw sensory and perceptual input

Boeckx (forthcoming) suggests that these four distinct features of human thought boil down to our ability to lexicalize concepts and combine them freely, echoing a proposal discussed in Uriagereka (1998, chap. 6).

Conclusion

To sum up, we have discussed three ways in which biolinguistics relates to information: (1) innate information (qua constraints) are needed to grow a language (Universal Grammar); (2) third-factor information entering into the nature of the language faculty makes it possible to reduce the amount of genetic information dedicated to language (Linguistic Minimalism); and (3) cross-modular information transactions are made possible by lexicalization ("language as the seat of humaniqueness").

References

Arp, Robert. 2008. *Scenario visualization: An evolutionary account of creative problem solving*. Cambridge, Mass.: MIT Press.

Baroody, Arthur J., and Ann Dowker. 2003. *The development of arithmetic concepts and skills: Constructing adaptive expertise*. Mahwah, N.J.: Lawrence Erlbaum and Associates.

Berwick, Robert C., and Amy S. Weinberg. 1984. Deterministic parsing and linguistic explanation. *Language and Cognitive Processes* 1 (2): 109–134.

Boeckx, Cedric. 2006. *Linguistic minimalism: Origins, concepts, methods, and aims*. Oxford: Oxford University Press.

Boeckx, Cedric. Forthcoming. Some reflections on Darwin's problem in the context of Cartesian biolinguistics. In *The Biolinguistizzenterprise: New perspectives on the evolution and nature of the human language faculty*, ed. Anna Maria Di Sciullo and Cedric Boeckx. Oxford: Oxford University Press.

Carruthers, Peter. 2002. The cognitive functions of language. *Behavioral and Brain Sciences* 25 (6): 657–673.

Carruthers, Peter. 2006. *The architecture of the mind: Massive modularity and the flexibility of thought*. Oxford: Oxford University Press.

Chomsky, Noam. 1956. "Three models for the description of language." *IRE Transactions on Information Theory: Proceedings of the Symposium on Information Theory* IT-2 (3): 113–124.

Chomsky, Noam. 1957. *Syntactic structures*. The Hague: Mouton.

Chomsky, Noam. 1959. Review of *Verbal behavior*, by B. F. Skinner. *Language* 35: 26–58.

Chomsky, Noam. 1965. *Aspects of the theory of syntax*. Cambridge, Mass.: MIT Press.

Chomsky, Noam. 1970. Remarks on nominalization. In *Readings in English transformational grammar*, ed. Roderick A. Jacobs and Peter S. Rosenbaum, 184–221. Waltham, Mass.: Ginn-Blaisdell Publishing.

Chomsky, Noam. 1973. Conditions on transformations. In *A festschrift for Morris Halle*, ed. Stephen Anderson and Paul Kiparsky, 232–286. New York: Holt, Rinehart, and Winston.

Chomsky, Noam. 1986. *Knowledge of language: Its nature, origin, and use*. New York: Praeger.

Chomsky, Noam. 1995. *The minimalist program*. Cambridge, Mass.: MIT Press.

Chomsky, Noam. 2000. *New horizons in the study of language and mind*. Cambridge: Cambridge University Press.

Chomsky, Noam. 2005. Three factors in language design. *Linguistic Inquiry* 36 (1): 1–22.

Chomsky, Noam. 2007. Approaching UG from below. In *Interfaces + recursion = language? Chomsky's minimalism and the view from syntax-semantics*, ed. Uli Sauerland and Hans-Martin Gärtner, 1–29. Berlin: Mouton de Gruyter.

Chomsky, Noam, and George A. Miller. 1963. Introduction to the formal analysis of natural languages. In *Handbook of mathematical psychology*, vol. 2, ed. R. Duncan Luce, Robert R. Bush and Eugene Galanter, 269–321. New York: Wiley.

Fodor, Jerry A. 1983. *The modularity of mind: An essay on faculty psychology*. Cambridge, Mass.: MIT Press.

Gallistel, Charles R. 2001. Mental representations, psychology of. In *International Encyclopedia of the Behavioral and Social Sciences*, ed. Neil J. Smelser and Paul B. Baltes, 9691–9695. Oxford: Elsevier.

Gallistel, Charles R. 2006. The nature of learning and the functional architecture of the brain. In *Psychological science around the world, vol. 1: Neural, cognitive, and developmental issues. Proceedings of the 28th International Congress of Psychology*, ed. Qicheng Jing, Mark R. Rosenzweig, Gery d'Ydewalle, Houcan Zhang, Hsuan-Chih Chen, and Kan Zhang, 63–71. Sussex: Psychology Press.

Gallistel, Charles R. 2007. Learning organs. In *Cahier no. 88: Noam Chomsky*, ed. Jean Bricmont and Julie Franck, 181–187. Paris: L'Herne.

Gallistel, Charles R., and Audrey E. Cramer. 1996. Computations on metric maps in mammals: Getting oriented and choosing a multi-destination route. *Journal of Experimental Biology* 199 (1): 211–217.

Gallistel, Charles R., and Adam Philip King. 2009. *Memory and the computational brain: Why cognitive science will transform neuroscience*. Malden, Mass.: Wiley-Blackwell.

Gould, Stephen Jay. 2002. *The structure of evolutionary theory*. Cambridge, Mass.: Belknap.

Halle, Morris. 1995. Feature geometry and feature spreading. *Linguistic Inquiry* 26 (1): 1–46.

Halle, Morris. 2002. *From memory to speech and back: Papers on phonetics and phonology, 1954–2002*. Berlin: Mouton de Gruyter.

Hauser, Marc. 2008. Humaniqueness. Paper presented at the AAAS Annual Meeting, Boston.

Humboldt, Wilhelm von. 1836/1988. *On language: The diversity of human language-structure and its influence on the mental development of mankind*. Trans. Peter Heath. Cambridge: Cambridge University Press.

Joshi, Aravind K. 1985. Tree adjoining grammars: How much context-sensitivity is required to provide reasonable structural descriptions? In *Natural language parsing: Psychological, computational, and theoretical perspectives*, ed. David R. Dowty, Lauri Karttunen, and Arnold M. Zwicky, 206–250. Cambridge: Cambridge University Press.

Kauffman, Stuart, Robert K. Logan, Robert Este, Randy Goebel, David Hobill, and Ilya Shmulevich. 2008. Propagating organization: An enquiry. *Biology and Philosophy* 23 (1): 27–45.

Lasnik, Howard. 2000. *Syntactic structures revisited: Contemporary lectures on classical transformational theory*. Cambridge, Mass.: MIT Press.

Lasnik, Howard. 2007. Up and down the Chomsky hierarchy. Paper presented at Syntactic Structures—A 50th Anniversary Celebration, Princeton, N.J. Available at http://www.ling.umd.edu//~lasnik/.

Lasnik, Howard, Juan Uriagereka, and Cedric Boeckx. 2005. *A course in minimalist syntax: Foundations and prospects*. Oxford: Blackwell.

Lenneberg, Eric H. 1967. *Biological foundations of language*. New York: Wiley & Sons.

Lewontin, Richard. 2000. *The triple helix: Gene, organism, and environment*. Cambridge, Mass.: Harvard University Press.

Miller, George A. 2003. The cognitive revolution: A historical perspective. *Trends in Cognitive Sciences* 7 (3): 141–144.

Mithen, Steven. 1996. *The prehistory of the mind: The cognitive origins of art, religion, and science*. London: Thames & Hudson.

Peirce, Charles S. 1934. Methods for attaining truth. In *Collected papers of Charles Sanders Peirce*, vol. 4, ed. Charles Hartshorne, Paul Weiss, and A. W. Burks, 399–422. Cambridge, Mass.: Harvard University Press. Press.

Pietroski, Paul M. 2005. Meaning before truth. In *Contextualism in philosophy: Knowledge, meaning, and truth*, ed. Gerhard Preyer and Georg Peter, 253–300. Oxford: Oxford University Press.

Pietroski, Paul M. Forthcoming. *Semantics without truth values*. Oxford: Oxford University Press.

Pullum, Geoffrey K., and James Rogers. 2006. Animal pattern-learning experiments: Some mathematical background. Ms., Radcliffe Institute for Advanced Study, Harvard University. Available at http://ling.ed.ac.uk/~gpullum/Monkey-Math.pdf.

Pylyshyn, Zenon W. 1984. *Computation and cognition: Toward a foundation for cognitive science*. Cambridge, Mass.: MIT Press.

Pylyshyn, Zenon W. 1999. Is vision continuous with cognition? The case for cognitive impenetrability of visual perception. *Behavioral and Brain Sciences* 22 (3): 341–423.

Shannon, Claude E. 1948. A mathematical theory of communication. *Bell System Technical Journal* 27 (3–4): 379–423, 623–656.

Shannon, Claude E., and Warren Weaver. 1949. *The mathematical theory of communication*. Urbana: University of Illinois Press.

Spelke, Elizabeth S. 2003. What makes us smart? Core knowledge and natural language. In *Language in mind: Advances in the study of language and thought*, ed. Dedre Gentner and Susan Goldin-Meadow, 277–311. Cambridge, Mass.: MIT Press.

Strawson, Peter F. 1950. On referring. *Mind* 59 (235): 320–344.

Tinbergen, Niko. 1964. *Social behavior in animals: With special reference to vertebrates*. Berlin: Springer.

Uriagereka, Juan. 1998. *Rhyme and reason: An introduction to minimalist syntax*. Cambridge, Mass.: MIT Press.

Uriagereka, Juan. 2008. *Syntactic anchors: On semantic structuring*. Cambridge: Cambridge University Press.

13

The Biology of Personality

Aurelio José Figueredo, W. Jake Jacobs, Sarah B. Burger, Paul R. Gladden, and Sally G. Olderbak

The biology of personality can be described at multiple, complementary levels of analysis: *descriptive, behavioral-genetic, neuroanatomical, neurochemical, situational,* and *behavioral-ecological.* None of these individual levels of analysis separately presents a complete picture of the biology of personality, but we provide a multilayered perspective by presenting a theoretically coherent and *consilient* (Wilson 1998), vertical integration of these various levels, linking proximate with ultimate levels of causation. Each of these levels of analysis can be thought of as members of a constitutive hierarchy, as described by Ernst Mayr:

In such a hierarchy the members of a lower level, let us say tissues, are combined into new units (organs) that have unitary functions and emergent properties.... At each level there are different problems, different questions to be asked, and different theories to be formulated. Each of these levels has given rise to a separate branch of biology; molecules to molecular biology, cells to cytology, tissues to histology, and so forth, up to biogeography and the study of ecosystems. (1982, 65)

Mayr is using the concept of emergence here in the *weak* (unproblematic) sense, in that the properties of a system are ultimately reducible to the products of the interactions among its constituents, as opposed to the *strong* sense, in which the properties of an integrated whole are *not* reducible to the synergistic effects of its component parts (Laughlin 2005).

We can gainfully express the increasing levels of complexity of the ascending systems, and of the interactions among the components involved, in terms of formal information theory (Shannon and Weaver 1949). As stated succinctly by Dretske: "Information theory identifies the amount of information associated with, or generated by the occurrence of an event (or the realization of a state of affairs), with the reduction in uncertainty, the elimination of possibilities, represented by that event of state of affairs" (1981, 4). Thus, we can conceive of Shannon-Weaver (SW) information, or reduction in uncertainty, as the degree of *constraint* that each level of hierarchical organization exerts upon another. Hence, through bottom-up causal interactions, the material properties of a lower level of organization can constrain those of the levels above it; through top-down causal interactions, the emergent properties of a higher level of organization can constrain the behavior of the levels below it. Constraint, then, may be a mutual and reciprocal process among these hierarchically organized levels. Thus, the SW information content of any system increases as a function of increasing causal interactions among levels of biological organization.

The use of a multilayered and vertically integrated analytic approach to such a hierarchy maximizes the SW information available to theorists in the construction of predictive models. By increasing the number of identified constraints upon a system, we enhance our capacity to predict and possibly influence the behavior of that system. Because Shannon and Weaver (1949) explicitly eschewed *semantics* (meaning) in their definition of information, this formulation is the most useful one for a consilient approach, because we can apply it across hierarchical levels of organization without attributing semantic meanings to the molecular structures of the genetic code, the chemical actions of neurotransmitters, the functioning of the brain, the affordances of situations, or the selective pressures of the ecology.

Further, in pursuing this vertical integration of analytic levels, we revisit the so-called person-situation debate (Fleeson 2004; Kenrick and Funder 1988; Rowe 1987) by exploring the adaptive functions of the temporal stability, consistency, and persistence of interindividual variation in personality traits across developmental, ecological, and evolutionary time. However, we also examine the countervailing adaptive functions of intraindividual behavioral flexibility in response to the classes of adaptive problem that were recurrent in the "mosaic" (compare Hutchinson 1959) of physical, biological, and social environments of human evolutionary adaptedness. Finally, we resolve this apparent dichotomy by applying evolutionary concepts from quantitative theoretical ecology to the problem of intraspecific variation.

The Description of Personality

How Many Personality Factors Are There?

Different levels of data aggregation can generate practically any desired number of personality factors (Zuckerman et al. 1993). Ostensibly established personality models yield different numbers of factors, including the *16 Primary Factors* (Cattell 1946), the *Big Five* (Costa and McCrae 1992), the *Gigantic Three* (Eysenck and Eysenck 1975), two *Metatraits* or *Big Two* factors (Digman 1997; DeYoung, Peterson, and Higgins 2001, 2005), and even a single higher-order *General Factor of Personality* (GFP) or *Big One* factor (Vásquez 2004; Figueredo et al. 2004; Musek 2007). Furthermore, there is evidence to indicate that these are not completely rival hypotheses or alternative models, but instead that the lower numbers of factors obtained by some models can often best be described as higher-order aggregates of the higher numbers of factors obtained by certain others (Musek 2007).

Higher-order constructs are better at predicting certain traits, such as delinquency and sensational interests, than individual personality dimensions (Cooper et al. 2003; Egan et al. 2005). Certain composites of traditional personality factors reflect the mating preferences and reproductive life history strategies of individuals (Buss 1989, 1991, 1997; Buss and Greiling 1999). Vásquez (2004) reported that a GFP can be found whether one uses the NEO Five-Factor Inventory (Costa and McCrae 1992) or the Zuckerman-Kuhlman Personality Questionnaire (Zuckerman 2002), and that this GFP correlated positively with the Mate Value Inventory, a measure of self-reported mate value, irrespective of the personality inventory used. Most recently, Figueredo et al. (2004) found that a GFP was highly positively correlated (both phenotypically and genetically) with a multivariate construct assessing a *slow* (more prosocial and higher-parental-investment) reproductive life history strategy. The GFP was also highly positively correlated (both phenotypically and genetically) with a Covitality construct (Weiss, King, and Enns 2002) that reflected higher levels of physical and psychological health. The GFP (Musek 2007) has also been independently associated with general affect, subjective well-being, satisfaction with life, self-esteem, and self-competence. We interpret these additional correlates as representing convergent indicators of a *slow* life-history strategy.

Thus, there is converging evidence that higher-order personality factors are meaningful and useful constructs that can relate to social deviance, a reproductive life-history strategy, or one's perceived mate value. These results do not speak directly to the long-standing debate over how many personality factors are "real," but instead specify an optimal degree of aggregation of personality data for the specific purposes of predicting a particular set of criterion variables. Nevertheless, as with the well-documented hierarchical structure of general intelligence (e.g., Carroll 1993; Musek 2007), lower-order factors are often more predictive of specific behaviors, such as performance on particular tasks, than the higher-order constructs. For example, immediately below lower-order personality factors in the hierarchy are personality *facets*, which are closer than the common factors to the behaviors expressed in specific situations.

In conceptualizing the problems inherent in dealing with hierarchically organized data, Wittmann (1988) has elaborated the principle of *Brunswik Symmetry*, which states that both the predictor and the criterion in any predictive validity study should be constructed at the same level of data aggregation. Wittmann has applied this principle to many different areas of scientific psychology with great success, including meta-analysis of psychotherapy outcome research, personality research, attitude research, intelligence and working memory capacity research, intelligence and complex problem solving, and program evaluation research. Thus, the information content inherent at different levels of hierarchical organization in personality ranges from the very specific to the very general, and the description that we select should reflect the level of data aggregation that is most appropriate (*Brunswik-Symmetrical*) to the phenomenon under investigation.

The Behavioral Genetics of Personality

Behavioral genetic researchers have attempted to estimate the relative magnitudes of genetic and environmental influences on personality (Rowe 1994; Harris 2006; Plomin 1989). The percentage of the variance among individuals that is due to genetic influence is referred to as the *heritability*. Heritability coefficients are not descriptive of any one individual but instead represent characteristics of the sampled population. Heritabilities can be referenced in the *narrow sense*, which accounts for only additive effects, or in the *broad sense*, which accounts for both additive and nonadditive effects. Additive effects result when the effects of different genes are equal to the sum of their individual effects. Nonadditive effects represent interactions among genes, such as *dominance* and *epistasis. Dominance* refers to the interaction of two alleles at the same locus on a homologous chromosome in which the effect of one gene dominates, or overshadows, that of the other. *Epistasis* is similar to dominance, but instead refers to alleles at different loci (Rowe 1994).

Heritability is often studied by comparing monozygotic (MZ) twins, who share 100 percent of their genes, with dizygotic (DZ) twins, who share 50 percent of their genes. If the genetic influence of a trait is solely of the additive type, then DZ twins (or other non-MZ siblings) should show a correlation of only about 0.50, whereas MZ twins should show a perfect correlation of 1.0 in the genetic influences upon their traits (Sherman et al. 1997). Falconer's Index (Falconer 1981) was designed to estimate the total genetic effects on any given trait:

$$h^2 = 2(r_{MZ} - r_{DZ})$$

This equation essentially states that heritabilities (h^2) can be estimated as equal to twice the difference between the monozygotic (r_{MZ}) and dizygotic (r_{DZ}) twin correlations, which are the phenotypic correlations within pairs of twins for the same traits. More sophisticated multivariate techniques, such as biometric structural equation models, are now available to estimate heritabilities, but they are essentially based on the same theoretical principle. To increase the generalizability of these estimates, numerous studies with large twin populations are often conducted (Plomin 1989).

The remainder of the phenotypic variance is generally attributed to environmental influences, and these environmental influences are further separated into shared and nonshared portions. *Shared environment* refers to the effects of having a similar environment growing up, such as the same family unit, and *nonshared environment* is typically understood to represent environmental experiences not shared by both individuals, but is in fact merely the residual (unaccounted for) variance. Because measurement error (typically around 15 to 25 percent of the variance) is generally not accounted for in twin studies, error is generally included with nonshared environment, thus inflating this estimate (Harris 2006).

Genes generally account for 45 percent of the variance in personality traits. Little, if any, of the variance is accounted for by shared environment; the remaining 55 percent is attributed to nonshared environment (Harris 2006). Heritability ranges from 41 to 67 percent for Extraversion, from 31 to 56 percent for Neuroticism (Saudino et al. 1999; Pedersen et al. 1988; Loehlin 1989; Plomin 1989), from 43 to 49 percent for Impulsivity, and from 23 to 53 percent for Monotony Avoidance (Saudino et al. 1999; Pedersen et al. 1988). When separated by sex, heritability of Neuroticism was 52 percent for males and 56 percent for females, and for Extraversion around 58 percent for males and 67 percent for females. Thus, the effects of sex on the heritability of personality are relatively negligible (Loehlin 1989). The heritabilities of personality traits, however, generally increase with age (Plomin and Nesselroade 1990).

Estimates of shared environmental influence on personality are generally quite low (Rowe 1994). Estimates of shared environment may range up to 7 percent for Extraversion and up to 10 percent for Neuroticism (Pedersen et al. 1988; Loehlin 1989; Plomin 1989); estimates of shared environment are about 0 percent for Impulsivity and about 5 percent for Monotony Avoidance (Pedersen et al. 1988). Because Falconer's Index was used to calculate the heritabilities for these traits, a possible explanation for these low estimates is that these traits are influenced not only by additive effects, but by nonadditive effects as well (Loehlin 1989).

Most behavioral genetic models estimate the complementary effects of genes and environment as purely additive, and do not adequately account for *gene-environment interactions*, in which an organism's level of genotypic expression is dependent on cues from the environment (West-Eberhard 2003). Harris (2006) hypothesized the existence of such unmeasured interactions as a possible explanation for the large effects attributed in additive behavioral genetic models to nonshared environment. She proposed that it is actually a set of three evolved psychological mechanisms, or systems, which interact with the genes to create differences among individuals. The first, called the *relationship system*, influences humans to make fine distinctions between individuals so they can act appropriately with each person with whom they are acquainted. The second, called the *socialization system*, influences humans to be like others within their culture. The third, called the *status system*, influences humans to maximize their individual behavioral differences to achieve status within their society.

Besides gene-environment interactions, the existence of *gene-environment correlations* also interferes with our ability to separately estimate the complementary effects of genes and environments. A gene-environment correlation refers to the reciprocal relationship created between an organism's genotype and environment, with the two acting upon each other. These gene-environment correlations are generated by three separate mechanisms: *selection*, in which an individual organism chooses or avoids certain environments; *evocation*, in which the organism unintentionally elicits predictable reactions from its environment; and *manipulation*, in which the organism uses specific tactics to intentionally alter its environment (Buss 1987). Collectively, these three behavioral mechanisms constitute three different forms of *niche-picking*.

To understand the role of genetics in the development of personality in terms of information theory, it is important to recall what genes actually do. In spite of the popular metaphor, the information content inherent in the genetic code is not a static "blueprint" for an organism (West-Eberhard 2003). Instead, genes code for amino-acid sequences that specify which enzymes will be synthesized. Enzymes are catalysts that influence (by either up- or down-regulating) the rates at which all chemical reactions within the cell occur. These genetically influenced metabolic effects eventually work their way through (by bottom-up causal interactions) the ascending levels of biological organization to produce phenotypic traits. Thus, genes indirectly *constrain* behavioral development through the regulation of biochemical processes, but do not *determine* it.

Environmental influences also constrain development in parallel ways (West-Eberhard 2003). Furthermore, environmental influences constrain the function of the genes themselves by altering gene expression and hence the production of enzymes. These gene-environment interactions involve a biochemical *epigenetic code* by which *histones* (the protein molecules around which strands of DNA are coiled) are modified to regulate gene action. Histones in the chromatin are chemically modified (and the associated genes either partially, or fully, activated or suppressed) by the biochemical environment within the cell, including the metabolic products of other genes, either within the same cell or elsewhere. For example, hormones produced in, and transported from, other cells may function to regulate gene expression.

Conversely, genes also set the thresholds for *developmental switches* which determine organismic responses to environmental as well as to other genetic triggers (West-Eberhard 2003). The microevolutionary process by which the genetically controlled thresholds for environmental influences are naturally selected is called *genetic accommodation*, of which Waddington's (1953) *genetic assimilation* is a special case exclusively denoting the *lowering* of a developmental switch response threshold. The entire coevolutionary process for setting the response parameters of the developmental switches, including both genetic and environmental influences, is called *threshold selection*.

Thus, the causal relations among genes, environments, and behavioral development are *dynamic*. Information from the genes combines with information from the environment to constrain behavioral development. Some of this complementary information is additive and some is multiplicative, as in gene-environment interactions by which environments modify gene expression or genes set the threshold for environmental influences. Other causal pathways for this synergistic information are indirect, as in gene-environment correlations by which genotypes select, evoke, or manipulate their environments. The heritability and environmentality calculations of behavioral genetics represent quantitative estimates of the relative *constraints* upon development imposed by these complementary sources of information.

The Neuroanatomy of Personality

A Caveat on the Neurological Localization of Psychological Constructs

In discussing the neurobiological correlates of personality traits within individuals and of the situational affordances of behavioral variability, we explore how information theory aids us in reducing our theoretical uncertainty regarding the organization of both brain and behavior, not how information theory might aid us in understanding causal interactions between brain and behavior.

Although the brain mediates behavior, there may not be a direct correspondence between behaviors instantiated in neural structure and neurochemistry and the structure of personality (Cacioppo and Berntson 2004). Even assuming that classification of neural functions by brain areas is appropriate, there is no a priori reason to expect a one-to-one correspondence between *anatomically* specified neural areas and *functionally* specified psychological constructs. To attempt to bridge this gap, we draw upon recent evidence that the Big Five personality factors can be aggregated to form two higher-order factors: Plasticity and Stability (Digman 1997; DeYoung, Peterson, and Higgins 2001, 2005).

Plasticity: Openness to Experience/Intellect and Extraversion

Plasticity is a higher-order personality factor that refers to the tendency to flexibly seek and explore novelty (DeYoung, Peterson, and Higgins 2001, 2005). Openness to Experience and Extraversion are correlated and together constitute Plasticity. Because they are not identical, however, there are likely to be shared and distinctive neurobiological features underlying each.

Openness to Experience is the only Big Five personality factor that is positively correlated with both *fluid intelligence* (raw mental ability) and *crystallized intelligence* (cultural knowledge), though more closely with crystallized than fluid intelligence (Loehlin et al. 1998). In light of this association and recent evidence that activity in the dorsolateral prefrontal cortex (DLPFC, particularly areas 9 and 46) is related to fluid intelligence (Duncan et al. 2000), DeYoung, Peterson, and Higgins (2005) hypothesized that neuropsychological tests previously shown to be associated with DLPFC functioning would predict some facets of Openness to Experience. As expected, DLPFC function (as indicated by a multivariate composite of several tests) was closely related to fluid intelligence ($r = 0.62$), and was also correlated with Openness to Experience ($r = 0.33$).

Further, when the NEO PI-R Openness to Experience subscale was broken down into its six individual facets, two (Ideas and Values) were associated with both DLPFC scores and fluid intelligence. The remaining four facets were related to crystallized intelligence and Extraversion. Aesthetics was related to DLPFC, but only marginally significantly related to fluid intelligence. Extraversion was associated with the Actions facet of Openness to Experience. As the authors noted, this is not surprising because self-report measures of Extraversion generally ask about behavior, and measures of Openness to Experience generally ask about mental traits. Openness is also associated with crystallized intelligence. Clearly, the DLPFC is not the "Openness to Experience module" and many other areas are involved in Openness to Experience, but the DLPFC does seem to be involved in flexible cognitive exploration.

Eysenck (1967) argued that introverts have greater neural arousal than extroverts due to differences in regulation in the ascending reticular activation system (ARAS). Introverts are easily stimulated, so small amounts of stimulation lead to physiological overstimulation and avoidance behavior in introverts. Extroverts are not easily stimulated, which leads them to seek out and explore their environments for physiological stimulation. Consistent with this theory, introverts respond with a larger change in brainwave activity when presented with low-frequency tones (Stelmack and Michaud-Achorn 1985). Further, introverts, when given a choice, choose lower levels of background noise stimulation and perform tasks as well as extroverts when allowed to perform at their chosen noise level; but when introverts receive extroverts' chosen noise level (overstimulation) and extroverts receive introverts' noise level (understimulation), performance in both groups declines (Geen 1984).

The behavioral approach system (BAS) and the behavioral inhibition system (BIS) are two descriptive behavioral systems that guide an organism in response to specific stimuli (Sutton and Davidson 1997). The BAS guides opportunistic behavior in response to perceived incentives, and the BIS guides avoidance behavior in response to perceived threats. Individuals with a dominant BAS are more likely to be extroverted and to be optimistic in situations that present incentives. Individuals with greater relative left resting prefrontal activity (as measured with an EEG) tend to have higher scores on self-report BAS scales, whereas individuals with greater relative right prefrontal activity tend to have higher scores on BIS scales (Sutton and Davidson 1997). This is consistent with the observation that patients with left prefrontal lobe damage often show signs of depression. Nevertheless, prefrontal activity asymmetry predicts responses on the BAS and BIS scales better than it predicts positive and negative affect.

Stability: Neuroticism, Conscientiousness, and Agreeableness
A dominant BIS partially reflects Neuroticism, which is implicated in avoidance behavior (Sutton and Davidson 1997; Deckersbach et al. 2006). Highly neurotic individuals are expected to be superior at detecting threatening stimuli to avoid them. The dorsal anterior cingulate cortex (dACC), thought to be involved in discrepancy detection, relates to Neuroticism (Deckersbach et al. 2006). Further, dACC reactivity, as measured by functional Magnetic Resonance Imaging (fMRI), predicts individual differences in accuracy of self-reported physiological arousal ($r = 0.86$) far better than self-reported Neuroticism ($r = 0.40$) (Eisenberger, Lieberman, and Satpute 2005). Hence, the direct measurement of brain activity may predict behavioral outcomes thought to be associated with personality traits better than self-report. Neuroticism also correlated negatively ($r = -0.74$) with rostral anterior cingulate cortex (rACC) activity.

Psychopathy is related to low Conscientiousness and low Agreeableness (Oltmanns and Emery 2000). Psychopaths exhibit deficits on neuropsychological tests that require contributions from the orbitofrontal cortex (OFC), but not on tests that require contributions from the DLPFC (Blair et al. 2006). This is consistent with reports that patients with "acquired psychopathy" have acquired damage to the OFC and exhibit deficits in social functioning and decision making (Damasio 1994). The OFC is known to be both anatomically and functionally connected to limbic areas such as the amygdala, and criminal psychopaths exhibit decreased emotion-related activity in the amygdala and other limbic structures relative to criminal nonpsychopaths and controls (Kiehl et al. 2001).

Activity in the OFC and dorsal amygdala also correlates negatively with Impulsivity; activity in the ventral amygdala, caudate, parahippocampal gyrus, and dorsal anterior cingulate gyrus correlates positively with Impulsivity (Brown et al. 2006). Further, self-reported Impulsivity predicts performance on tests related to OFC functioning (Spinella 2004). Although Impulsivity is expected to correlate negatively with Conscientiousness, comparisons of OFC-lesioned patients, non-OFC-lesioned controls, and normals detected no differences in Conscientiousness among these groups.

The Neurochemistry of Personality

The Biochemical Bases of Personality Traits

We now move away from descriptive accounts of personality traits and toward explanatory accounts of both gene-based strategies and context-based tactics. Although we may not express it directly, in all cases we seek reductions in uncertainty (either in the system under consideration or in our understanding of it), and adhere to a selectionist view of how that reduction is achieved (Edelman 1987; Chamberlin 1897; Platt 1964).

The neurochemistry of personality parallels the neuroanatomy in that there is no simple and straightforward one-to-one correspondence between the action of specific neurotransmitters and particular personality traits. We should not expect anything even remotely resembling the Humoral Theory of Temperament originally proposed by Hippocrates (460–377 BCE) and later popularized by Claudius Galen (129–216 CE), who applied the theory to diagnose and treat illness. Nevertheless, there is an identifiable, if complex, pattern of neurochemical influence on personality traits.

For example, Zuckerman (1991, 1995) proposed a structural model in which three monoamine neurotransmitters (dopamine, serotonin, and norepinephrine) underlie three behavioral mechanisms (approach, inhibition, and arousal), which, in turn, jointly underlie at least three major personality traits (Extraversion/Sociability, Psychoticism/Impulsive Unsocialized Sensation Seeking, and Neuroticism/Anxiety). In this model, dopamine underlies Approach, derotonin underlies Inhibition, and norepinephrine underlies Arousal. After that, however, things get more complicated due to multiple interactions both among the neurochemicals and among the behavioral mechanisms.

First, the three monoamine neurotransmitters are themselves influenced by various other neurochemicals. Dopamine is antagonistically affected by monoamine oxidase (MAO Type B). MAO Type B is itself antagonistically affected by gonadal hormones (such as testosterone), which also directly and positively affect both Approach and Extraversion/Sociability. Norepinephrine is agonistically affected by dopamine-beta-hydroxylase (DBH), but antagonistically affected by endorphins and gamma-aminobutyric acid (GABA). GABA also directly and negatively affects Neuroticism/Anxiety. Furthermore, the monoamine neurotransmitters themselves interact: dopamine and serotonin reciprocally influence each other antagonistically; serotonin and norepinephrine also reciprocally influence each other antagonistically.

Second, the three monoamine neurotransmitters then influence three corresponding behavioral mechanisms: dopamine positively affects Approach, serotonin positively affects Inhibition, and norepinephrine positively affects Arousal. However, Approach, Inhibition, and Arousal themselves interact: Approach and Inhibition reciprocally influence each other negatively, whereas Arousal positively affects Inhibition.

Third, and finally, the three behavioral mechanisms directly influence the three personality traits: Approach positively affects both Extraversion/Sociability and Psychoticism/Impulsive Unsocialized Sensation Seeking; Inhibition negatively affects Psychoticism/Impulsive Unsocialized Sensation Seeking; and Arousal negatively affects Psychoticism/Impulsive Unsocialized Sensation Seeking and positively affects Neuroticism/Anxiety. This model is by no means exhaustive, but illustrates how personality might be mediated by neurotransmitter systems.

In addition to producing temporally stable behavioral patterns in personality, however, these neurochemical systems must have evolved in such a way as to support sufficient behavioral flexibility to meet the differing adaptive demands presented by transient situations that the organism might encounter. To this end, the organism must be able to apprehend feedback from the environment regarding the effectiveness of any behavioral tactics used upon meeting adaptive demands. Recent work investigating the functioning of neurotransmitter systems underlying the behavioral mechanisms of Arousal and Reward proves promising in this respect; we highlight some of that work here.

Neurotransmitter Systems Underlying Arousal and Reward
The dopaminergic, locus coeruleus-norepinephrine (LC-NE), cholinergic, and serotonergic systems have ascending projections that distribute widely into the neocortex. Rather than merely inhibiting and exciting

receptors on postsynaptic neurons directly, the neurotransmitters associated with these systems also modulate the effects of other neurotransmitters and, by extension, modulate basic psychological functions. Aston-Jones and Cohen (2005) point to the mediation of Arousal (traditionally associated with the LC-NE system) and the signaling of Reward (traditionally associated with the dopaminergic system) as prime examples of this neuromodulation. They argue that we have to move beyond the simple association of particular neurotransmitters with basic, pervasive psychological functions (such as arousal) if we are to understand the workings of these neurotransmitter systems in the brain: "Whereas functions such as reward and arousal have intuitive appeal, they have often escaped precise characterization, at both the neural and the computational levels" (Aston-Jones and Cohen 2005, 405).

Recent research has refined our thinking about the role of these and other neurotransmitter systems in the regulation of behavior. Because much of this research has focused on the LC-NE and dopaminergic systems, we will focus on these as an example of ways the brain may have evolved to allow an organism to obtain feedback from specific environments regarding how effectively behavioral tactics meet adaptive demands.

LC-NE Functioning: Moving Beyond Basic Arousal

In a recent review, Aston-Jones and Cohen described a complex and specialized role for the LC-NE system (2005). The LC-NE system is critical for the optimization of behavioral performance (Aston-Jones and Cohen 2005). LC neurons exhibit distinct kinds of activity: *phasic* and *tonic*. At low levels of tonic activity (or discharge), the organism's performance on tasks is poor (the organism is often characterized as being drowsy or nonalert). At high levels of tonic LC activity, performance is also observed to be poor. At intermediate levels, activity performance is optimal. In fact, following the presentation of goal-relevant stimuli, performance is observed to be optimal with moderate tonic activity and prominent phasic activity. In this way, the overall phasic and tonic activity of the LC neurons resembles an inverted U: the Yerkes-Dodson relationship between performance and arousal. This is consistent with the fact that norepinephrine has long been implicated in arousal.

Aston-Jones and Cohen (2005) point out that, in monkeys, the LC receives input from the OFC and the anterior cingulate (ACC). They have proposed that phasic LC activation is critical in helping an organism exploit resources available to them; it provides environmental feedback when a given strategy is no longer effective in solving adaptive demands. In other words, when a given task or strategy does not appear to be effective, the LC neurons exhibit a tonic activity. The organism is often described as being drowsy and nonalert, as noted previously, or as otherwise disengaging from the task at hand and becoming highly distractible. Although these responses may seem problematic, in the long term, they are highly adaptive. While performance on the current task (or using the current strategy) suffers, the organism may explore other, potentially more effective, means of addressing adaptive demands; the organism will try something new. On the other hand, when an organism's behavior effectively solves presented adaptive demands, LC neurons exhibit bursts of activity described as phasic. The likelihood increases that the organism will then use that strategy again in the future (in similar situations). Hence, activity in LC neurons provides information (reduces uncertainty) regarding the effectiveness of specific behavioral tactics in relation to specific adaptive problems.

Dopamine: Moving Beyond Basic Models of Reinforcement Signaling
Researchers have also spent time refining our understanding of the dopaminergic system and its relation to behavioral control (see, e.g., Montague, Hyman, and Cohen 2004; McClure, Daw, and Montague 2003; Pagnoni et al. 2002). These researchers suggest that the LC-NE and dopaminergic systems operate in synergy.

Dopamine release has long been associated with reward-based learning. Recent research suggests, however, that dopamine release "does not signal reward per se but rather mediates a learning signal that allows the system to predict better when rewards are likely to occur and thereby contribute to the optimization of reward-seeking behaviors" (Aston-Jones and Cohen 2005, 405). Dopaminergic projections are most common in the DLPFC, and research has shown that dopamine neurons of the substantia nigra and of the ventral tegmental area "show phasic changes in spike activity that correlate with the history of reward delivery" (Montague, Hyman, and Cohen 2004, 761). Thus, the brain has the means to instantiate an organism's experiential history, and this history serves to set the occasion for an organism's behavior in extant contexts.

It appears that dopamine neurons exhibit *pause* and *burst* responses that are critical in facilitating the organism's evaluation of the effectiveness of current behavioral tactics at solving adaptive problems. More specifically, it appears that the organism's prior experiential history with a given tactic has been instantiated in the brain in such a way that current use of this tactic is evaluated in relation to past effectiveness. When a given tactic is working better than it has in the past, dopamine neurons exhibit a burst of activity (a *positive* reward-prediction error). On the other hand, when a given tactic is not working as well as it has in the past, dopamine neurons exhibit a pause in activity (a *negative* reward-prediction error). When the activity of these neurons is close to baseline, then, the tactic appears to be as effective as it has been in the past at meeting the adaptive demands. "This verbal interpretation of dopaminergic activity belies the sophistication of the underlying neural computations" (Montague, Hyman, and Cohen 2004, 761), but nonetheless presents a basic working model of these admittedly complex phenomena.

These findings indicate that, in addition to generating temporally stable behavioral patterns in personality, neurotransmitter systems have evolved that are exquisitely responsive to the differing adaptive demands presented by transient situations. These mechanisms permit the organism to obtain feedback from the environment regarding the effectiveness of employed behavioral tactics at meeting adaptive demands and then to shift tactics in response to such feedback in case the current tactics are no longer working.

The Contingency of Personality

Traits and Situations

Approximately forty years ago, Walter Mischel (1968) stunned personality theory, and personality theorists, with a literature review demonstrating that we cannot use well-measured personality traits to predict *situationalized* behavior consistently or precisely. Expressed in terms of information theory, the application of theoretically based personality traits does *not* sufficiently decrease theoretical measures of behavioral uncertainty inherent in contextualized behavior: intraindividual variability in behavior continued to surprise personality theorists. So began the misnamed "person-situation debate" that dominated personality theory during that time (Fleeson 2004; Kenrick and Funder 1988; Rowe 1987).

At the center of this debate were demonstrations that, although relatively invariant internal dispositions (*traits*) are detectable when taken as an aggregate, momentary behavior varies appreciably as a function of subtle changes in situations. This is no great surprise from an evolutionary perspective: an organism would not long survive nor reproduce if it possessed trait-anchored behavior insensitive to extant adaptive demands. Consistent with this intuition, Walter Mischel, Yuichi Shoda, and their colleagues (Mischel and Shoda 1995; Mischel, Shoda, and Mendoza-Denton 2002; Mischel, Shoda, and Smith 2004) demonstrated stable ideographic patterns of behavior across identifiable *classes* of situations, a phenomenon they name *behavioral signatures*. A behavioral signature reflects *within-situation class* consistencies in behavioral tactics across time. Put another way, personality theory has identified *two* loci of behavioral stability in personality: one at the level of stable individual differences in overall behavior (traits) and another at the level of stable situation-specific behavioral patterns.

Hence, individuals exhibit relatively invariant behavioral tactics *within* classes of situations and a higher-level set of invariant behavioral strategies that are evident, on the mean, *across* classes of contexts. Because formal information theory permits us to express uncertainty through the variance of a measure, we may say that situations constrain behavioral tactics (the behaviors actually exhibited in a situation), thereby reducing behavioral variability within the situation. Traits, on the other hand, impose another set of constraints on behavioral strategies (the general goals and values that guide behavioral tactics), reducing variability in strategies that are applied to specific situations.

Thus far, the field has focused on identifying internal cognitive and affective personality characteristics such as encodings, expectancies and beliefs, emotions, goals, and values, and competencies and self-regulatory plans that may mediate personality and situations; the field has invested little effort in classifying the various situations yielding a behavioral signature. Here, we apply evolutionary thinking as a step toward a properly conceptualized and measured taxonomy of the functional aspects of situations that draw out specific behavioral tactics and strategies (Funder 2006).

The Taxonomy of Contexts

Any first steps in systematics require a theory-driven statement of its general principles. Hence, we offer a classification system in terms of the source of adaptive problems.

An *environment* is the full set of adaptive problems an individual of a given species faces across development, independent of time (that is, the full set of adaptive problems that an individual faced, is facing, and will face). A *context* is the subset of extant adaptive problems that the individual is facing. A *setting* is a subset of the context: the adaptive problems originating from the abiotic (nonliving) environment. A *situation* is also a subset of the context: in contrast to a setting, however, a *situation* consists of the subset of adaptive problems originating from biotic (living) sources. An *interspecies* (between-species) *situation* consists of the subset of adaptive problems originating from biotic (living) sources other than the species in question. An *intraspecies* (within-species) *situation* consists of the subset of adaptive problems originating from intraspecies competition and/or cooperation. Most situations relevant to personality variation are of the intraspecies variety. Unfortunately, in the so-called person-situation debate, what have been traditionally referred to as *situations* are more properly understood as *contexts*.

Four major categories of intraspecies situations exist (Figueredo et al. 2009): intrasexual situations composed of the adaptive problems that (1) a male presents to another male and (2) a female presents to another female; intersexual situations composed of the adaptive problems that (3) a male presents to a female and (4) a female presents to a male. These adaptive problems may be competitive, cooperative, or any combination of the two. The form of the adaptive problems offered by intraspecies situations will be determined by factors such as the degree of perceived relatedness and the developmental stages of the participants (Hamilton 1964; Trivers 1971).

Behavior within any given context may produce conflicting outcomes. For example, certain behaviors within the confines of an intrasexual situation involving two males may reap consequential immediate costs but simultaneously reap consequential benefits in an encompassing intraspecies situation (e.g., injurious fighting may produce reproductive gains). Conversely, certain behaviors within the confines of, for example, an intersexual situation may reap consequential immediate benefits but simultaneously reap consequential costs in an encompassing context (e.g., immediate sexual activity produces immediate gains but consequential opportunity costs in the long term). Following Baum (2005), we label those contexts that involve short-term gains but substantial long-term costs *contingency traps* and those contexts that involve short-term costs but long-term gains *contingency augmentations*.

One may characterize the environment, especially intraspecies situations, along a *strong-to-weak* dimension: in *strong* environments, one must solve apparent adaptive problems immediately and efficiently. In this case, there will be little evidence of stable individual differences in overall behavior (traits); instead, one will observe stable situation-specific behavioral patterns. In *weak* environments, there are few important extant adaptive problems. In this case, there will be consistent evidence of stable individual differences in overall behavior (traits) and few idiosyncratic situations-specific behavioral patterns.

An individual, by definition, cannot act on unsensed, or unrecognized, adaptive demands. Hence, following Isaac Newton (1642–1727 CE), we contrast the *apparent* with the *true* (Newton 1729). The true context, setting, and situation are defined as the *full* extant state of adaptive problems that an individual is facing from these sources; the apparent context, setting, and situation are defined as the *apperceived* and ordered psychological manifestation of the extant adaptive problems that an individual is facing from these sources as produced by the senses and other psychological processes. An *experiential history* is the full set of adaptive problems that an individual has faced. The extant physical, psychological, and behavioral structure of the organism bears traces of the experiential history of the *current organism,* which produces the *perspective* that the organism takes on in its current context. The *perspective* is the subjective importance of the extant adaptive problems constituting a full environment. The current organism matches behavior to true adaptive problems by sensing, classifying, and differentially responding to apparent adaptive problems. This process is called *discrimination.* The apparent environment within which an organism responds is a product of its evolutionary, developmental, and experiential history. Hence, context discrimination sets the stage for various behavioral tactics. In so doing, a context reduces behavioral uncertainty.

Given that, two questions become immediately obvious. First, how do contexts come to reduce behavioral uncertainty, and second, where does the information, defined as a reduction in uncertainty, reside?

How Do Contexts Come to Reduce Behavioral Uncertainty?
To address this aspect of the question, we turn to behavioral psychology, which has discovered many of the laws that describe the experiential history necessary to produce perspective and discrimination.

One way of expressing uncertainty formally is through the variance of a distribution. Distributions without much variance (e.g., *leptokurtic* distributions) do not contain much uncertainty and are highly informative. In contrast, distributions with a great deal of variation (e.g., *platykurtic* distributions) contain a great deal of uncertainty and are somewhat uninformative.

Cast this way the question becomes how contexts come to reduce behavioral variability across development in time. To greatly oversimplify, the behaviorists claim that a history of behavioral *consequences* (e.g., reinforcement, punishment, or penalty) reduce behavioral uncertainty by eliminating both maladaptive classes of behavior through penalty or punishment and nonproductive classes of behavior through extinction—leaving only behavioral classes that are appropriately responsive to the adaptive problems that form the apparent context. According to this view, a context (i.e., the success and failure of various behavioral classes to solve adaptive problems presented by the extant abiotic setting and biotic situation) reduces behavioral variability by setting conditions that select against classes of behavior that are either maladaptive or neutral with regard to the extant adaptive problems (see Baum 2005, chaps. 4 and 6 for an introduction, Mazur 2002 for an advanced discussion, and Skinner 1981; see also Edelman 1987, 1988, and Edelman and Changeux 2000). This view of behavioral selection and its "fit" with contexts make it clear that solving adaptive problems changes both the behavior of the individual (through, e.g., reinforcement or punishment) and the context—which, in its turn, presents new adaptive problems to be solved.

Personality psychology and developmental psychology have made it clear that individual differences in specific and temporally persistent behavioral traits exist and serve as raw material upon which the laws discovered by behavioral psychology act. Behavioral genetics and related fields have discovered many of the laws relating genetic contributions to these traits and have revealed that traits exhibited by the current organism produce a tendency for it to be *absorbed* by adaptive problems presented by some contexts; to be *rebuffed* by adaptive problems presented by others; and to change the true context, apparent context, or both to match the traits and perspectives of the current organism. These forms of *niche picking* (including *niche rejection* and *niche building*) appear to be a product of the innate behavioral dispositions, preferences, and abilities in combination with the experientially acquired behaviors of the organism and the adaptive problems that it encounters in the environment.

Where Does Contextual Information Reside?
To answer the second of these questions adequately, we must think carefully. Common sense tells us that the information is internalized or embodied somewhere in the individual's brain. Although embodiment in the brain makes sense in terms of the various informal notions of "information" (e.g., meaning, knowledge, instruction, or representation) that

are often used in cognitive psychology, embodying information, as defined formally, does not.

Hence, to remain true to a formal definition of SW information, we descriptively claim that classes of contexts embody rules (expressed in English as "if-then" rules and expressed more formally in conditional logic). In so doing, a context serves as *discriminative cue* for specific behavioral tactics or, more accurately, as a force that selects against all but those behavioral tactics that have produced reinforcement or avoided punishment either in the organism's evolutionary or developmental history.

The Behavioral Ecology of Personality

Return to Santa Rosalia

More than half a century ago, Hutchinson (1957) introduced several revolutionary concepts into evolutionary ecology, transforming it into a quantitative science. Although personality theorists generally do not obtain the kind of precise quantitative data that have been collected by evolutionary ecologists, these seminal concepts are critical to a complete understanding of the maintenance of individual differences in behavioral dispositions in both human and nonhuman populations in the face of natural selection. Hutchinson (1959) asked, "Why are there so many kinds of animals?" and the answer may be the same for why there are so many kinds of people.

We begin with the multidimensional conceptualization of the *ecological niche* (Hutchinson 1978). An ecological niche is defined as a hypervolume in multidimensional hyperspace, in which each of the hyperspatial dimensions is one of the parameters describing the biotic (living) or abiotic (nonliving) factors in the ecology of the species. Abiotic factors include habitable ranges along spatial dimensions such as latitude and altitude (or depth), as well as along nonspatial physical parameters such as temperature and humidity. Biotic factors include the predators, prey, hosts, and parasites with which a given species interacts. Biotic factors also include the social environment, for example, cooperation and competition from conspecifics. The *fundamental niche* of a species is the hypervolume defined by the outer limits of each of these dimensions within which members of a given species can survive.

The *realized niche* of a species is that portion of the *fundamental niche* that the population actually inhabits. The *realized niche* may be more constrained than the *fundamental niche* due to competition between species, where the *fundamental niches* of two species overlap (Pianka 1961). This concept has been extended to the circumscribed *realized niches* of individuals within a single species (Putnam and Wratten 1984), where competition within species constrains the *realized niches* of different individuals to distinct regions within the broader *fundamental niche* of the population. Thus, the total niche breadth of a generalist species may actually be composed of narrower niches occupied by individual specialists.

Just as biotic factors render the description of an ecological niche more complex than abiotic factors alone, social factors, as a subset of biotic factors, generate even more multidimensionality in the niche space of individuals. In social species, diverse social niches are also nested within the *fundamental niche* hypervolume of the species. The *realized niches* of different individuals may therefore be constrained within these diverse social niches by competition within species. Because competition is exacerbated within social species, Figueredo (1995; Figueredo et al. 2005a) proposed that individual differentiation should be greater in social than in solitary species, and noted that the overwhelming majority of nonhuman animal species for which systematic variation in personality has been documented in the literature have been social.

Both Wilson (1994) and Figueredo (1995; Figueredo et al. 2005a) have suggested that the diversification of individual traits to fit different social niches might be ultimately due to frequency-dependent selection, where social competition drives individuals into different social niches and filling these diverse niches offers partial release from competitive pressure from conspecifics. Thus, intraspecific *niche-splitting*, the fragmentation of the ecological hyperspace into more specialized niches, leads to intraspecific *character displacement*, the differentiation of individual traits to adapt to these diversified niches. The cost of deviating from the species-typical optimum is compensated by the benefit of competitive release. This centrifugal displacement of individual traits creates an *ideal free distribution* of alternative behavioral phenotypes, where the balance of costs and benefits are equalized among different individuals. Such behavioral variation generates individual differences in the relative effectiveness of adopting different adaptive strategies within complex social groups (Buss 1991, 1997; Buss and Greiling 1999). Thus, different individuals become better suited to particular social niches that others would not be as well-suited for, where greater behavioral specialization renders the various personality characteristics optimal under differing local conditions (MacDonald 1995, 1998).

Environments that are variable or heterogeneous pose special adaptive problems, and complex social environments are both. Ecological contingencies that are variable over evolutionary time are expected to select organisms that are phenotypically plastic enough to adapt by means of learning over developmental time (Figueredo, Hammond, and McKiernan 2006). Such adaptive developmental plasticity depends critically on the existence of reliable and valid cues that signal which alternative phenotype is optimal under each set of localized conditions (West-Eberhard 2003); in the absence of such reliable and valid cues, the adaptive solution to environmental variability is the production of genetically diverse individuals that are dispersed along the expected distribution of locally optimal trait values. However, ecological cues are typically neither completely reliable and valid nor completely unreliable and invalid, but are instead characterized by some *ecological validity* coefficient ranging between zero and one (Figueredo, Hammond, and McKiernan 2006). Under such intermediately stochastic conditions, organisms should evolve a combination of genetic diversity and developmental plasticity to collectively fill the available ecological niche space. The partial heritability and partial environmentality of personality variation actually observed conforms to these evolutionary ecological predictions: personality traits are heritable in humans by a margin of 0.30 to 0.50 (e.g., MacDonald 1995).

What all these evolutionary-ecological models have in common is that the environment itself is a source of information that constrains behavioral development over both evolutionary and developmental time (West-Eberhard 2003). This perspective was always inherent in Darwin's (1859) theory of natural selection, in which the ecology ultimately shapes organismic traits through the differential survival and reproduction of heritably variable individuals within their respective portions of the species-typical niche space. Any selectionist model is essentially a model of ecological *constraint*, and hence of SW information. Furthermore, evolving and developing organisms make strategic use of the information available in their environments where such information is available, as in the ecological affordance of reliable and valid cues to the relative fitness of alternative behavioral phenotypes. These are cases of top-down causal interactions within the biological hierarchy.

Nevertheless, we also see that individuals and their traits reciprocally interact with their ecology and therefore affect the selective pressures upon other individuals, as in the case of frequency-dependent selection. For example, the essential difference between the narrower *realized niche* and the broader *fundamental niche* of either the individual or of the population as a whole is entirely due to competitive pressures from other individuals of the same or other species. These are further cases of bottom-up causal interactions within the biological hierarchy. Thus, information from individual variation and information from environmental variation are dynamically complementary in mutually constraining each other as well as synergistically constraining behavioral development.

Complementarity of Traits and Contexts

Temporally persistent genetic biases in behavioral dispositions might at first seem maladaptive, because they appear to constrain individuals from being able to exploit the full range of possibilities inherent in the different contexts that they might encounter (e.g., Mischel, Shoda, and Smith 2004). Such apparent paradoxes have led to the controversies in personality psychology described previously. However, an evolutionary ecological perspective reveals how individual differences in behavioral predispositions and contextual selection are mutually complementary rather than antagonistic factors.

A good example of this complementary relationship from behavioral ecology is the relationship between microhabitat choice and either cryptic or conspicuous coloration. Behavioral choice of microhabitat (a discrete portion of a heterogeneous ecological niche) is particularly important among species in which cryptic coloration is a fixed individual trait, unlike that of the proverbial chameleon that is able to change color to match its background. In visually heterogeneous environments, optimally cryptic coloration may be either a monomorphic compromise between differing microhabitats or a diversification of distinct morphs, each entirely adapted to only one of them. The latter adaptation is facilitated when individuals belonging to each distinct morph evolve a behavioral preference for selecting the microhabitat where it has better background-specific crypsis (Merilaita, Tuomi, and Jormalainen 1999). For example, microhabitat selection by different color morphs and sexes of the pygmy grasshopper (*Tetrix undulate*) is controlled by a tradeoff between thermal benefits and predator avoidance: differently colored individuals disproportionately select areas of different substrate types and surface temperatures depending on varying risks of predation (Ahnesjö and Forsman 2006). Cryptic caridean shrimp (*Tozeuma carolinense*) also shift among microhabitats within marine seagrass meadows in response to predators and execute substrate-specific predator avoidance behaviors in the appropriate context (Main 1987). Male Trinidadian guppies (*Poecilia reticulata*) of high carotenoid ornamentation, which is preferred by females but incurs higher risks of predation, are sexually selected by females to display at different water velocities and depths within different microhabitats depending on relative risk of predation (Kodric-Brown and Nicoletto 2005): by preferring male courtship displays performed in higher-risk areas, females select for "honest" male signals (ecologically valid fitness indicators).

In nature, temporally persistent traits (such as fixed cryptic or conspicuous coloration) and transitory situations (such as conditions within localized microhabitats) are therefore not orthogonal dimensions, but are correlated and mutually reinforcing. The conditions within localized microhabitats naturally and sexually select for certain traits; individuals possessing conditionally adaptive traits behaviorally self-select into appropriate microhabitats. Thus, situations (*socioecological microniches*) shape context-specific traits over evolutionary time, and individuals possessing those traits gravitate toward their native situations within developmental time. Behavioral ecology, therefore, resolves the person-situation dichotomy.

As in the evolutionary ecological models, information from individual traits and information from environmental situations are dynamically complementary: situations set the occasion for the contingent expression of individual traits and individuals possessing different adaptive traits strategically self-select into situations in which these traits are most effective. As with the nature-nurture debate, the person-situation dichotomy is an illusion.

Evidence for Human Niche-Picking

Scarr (1996) has reviewed behavioral genetic evidence that humans select, evoke, and manipulate their own social environments. For example, studies comparing MZ and DZ twins, first-degree relatives, and adopted siblings have found heritabilities of 15 to 35 percent in both twin and nontwin perceptions of their family environments of origin. The similarity in family perceptions by adult MZ twins who were raised apart was as high as those of DZ twins who were raised together. Furthermore, when asked about how they raised their own children, the similarities in rearing practices of MZ twins raised together were as high as those of MZ twins raised apart, but were higher than the similarities of DZ twins raised either together or apart. In general, heritabilities ranged from 12 to 40 percent for child rearing practices. Overall, approximately 25 percent to 50 percent of the variance in family perceptions was accounted for by shared genes.

In addition to twin similarities in perceptions of their family environment, which could possibly be due to shared reporting bias (although a necessarily heritable one), there has been further evidence supporting the existence of such gene-environment correlations using behavioral observations (Scarr 1996). In one of the largest studies of genetic influences on family environment, researchers videotaped and coded the positive and negative interactions of adolescents with their biological family members, and found that the behavior of the adolescents in these interactions had heritabilities of up to 64 percent and that the behavior of their parents in these interactions had heritabilities of up to 38 percent. This pattern suggests that the mechanism of evocation was operating because the genetic traits of adolescents were more salient than those of their parents in influencing these familial interactions. Although humans cannot select their own family members, they do have more freedom to select their peer relationships. Researchers found evidence for selection as well as evocation in the latter. Adolescents' ratings of their relationships with peers and teachers showed heritabilities ranging from 31 to 38 percent. Perhaps more significantly, parental ratings of the phenotypic (and perhaps genotypic) traits of the adolescents' peers were also influenced by the genetic traits of the adolescents themselves, with heritabilities ranging from 62 to 73 percent for peer popularity, from 73 to 85 percent for peer college orientation, from 49 to 70 percent for peer delinquency, and from 72 to 74 percent for peer substance abuse. This indicates that adolescents are selecting their peers based largely on their own genetic traits, and perhaps based on the genetic traits of the peers themselves.

According to *Genetic Similarity Theory* (Rushton 1989), individuals seek out genetically similar romantic and social partners by *phenotype matching*, as indicated by the high degree of phenotypic similarity on a variety of traits shown by both friends and romantic partners. Both friends and lovers assort more strongly on more heritable characteristics, indicating that phenotypic similarity may be serving as a proxy for genotypic similarity. A recent twin study (Rushton and Bons 2005) examined the genetic contribution to people's preference for similarity to themselves of spouses and best friends in personality, attitudes, and demographics. Surprisingly, spouses and best friends were as similar to each other as DZ twins. Furthermore, both spouses and best friends selected by MZ twins were more similar to each other than spouses and best friends selected by DZ twins. Preferences for similarity in both spouses and friends were found to have a heritability of 34 percent. Again, phenotypic similarity to partners was higher for more heritable traits.

Thus, what are arguably the most important long-term social relationships in a person's life—those with one's parents, one's own children, one's teachers, one's peers, one's best friends, and one's romantic partners—are largely selected, evoked, and manipulated by a person's own genetic traits. An individual's traits therefore shape the social environment at least as much as the environment shapes them. This is true not only for long-term relationships, but also even for transient situations. A fairly recent observational study (Mehl and Pennebaker 2003) used portable recording devices to sample naturally occurring conversation at intervals of approximately twelve minutes. This study found a remarkable amount of temporal stability in behavior within individuals. This was attributed to the finding that individuals possessing different personality traits gravitated toward the different social niches that seemed to suit their personalities best. Temporal stability in a person's *traits* translated naturally, by means of niche-picking mechanisms, into temporal stability in the *situations* that the person habitually inhabits.

Personality traits, therefore, systematically bias the adaptive strategies of individuals. Extroverts exhibit less sexually restricted reproductive strategies (Eysenck 1976), neurotics exhibit more inhibited and avoidant behavioral strategies (MacDonald 1995), conscientious individuals exhibit more long-term adaptive strategies (e.g., desire for control, dependability, and behaviors reflecting those preferences; Friedman et al. 1993; Schwartz et al. 1995), and impulsive sensation-seekers exhibit more risk-taking strategies (Zuckerman and Kuhlman 2000).

In summary, by *constraining* behavior, personality traits add SW information to the system. Because an organism cannot be adapted to *all possible* environments, niche-picking permits organisms to specialize in particular socioecological niches. Constraint should not be viewed as maladaptive, because biased adaptive strategies can evolve to be expressed contingently in the appropriate environmental context (West-Eberhard 2003). Niche-picking unites bottom-up and top-down causal interactions combining information from individual and environmental variation to achieve an optimal match of persons and situations.

Personality and Life-History Strategy
Individual differences in personality also correlate with a diverse array of life-history traits, including longevity and fertility (see Figueredo et al.

2005b, for a review). For example, longitudinal studies (Friedman et al. 1993; Schwartz et al. 1995) indicate that Conscientiousness is associated with greater longevity in both men and women, whereas Neuroticism is associated with a greater lifelong risk of mortality in men. A retrospective twin study of postmenopausal women (Eaves et al. 1990) found that women with either (1) high Neuroticism and low Extraversion or (2) low Neuroticism and high Extraversion had the highest completed fertility.

Another recent line of research has related human personality to multivariate constructs composed of a diverse array of cognitive and behavioral indicators of life-history strategy rather than to individual life-history traits (see Figueredo et al. 2006 for a review). One recent study (Figueredo et al. 2005b) constructed a single common factor (K) for a slow life-history strategy that loaded positively on both attachment to and investment from the biological father and adult romantic partner attachment, and loaded negatively on attachment to and investment from father figures other than the biological father, mating effort, Machiavellianism, and risk taking. The three higher-order personality factors found in this study were also multivariate constructs developed from three major personality inventories, the NEO-FFI (Costa and McCrae 1992), the EPQ (Eysenck and Eysenck 1975), and the ZKPQ (Zuckerman 2002), replicating results previously obtained by Zuckerman, Kuhlman, Joireman, Teta, and Kraft (1993). *Big N* (for Neuroticism) loaded positively on NEO-FFI Neuroticism, EPQ Neuroticism, and ZKPQ Neuroticism/Anxiety; *Big E* (for Extraversion) loaded positively on NEO-FFI Extraversion, EPQ Extraversion, and ZKPQ Sociability; *Big P* (for Psychoticism) loaded negatively on NEO-FFI Conscientiousness and NEO-FFI Agreeableness, but loaded positively on EPQ Psychoticism, ZKPQ Impulsivity/Sensation Seeking, and ZKPQ Aggression/Hostility. Slow life history strategy (K) correlated significantly with both Big N ($r = -0.24$) with Big P ($r = -0.67$), and approaching significance with Big E ($r = 0.12$). The high negative correlation of Big P with K supported Zuckerman and Brody's (1988) prediction that Psychoticism should be more relevant to life history strategy than either Neuroticism or Extraversion.

In an even more recent, secondary analysis (Figueredo et al. 2007) of the MIDUS survey data (Brim et al. 2000), a comparable, but more extensive, common factor for slow life-history strategy was constructed that loaded positively on mother relationship quality, father relationship quality, marital relationship quality, children relationship quality, family support, altruism toward kin, friends' support, altruism toward non-kin, close relationship quality, communitarian beliefs, religiosity, financial status, health control, agency, advice seeking, foresight/anticipation, insight into past, primary control/persistence, flexible/positive reappraisal, and self-directedness/planning. This expanded *K-Factor* correlated positively ($r = 0.66$) with a higher-order *General Factor of Personality* (GFP) that loaded positively on Openness to Experience, Conscientiousness, Extraversion, and Agreeableness, and loaded negatively on Neuroticism. In the same study, this GFP also correlated positively ($r = 0.36$) with a Covitality factor (see Weiss, King, and Enns 2002) that loaded positively on subjective well-being, positive affect, and general health, and loaded negatively on both negative affect and medical symptoms. Using the otherwise comparable but genetically informative MIDUS Twin sample, a related study (Figueredo et al. 2004), found that the *genetic* correlations of this GFP ($h^2 = 0.59$) were even higher with both the K-Factor ($r = 0.78$; $h^2 = 0.65$) and the Covitality factor ($r = 0.70$; $h^2 = 0.52$).

These genetic correlations suggest the same set of *pleiotropic genes* (genes with multiple phenotypic effects) influence multiple life-history traits and that life-history strategy as a whole may be predominantly under the control of regulatory genes that coordinate the expression of an entire array of life-history traits (Bailey 1998). Presumably, common genetic control is necessary to integrate these individual tactical elements into a coherent and internally consistent reproductive strategy. The existence of such higher-order regulatory genes, however, would not rule out adaptive interaction with the environment (Belsky 2000). It is quite probable that the expression of these regulatory genes is conditional and subject to environmental triggers.

Nevertheless, this genetic connection between the K-Factor and the GFP has implications for the principle of Brunswik Symmetry (Wittmann 1988). First, the genetic correlations among traits *within* each multivariate construct suggest that the higher-order levels of aggregation proposed as having merely heuristic utility may be neither intellectual conveniences nor statistical artifacts, but instead reflect a genuine functional integration of adaptive traits at the genetic level. Second, the genetic correlations *among* the higher-order constructs (the K-Factor and the GFP) suggest that a correct application of Brunswik Symmetry identifies the level of biological organization at which the aggregates are functionally linked. In short, the heuristically optimal structure of our subjective descriptions of these multivariate constructs reflects the structure of the SW information objectively contained within the biological system. Evidently, individual differences in personality are intimately connected with individual variation in human life-history strategy. Because of the central importance of life-history strategy to inclusive fitness, it is quite possible that any natural or sexual selection on human personality traits is ultimately driven by selection on life history, which is more directly relevant to survival and reproduction in different environments.

Conclusion

We have integrated a theoretically coherent and consilient biology of personality across the descriptive, behavioral-genetic, neuroanatomical, neurochemical, situational, and behavioral-ecological levels of analysis, linking proximate with ultimate levels of causation. We have applied this integration to explain the adaptive significance of both the consistency and the variability of personality both within and between individuals. In doing so, we have described the *reciprocal interaction* between genes and environment in the evolution and development of personality. It is an axiom of natural selection that environments ultimately shape the traits of individuals, over both evolutionary and developmental time, but it is also a fact of tragic destiny that the traits of individuals select, evoke, and manipulate the environments in which they evolve and develop. Again, the behavioral development and expression of personality requires the input of synergistic information from both genetic and environmental sources.

Throughout these multilayered descriptions, we have integrated our treatment of personality at all these hierarchically nested levels of analysis using a single operational definition of information: SW information as *constraint*. Thus, successive levels of analysis reflect levels of biological organization that collectively constrain the emerging properties of the evolving and developing system. The concept of constraint does not carry a negative connotation. By selecting against the diverse array of maladaptive outcomes that would otherwise be possible, constraint selectively guides the evolution and development of personality toward ecologically adaptive outcomes. The incremental input of information at every successive level of the hierarchy of biological organization also constrains the system to be more systematically predictable, and this enhanced predictability amply justifies the scientific use of a multilayered heuristic approach to the biology of personality.

References

Ahnesjö, Jonas, and Anders Forsman. 2006. Differential habitat selection by pygmy grasshopper color morphs: Interactive effects of temperature and predator avoidance. *Evolutionary Ecology* 20 (3): 235–257.

Aston-Jones, Gary, and Jonathan D. Cohen. 2005. An integrative theory of locus coeruleus-norepinephrine function: Adaptive gain and optimal performance. *Annual Review of Neuroscience* 28: 403–450.

Bailey, J. Michael. 1998. Can behavior genetics contribute to evolutionary behavioral science? In *Handbook of evolutionary psychology: Ideas, issues, and applications*, ed. Charles B. Crawford and Dennis L. Krebs, 211–233. Mahwah, N.J.: Lawrence Erlbaum Associates.

Baum, William M. 2005. *Understanding behaviorism: Science, behavior, and culture*. Malden, Mass.: Blackwell Publishing.

Belsky, Jay. 2000. Conditional and alternative reproductive strategies: Individual differences in susceptibility to rearing experiences. In *Genetic influences on human fertility and sexuality: Theoretical and empirical contributions from the biological and behavioral sciences*, ed. Joseph Lee Rodgers, David C. Rowe, and Warren B. Miller, 127–146. Boston: Kluwer.

Blair, Karina, Chris Newman, Derek G. V. Mitchell, Rebecca Richell, Alan Leonard, John Morton, and Robert J. Blair. 2006. Differentiating among prefrontal substrates in psychopathy: Neuropsychological test findings. *Neuropsychology* 20 (2): 153–165.

Brim, Orville G., Paul B. Baltes, Larry L. Bumpass, Paul D. Cleary, David L. Featherman, William R. Hazzard, Ronald C. Kessler, et al. 2000. *National survey of midlife development in the United States (MIDUS), 1995–1996*. Available at http://www.icpsr.umich.edu/cocoon/ICPSR/STUDY/02760.xml.

Brown, Sarah M., Stephen B. Manuck, Janine D. Flory, and Ahmad R. Hariri. 2006. Neural basis of individual differences in impulsivity: Contributions of corticolimbic circuits for behavioral arousal and control. *Emotion* 6 (2): 239–245.

Buss, David M. 1987. Selection, evocation, and manipulation. *Journal of Personality and Social Psychology* 53 (6): 1214–1221.

Buss, David M. 1989. Sex differences in human mate preferences: Evolutionary hypotheses tested in 37 cultures. *Behavioral and Brain Sciences* 12 (1): 1–49.

Buss, David M. 1991. Evolutionary personality psychology. *Annual Review of Psychology* 42 (January): 459–491.

Buss, David M. 1997. Evolutionary foundations of personality. In *Handbook of personality psychology*, ed. Robert Hogan, John A. Johnson, and Stephen R. Briggs, 317–344. London: Academic Press.

Buss, David M., and Heidi Greiling. 1999. Adaptive individual differences. *Journal of Personality* 67 (2): 209–243.

Cacioppo, John T., and Gary G. Berntson, eds. 2004. *Social neuroscience.* New York: Psychology Press.

Carroll, John B. 1993. *Human cognitive abilities: A survey of factor-analytical studies.* New York: Cambridge University Press.

Cattell, Raymond B. 1946. *Description and measurement of personality.* New York: World Book.

Chamberlin, T. C. 1897. The method of multiple working hypotheses. *Journal of Geology* 5 (8): 837–848.

Cooper, M. Lynne, Phillip K. Wood, Holly K. Orcutt, and Austin W. Albino. 2003. Personality and the predisposition to engage in risky or problem behaviors during adolescence. *Journal of Personality and Social Psychology* 84 (2): 390–410.

Costa, Paul T., Jr., and Robert R. McCrae. 1992. *Revised NEO personality inventory (NEO PI-R) and NEO five-factor inventory (NEO-FFI): Professional manual.* Odessa, Fla.: Psychological Assessment Resources.

Damasio, Antonio R. 1994. *Descartes' error: Emotion, reason, and the human brain.* New York: Putnam.

Darwin, Charles. 1859. *The origin of species by means of natural selection.* London: John Murray.

Deckersbach, Thilo, Karen K. Miller, Anne Klibanski, Alan Fischman, Darin D. Dougherty, Mark A. Blais, David B. Herzog, and Scott L. Rauch. 2006. Regional cerebral brain metabolism correlates of neuroticism and extraversion. *Depression and Anxiety* 23 (3): 133–138.

DeYoung, Colin G., Jordan B. Peterson, and Daniel M. Higgins. 2001. Higher-order factors of the big five predict conformity: Are there neuroses of health? *Personality and Individual Differences* 33 (4): 533–552.

DeYoung, Colin G., Jordan B. Peterson, and Daniel M. Higgins. 2005. Sources of openness/intellect: Cognitive and neuropsychological correlates of the fifth factor of personality. *Journal of Personality* 73 (4): 825–858.

Digman, John M. 1997. Higher-order factors of the big five. *Journal of Personality and Social Psychology* 73 (6): 1246–1256.

Dretske, Fred I. 1981. *Knowledge and the flow of information.* Cambridge, Mass.: MIT Press.

Duncan, John, Rüdiger J. Seitz, Jonathan Kolodny, Daniel Bor, Hans Herzog, Ayesha Ahmad, Fiona N. Newell, and Hazel Emslie. 2000. A neural basis for general intelligence. *Science* 289 (5478): 457–460.

Eaves, Lindon J., Nicholas G. Martin, Andrew C. Heath, John K. Hewitt, and Michael Neale. 1990. Personality and reproductive fitness. *Behavior Genetics* 20 (5): 563–568.

Edelman, Gerald M. 1987. *Neural Darwinism: The theory of neuronal group selection.* New York: Basic Books.

Edelman, Gerald M. 1988. *Topobiology: An introduction to molecular embryology.* New York: Basic Books.

Edelman, Gerald M., and Jean-Pierre Changeux, eds. 2000. *The brain.* Edison, N.J.: Transaction Publishers.

Egan, Vincent, Aurelio José Figueredo, Pedro Wolf, Kara McBride, Jon Sefcek, Geneva Vásquez, and Kathy Charles. 2005. Sensational interests, mating effort, and personality: Evidence for cross-cultural validity. *Journal of Individual Differences* 26 (1): 11–19.

Eisenberger, Naomi I., Matthew D. Lieberman, and Ajay B. Satpute. 2005. Personality from a controlled processing perspective: An fMRI study of neuroticism, extraversion, and self-consciousness. *Cognitive, Affective, and Behavioral Neuroscience* 5 (2): 169–181.

Eysenck, Hans J. 1967. *The biological basis of personality.* Springfield, Ill.: Charles C. Thomas.

Eysenck, Hans J. 1976. *Sex and personality.* London, U.K.: Open Books.

Eysenck, Hans J., and Sybil B. G. Eysenck. 1975. *Manual of the Eysenck Personality Questionnaire.* London: Hodder & Stoughton.

Falconer, Douglas S. 1981. *Introduction to quantitative genetics.* London: Longmans Green.

Figueredo, Aurelio José. 1995. The evolution of individual differences. Available at http://www.sciencedirect.com.

Figueredo, Aurelio José, Kenneth R. Hammond, and Erin C. McKiernan. 2006. A Brunswikian evolutionary developmental theory of preparedness and plasticity. *Intelligence* 34 (2): 211–227.

Figueredo, Aurelio José, Geneva Vásquez, Barbara Hagenah Brumbach, and Stephanie M. R. Schneider. 2004. The heritability of life history strategy: The K-factor, covitality, and personality. *Social Biology* 51 (3–4): 121–143.

Figueredo, Aurelio José, Geneva Vásquez, Barbara Hagenah Brumbach, and Stephanie M. R. Schneider. 2007. The K-factor, covitality, and personality: A psychometric test of life history theory. *Human Nature* 18 (1): 47–73.

Figueredo, Aurelio José, Paul R. Gladden, Geneva Vásquez, Pedro Wolf, and Daniel N. Jones. 2009. Evolutionary theories of personality. In *Cambridge handbook of personality psychology: Part IV. biological perspectives*, ed. Phillip J. Corr and Gerald Matthews, 265–274. Cambridge, UK: Cambridge University Press.

Figueredo, Aurelio José, Jon A. Sefcek, Geneva Vásquez, Barbara Hagenah Brumbach, James E. King, and W. Jake Jacobs. 2005a. Evolutionary personality psychology. In *The handbook of evolutionary psychology*, ed. David M. Buss, 851–877. Hoboken, N.J.: Wiley.

Figueredo, Aurelio José, Geneva Vásquez, Barbara Hagenah Brumbach, Jon A. Sefcek, Beth R. Kirsner, and W. Jake Jacobs. 2005b. The K-Factor: Individual differences in life history strategy. *Personality and Individual Differences* 39 (8): 1349–1360.

Figueredo, Aurelio José, Geneva Vásquez, Barbara Hagenah Brumbach, Stephanie M. Schneider, Jon A. Sefcek, Ilanit R. Tal, Dawn Hill, Christopher J.

Wenner, and W. Jake Jacobs. 2006. Consilience and life history theory: From genes to brain to reproductive strategy. *Developmental Review* 26 (2): 243–275.

Fleeson, William. 2004. Moving personality beyond the person-situation debate: The challenge and the opportunity of within-person variability. *Current Directions in Psychological Science* 13 (2): 83–87.

Friedman, Howard S., Joan S. Tucker, Carol Tomlinson-Keasey, Joseph E. Schwartz, Deborah L. Wingard, and Michael H. Criqui. 1993. Does childhood personality predict longevity? *Journal of Personality and Social Psychology* 65 (1): 176–185.

Funder, David C. 2006. Towards a resolution of the personality triad: Persons, situations, and behaviors. *Journal of Research in Personality* 40 (1): 21–34.

Geen, Russell G. 1984. Preferred stimulation levels in introverts and extroverts: Effects on arousal and performance. *Journal of Personality and Social Psychology* 46 (6): 1303–1312.

Hamilton, W. D. 1964. The genetical evolution of social behavior. *Journal of Theoretical Biology* 7 (1): 1–52.

Harris, Judith Rich. 2006. *No two alike: Human nature and human individuality.* New York: W. W. Norton.

Hutchinson, G. Evelyn. 1957. Concluding remarks. *Cold Spring Harbor Symposia on Quantitative Biology* 22 (2): 415–427.

Hutchinson, G. Evelyn. 1959. Homage to Santa Rosalia, or why are there so many kinds of animals? *American Naturalist* 93 (870): 145–159.

Hutchinson, G. Evelyn. 1978. *An introduction to population ecology.* New Haven, Conn.: Yale University Press.

Kenrick, Douglas T., and David Funder. 1988. Profiting from controversy: Lessons from the person-situation debate. *American Psychologist* 43 (1): 23–34.

Kiehl, Kent A., Andra M. Smith, Robert D. Hare, Adrianna Mendrek, and Bruce B. Forster, Johann Brink, and Peter F. Liddle. 2001. Limbic abnormalities in affective processing by criminal psychopaths as revealed by functional magnetic resonance imaging. *Society of Biological Psychiatry* 50 (9): 677–684.

Kodric-Brown, Astrid, and Paul Nicoletto. 2005. Courtship behavior, swimming performance, and microhabitat use of Trinidadian guppies. *Environmental Biology of Fishes* 73 (3): 299–307.

Laughlin, Robert B. 2005. *A different universe: Reinventing physics from the bottom down.* New York, NY: Basic Books.

Loehlin, John C. 1989. Partitioning environmental and genetic contributions to behavioral development. *American Psychology* 44 (10): 1285–1292.

Loehlin, John C., Robert R. McCrae, Paul T. Costa, and Oliver P. John. 1998. Heritability of common and measure-specific components of the big five personality factors. *Journal of Research in Personality* 32 (4): 431–453.

MacDonald, Kevin. 1995. Evolution, the five-factor model, and levels of personality. *Journal of Personality* 63 (3): 525–567.

MacDonald, Kevin. 1998. Evolution, culture, and the five-factor model. *Journal of Cross-Cultural Psychology* 29 (1): 119–149.

Main, Kevan L. 1987. Predator avoidance in seagrass meadows: Prey behavior, microhabitat selection, and cryptic coloration. *Ecology* 68 (1): 170–180.

Mayr, Ernst. 1982. *The growth of biological thought: Diversity, evolution, and inheritance.* Cambridge, Mass.: Harvard University Press.

Mazur, James E. 2002. *Learning and behavior.* Englewood Cliffs, N.J.: Prentice-Hall.

McClure, Samuel M., Nathaniel D. Daw, and P. Read Montague. 2003. A computational substrate for incentive salience. *Trends in Neurosciences* 26 (8): 423–428.

Mehl, Matthias R., and James W. Pennebaker. 2003. The sounds of social life: A psychometric analysis of students' daily social environments and natural conversations. *Journal of Personality and Social Psychology* 84 (4): 857–870.

Merilaita, Sami, Juha Tuomi, and Veijo Jormalainen. 1999. Optimization of cryptic coloration in heterogeneous habitats. *Biological Journal of the Linnean Society. Linnean Society of London* 67 (2): 151–161.

Mischel, Walter. 1968. *Personality and assessment.* New York: Wiley.

Mischel, Walter, and Yuichi Shoda. 1995. A cognitive-affective system theory of personality: Reconceptualizing situations, dispositions, dynamics, and invariance in personality structure. *Psychological Review* 102 (2): 246–268.

Mischel, Walter, Yuichi Shoda, and Rodolfo Mendoza-Denton. 2002. Situation-behavior profiles as a locus of consistency in personality. *Current Directions in Psychological Science* 11 (2): 50–54.

Mischel, Walter, Yuichi Shoda, and Ronald E. Smith. 2004. *Introduction to personality: Toward integration.* Hoboken, N.J.: Wiley & Sons.

Montague, P. Read, Steven E. Hyman, and Jonathan D. Cohen. 2004. Computational roles for dopamine in behavioral control. *Nature* 431 (7010): 760–767.

Musek, Janek. 2007. A general factor of personality: Evidence for the big one in the five-factor model. *Journal of Research in Personality* 41 (6): 1213–1233.

Newton, Sir Isaac. 1729. *The mathematical principles of natural philosophy.* Trans. Andrew Motte. London: Knight and Compton.

Oltmanns, Thomas F., and Robert E. Emery. 2000. *Abnormal psychology.* Englewood Cliffs, N.J.: Prentice Hall.

Pagnoni, Giuseppe, Caroline F. Zink, P. Read Montague, and Gregory S. Berns. 2002. Activity in human ventral striatum locked to errors of reward prediction. *Nature Neuroscience* 5 (2): 97–98.

Pedersen, Nancy L., Robert Plomin, Gerald E. McClearn, and Lars Friberg. 1988. Neuroticism, extraversion, and related traits in adult twins reared apart and reared together. *Journal of Personality and Social Psychology* 55 (6): 950–957.

Pianka, Eric R. 1961. *Theoretical ecology: Principles and applications*. Oxford: Blackwell Science.

Platt, John R. 1964. Strong inference. *Science* 146 (3642): 347–353.

Plomin, Robert. 1989. Environment and genes: Determinants of behavior. *American Psychologist* 44 (2): 105–111.

Plomin, Robert, and John R. Nesselroade. 1990. Behavioral genetics and personality change. *Journal of Personality* 58 (1): 191–216.

Putnam, Rory, and Stephen D. Wratten. 1984. *Principles of ecology*. London: Chapman and Hall.

Rowe, David C. 1987. Resolving the person-situation debate: Invitation to an interdisciplinary dialogue. *American Psychologist* 42 (3): 218–227.

Rowe, David C. 1994. *The limits of family influence: Genes, experience, and behavior*. New York: Guilford Press.

Rushton, J. Philippe. 1989. Genetic similarity, human altruism, and group selection. *Behavioral and Brain Sciences* 12 (3): 503–559.

Rushton, J. Philippe, and Trudy Ann Bons. 2005. Mate choice and friendship in twins: Evidence for genetic similarity. *Psychological Science* 16 (7): 555–559.

Saudino, Kimberly J., Jeffrey R. Gagne, Julie Grant, Anna Ibatoulina, Tatinana Marytuina, Inna Ravich-Scherbo, and Keith Whitfield. 1999. Genetic and environmental influences on personality in adult Russian twins. *International Journal of Behavioral Development* 23 (2): 375–389.

Scarr, Sandra. 1996. How people make their own environments: Implications for parents and policy makers. *Psychology, Public Policy, and Law* 2 (2): 204–228.

Schwartz, Joseph E., Howard S. Friedman, Joan S. Tucker, Carol Tomlinson-Keasey, Deborah L. Wingard, and Michael H. Criqui. 1995. Sociodemographics and psychosocial factors in childhood as predictors of adult mortality. *American Journal of Public Health* 85 (9): 1237–1245.

Shannon, Claude E., and Warren Weaver. 1949. *The mathematical theory of communication*. Urbana, Ill.: University of Illinois Press.

Sherman, Stephanie L., John C. DeFries, Irving I. Gottesman, John C. Loehlin, Joanne M. Meyer, Mary Z. Pelias, John Rice, and Irwin Waldman. 1997. Recent developments in human behavioral genetics: past accomplishments and future directions. *American Journal of Human Genetics* 60 (6): 1265–1275.

Skinner, B. F. 1981. Selection by consequences. *Science* 213 (4507): 501–504.

Spinella, Marcello. 2004. Neurobehavioral correlates of impulsivity: Evidence of prefrontal involvement. *International Journal of Neuroscience* 114 (1): 95–104.

Stelmack, Robert M., and Aurelda Michaud-Achorn. 1985. Extraversion, attention, and habituation of the auditory evoked response. *Journal of Research in Personality* 19 (4): 416–428.

Sutton, Steven K., and Richard J. Davidson. 1997. Prefrontal brain asymmetry: A biological substrate of the behavioral approach and inhibition systems. *Psychological Science* 8 (3): 204–210.

Trivers, Robert L. 1971. The evolution of reciprocal altruism. *Quarterly Review of Biology* 46 (1): 35–57.

Vásquez, Geneva. 2004. Female personality, risk, and mate selection. Unpublished master's thesis, Department of Psychology, University of Arizona.

Waddington, C. 1953. Genetic assimilation of an acquired character. *Evolution: International Journal of Organic Evolution* 7 (6): 118–126.

Weiss, Alexander, James E. King, and R. Mark Enns. 2002. Subjective well-being is heritable and genetically correlated with dominance in chimpanzees (*Pan troglodytes*). *Journal of Personality and Social Psychology* 83 (5): 1141–1149.

West-Eberhard, Mary Jane. 2003. *Developmental plasticity and evolution*. New York: Oxford University Press.

Wilson, David Sloan. 1994. Adaptive genetic variation and human evolutionary psychology. *Ethology and Sociobiology* 15 (4): 219–235.

Wilson, Edward O. 1998. *Consilience: The unity of knowledge*. New York: Alfred A. Knopf.

Wittmann, W. W. 1988. Multivariate reliability theory: Principles of symmetry and successful validation strategies. In *Handbook of multivariate experimental psychology*, ed. J. Nesselroade and R. Cattell, 505–560. New York: Plenum Press.

Zuckerman, Marvin. 1991. *Psychobiology of personality*. Cambridge: Cambridge University Press.

Zuckerman, Marvin. 1995. Good and bad humors: Biochemical bases of personality and its disorders. *Psychological Science* 6 (6): 325–332.

Zuckerman, Marvin. 2002. The Zuckerman-Kuhlman Personality Questionnaire (ZKPQ): An alternative five-factorial model. In *Big Five Assessment*, ed. Boele De Raad and Marco Perugini, 377–396. Cambridge, Mass.: Hogrefe and Huber.

Zuckerman, Marvin, and Nathan Brody. 1988. Oysters, rabbits and people: A critique of *Race differences in behavior* by J. P. Rushton. *Personality and Individual Differences* 9 (6): 1025–1033.

Zuckerman, Marvin, and D. Michael Kuhlman. 2000. Personality and risk-taking: Common biosocial factors. *Journal of Personality* 68 (6): 999–1029.

Zuckerman, Marvin, D. Michael Kuhlman, Jeffrey Joireman, Paul Teta, and Michael Kraft. 1993. A comparison of three structural models for personality: The big three, the big five and the alternative five. *Journal of Personality and Social Psychology* 65 (4): 757–768.

Contributors

Robert Arp OntoReason, LLC
David Attewell Department of Experimental Psychology, University of Bristol
Roland Baddeley Department of Experimental Psychology, University of Bristol
Cedric Boeckx Department of Linguistics, Harvard University
Luciano Boi Ecole des Hautes Etudes en Sciences Sociales, Centre de Mathématiques
Nicolas J. Bullot Department of Philosophy, University of Toronto at Mississauga
Sarah B. Burger Department of Psychology, University of Arizona
María Cerezo Department of Philosophy, University of Navarra
Yaşar Demirel Department of Chemical and Biological Engineering, University of Nebraska, Lincoln
Charbel El-Hani Research Group in History, Philosophy, and Biology Teaching, Institute of Biology, Federal University of Bahia, Brazil.
Claus Emmeche Center for Philosophy of Nature and Science Studies, Faculty of Sciences, University of Copenhagen, Denmark.
Aurelio José Figueredo Department of Psychology, University of Arizona
Paul R. Gladden Department of Psychology, University of Arizona
Benoit Hardy-Vallée Department of Philosophy, University of Waterloo
W. Jake Jacobs Department of Psychology, University of Arizona
Kalevi Kull Department of Semiotics, University of Tartu, Tartu, Estonia
Natalia López-Moratalla Department of Biochemistry, University of Navarra
Alfredo Marcos Department of Philosophy, University of Valladolid
Alvaro Moreno Department of Logic and Philosophy of Science, University of the Basque Country
Sally G. Olderbak Department of Psychology, University of Arizona
Rebecca A. Pyles Department of Biological Sciences, East Tennessee State University

João Queiroz Research Group in History, Philosophy, and Biology Teaching, Institute of Biology, Federal University of Bahia, Brazil

Kepa Ruiz-Mirazo Department of Logic and Philosophy of Science, University of the Basque Country

Niall Shanks History and Philosophy of Science, Wichita State University

George Terzis Department of Philosophy, Saint Louis University

Juan Uriagereka Department of Linguistics, University of Maryland

Benjamin Vincent School of Psychology, University of Dundee

Index

Abramson, Norman, 58, 65–66
Adenosine triphosphate, xvi, 25–26, 36–39, 41–43, 46–47, 214–216, 222–223, 238
Affect, 272–276
Alternative splicing, xxv–xxvi, 108–109, 185
Amygdala, xxvix, 273–276, 380–381
Antibodies (immunity), 139–145
Antigens (immunity), xx–xxi, 139–147
Aristotelian causation, xxiv, xxvi, xxxvi, 162–166, 192–193
Aristotle, 67
Ascending reticular activation system (ARAS), 379
Aston-Jones, Gary, 383–384
ATP. *See* Adenosine triphosphate
Attention
  automatic versus controlled, 319, 321
  covert versus overt, xxxii, 322–325, 327–328, 334
  epistemic, 329–339
  perceptual, xxxii, 310–313, 318–335
  unitary versus divided, 320–321
Attentional constitutional principle (ACP), 309–313, 322–325, 339
Autopoiesis, 7, 18, 160

Baddeley, Roland J., xxxi, 294–296, 298–303
Ballard, Dana H., 328, 331, 334, 336–337

Ballestar, Esteban, 217, 225, 228
Bar-Hillel, Yehoshua, 80–81
Barwise, Jon, 62, 65, 67, 80
B cell antigen receptor (BCR), 114–122
B cells (immunity), xxii, 140–147
Behavioral ecology, 265–267
Bernard, Claude, 132
Bickhard, Mark H., xxiii, 13, 159–160, 169
Bioinformation, 55–83
Bioinformational equivalence, 60
Biolinguistics, 353–354
Biological decision making, 258–264
Biological Identity, 182–183, 189–193, 195, 198
Biology
  functional versus historical, 152–154
  systems, 233–235, 241
Biosemiotics, 91–95, 99–100, 102–103, 106–113
Bogdan, Radu J., 314–316
Boltzmann, Ludwig, 28, 75–78
Broadbent, Donald E., 317–319, 330–331, 334–335
Brunswik symmetry, 374, 398–399
Bullot, Nicolas J., 311–312, 325–326
Buss, David M., 373, 377, 391

Campbell, John, 311, 324, 331
Cancer, 147–152, 224–229

Cells
  eukaryotic, 35–36, 168, 206,
    210–212, 215, 219, 235, 239
  prokaryotic, 35–36
Cellular organization, structure of,
    162–166
Center-surround (retinal)
    organization, 292–293, 301–303
Chemoton, xiv, 10
Cholinergic system, 382–383
Chomsky, Noam, xxxiii–xxxiv,
    353–359, 361–364
Chomsky hierarchy (of automata),
    357–358, 363
Chromatin, xxvi–xxvii, 212–216,
    222–224, 227–228, 235–238
  code, 208, 212–213, 235–237
  compaction, 206–207, 212–215,
    221–222
  remodeling, 206–208, 212–218,
    222–226, 235–238, 241
Chromosome, xxvi–xxvii, 205–209,
    210–211, 218–229, 236, 237, 241,
    245–246
Cohen, Jonathan D., 255, 261, 267,
    269, 270–271, 383–385
Cosmides, Leda, 258–259, 272, 278,
    335–336
CpG island, xxvi, 224–225
Cremer, Thomas, 207, 210, 220, 236,
    238

Damasio, Antonio, xxix,
    272–273
Darwin, Charles, 7–10, 135–139,
    142–146, 148–154, 157–158,
    256–257
Darwinism
  ontogenetic, xxi–xxii, 135–137,
    142–154
  phylogenetic, xxi–xxii, 135–137,
    153–154
Davidson, Donald, 254, 277, 278
Dawkins, Richard, 136, 154,
    258–259
Decoupling, xiv, xxiii–xxiv, 14–15,
    159–160, 162–170

Definition
  descriptive, xiii, 5
  essentialist, xiii, 5–8
Deictic theory of vision, 336–339
Dennett, Daniel C., 256, 277, 278
DeYoung, Colin G., 373, 378–379
Disease, 153, 218–221, 224–226,
    229–230, 238
DNA, 34, 82, 106–113, 162–166
  codes for amino-acid sequences, xx,
    106–114
  decoupled from an organism's
    enzyme-controlled reactions, xiv,
    xxiv, 14, 165–168
  formal cause of protein synthesis,
    xxiv, 162–168
  insufficient to explain an organism's
    individuality, xxv, 180–189
  one of a number of layers of
    information, xxvii, 219–224
  part of a stable, yet flexible,
    chromatin package, xxvi–xxvii,
    211–215
Dopaminergic system, xxix, xxxv,
    267–271, 382–385
Dorsal anterior cingulated cortex
    (dACC), 380–381
Dorsolateral prefrontal cortex
    (DLPFC), 273–275, 379–380, 384
Dretske, Fred I., 62, 66–67, 69, 81,
    315–317, 329, 371–372

Early visual system, xxx, 291–297
Ecological niche, 265, 390–393
Economy of nature, xxix, 256–259
Ecopoiesis, 17
Eigen, Manfred, 8, 13, 160
El-Hani, Charbel, Niño, 60, 61,
    63–64, 68, 94–95, 105, 107, 113,
    122
Embryonic development, 183–185,
    189–200, 211–215, 217–218,
    226–229, 236–237, 238, 242
Emergent properties, 187, 233–235,
    372
Emmeche, Claus, 7, 34, 63–64, 68,
    95–96, 105, 107

Energy constraints, xxx–xxxi, 262–264, 265, 298–302
Enthalpy, 28
Entropy, 28–29, 30, 33, 75–76, 77–78 (*see also* Thermodynamics)
Epigenetics, xxxiv, 180–189, 211–215, 216–219, 224–226, 235–237, 361–363
 code, 186–187, 37
 information, xxv–xxvii, 177–178, 181–189, 192, 197–198, 224
 mechanisms, xxv–xxvi, xxvii, 179, 183–186, 194, 207, 211, 219, 226 (*see also* Alternative splicing; Methylation/Demethylation)
Epistemology, 309–312, 323–324, 328–339
Esteller, Manel, 217, 225, 228
Evans, Gareth, 309, 311, 324
Evolution, 7–11, 14–19, 43, 60–61, 80, 93–94, 135–137, 142, 152–154, 157–162, 165–168, 183, 211
 adaptations, 135–136
 open-ended, xxiii, 10, 11, 14–19, 160–162
Exergy, 26, 28, 45
Eysenck, Hans J., 373, 379, 396, 397

Falconer's index, 375–376
Feinberg, Gerald, 6–7
Fick's law (equation), 30
Field, David J., 296, 298–299
Figueredo, Aurelio José, 373, 387, 391–392, 396–398
Floridi, Luciano, 59–60, 261
Form (Peircean), xix–xx, 104–106
Foster, David A., 296–297

Gánti, Tibor, 9–10
Gatlin, Lila L., 68–69
Gaussian function, 293
Gene coding, 106, 143, 377
Gene expression, 106, 112–114, 120–121, 184–185, 206–207, 210–212, 216–229, 236, 262, 377–378 (*see also* Epigenetics)

Genome, 206–208, 210–229, 231, 235–240
Gibbs free energy, xv–xvi, 28, 30, 32, 33
Gibson, James J., 314, 317–318, 319, 321, 322, 327, 334–335
Glimcher, Paul W., 255, 267, 271
Goal orientation, 259–260, 265–271
Greaves, Mel, 148–149, 152
Griffiths, Paul, 61, 65, 83, 96, 107, 272, 273

Harris, Judith Rich, 374–376
Hauser, Marc, 255, 366
Hebb's rule, 293–294
Heritability (genetics), 374–378, 394–396
Higgins, Daniel M., 373, 378–379
Histones, 206, 210–217, 219, 222–227, 235–236, 377
Hubel, David H., xxx, 289–290
Humaniqueness, 366
Hutchinson, G. Evelyn, 372, 390–391
Hyman, Steven E., 384–385

Identity, biological, 178–179, 182–183, 189–193, 194–195, 198
Immune system, 133–147
 adaptive immune system, 138–147
 analogies with insect collectives, 133–135
 autoimmune disorders, 145, 147
 innate immune system, 138
Immunoecology, 131
Immunoinformatics, 132
Immunological stigmergy, 134, 141
Individual (agents, objects), 310–312, 315–316, 335–336
Individuality, biological, 178–179, 182–183, 189–194, 196–199
Information
 causal versus semantic, xxxi–xxxii, 314–316, 319
 constraint, xxxiv, 356–357, 372, 378, 386, 392–393, 396, 399
 definition of, 34, 56–67

Information (cont.)
  environmental, 317–318
  epigenetics (*see* Epigenetics, information)
  evaluative, 256, 263–264, 271, 276–277
  formal-mathematical, 316–317
  functional, xvi, 34, 59–60, 62–69,
  genetic, 106–113
  historic meanings, xvii, 56–57
  measurement of, xviii–xix, 68–75
  problems concerning, xviii, 59–60
  reduction of uncertainty, xxxiii, xxxiv, 261, 354–355, 371–372, 378, 381–382, 384, 388–389
  as a triadic relation, 62–68, 78–79, 81–82, 98–106, 109–111, 117–119
Information theory, 43–45, 60–62, 289–305, 354, 371–372, 378, 385–386 (*see also* Shannon, Claude E.; Weaver, Warren)
Introversion/extroversion, 379–380, 396
Israel, David J., 315–316, 330

Jablonka, Eva, 60, 96
James, William, 313–314, 321, 327
Joyce, Gerald, 8

Kahneman, Daniel, 277, 319, 320
Kauffman, Stuart, 14, 356
Keller, Evelyn Fox, 3, 11, 13, 96, 123, 158
Kolmogorov, Andrey N., 59, 62, 79–80

Land, Michael F., 338–339
Language
  dependence on epigenetic factors, 361–363
  innate informational constraints, 354–356
  transcends modularity, xxxiv, 364–365
Lateral geniculate nucleus (LGN), xxx, 291–292

Life, definition of, xiii–xv, 5–12, 16–19
Living systems (*see also* Thermodynamics)
  as nonequilibrium dissipative structures, 27, 29–30, 34–35, 47–48
  self-development, 180–184, 189–194, 196–198
  use of energy coupling, 28–29, 35–37, 43–45
  "well-informed" character, 26, 29, 34, 43–45, 48
Locus coeruleus-norepinephrine (LC-NE) system, xxxv, 382–385
Luisi, Pier Luigi, 8–9, 16

Mathematical theory of communication, 314–318
Maturana, Humberto R., 3, 7, 10, 158, 160
Maynard Smith, John, 8–9, 61, 110–111, 160, 177, 178, 261
Maxwell's Demon, 76
Mayr, Ernst, xxxiv, 92, 152, 169, 371
McDowell, John, 311, 324
Message, 63, 66, 69–74
Metabolic constraints
  on level of neural activity, 298–299
  on number of neurons, 298–303
Methylation/Demethylation, xxv, 184–186, 187, 191, 195, 198, 206, 211–212, 213, 214, 216–218, 219, 222, 224, 225–226, 227, 228–229
Miller, George A., 314, 319, 331, 340
Millikan, Ruth Garrett, 64–66, 81, 263
Mischel, Walter, 385–386, 393
Model-Pickup-Model-Drop (MPMD), 337–338
Montague, P. Read, 255, 261, 263, 267, 269, 270, 271, 384–385
Moreno, Alvaro, 67, 160, 161, 164, 165, 167, 168
Monozygotic twins (MZTs), 196–200, 375

Natural selection, xxviii, 8, 13, 15, 68, 80, 138, 150, 158, 160–161, 256, 258–259, 277–278, 392
Neisser, Ulric, 314, 319
Nesse, Randolph M., 137–138, 149, 153
Neuroeconomics, 267–275
Nucleic acids, 34
Nucleosome, 212–216, 222–224, 235–236

Olshausen, Bruno A., 290, 298–299
Onsager's reciprocal rules, 31, 38
Optimal foraging theory (OFT), xxiii–xxix, 265–267
Orbitofrontal cortex (OFC), 273, 380–381, 383–384
Orosz, Charles, 131–133
Oxidative phosphorylation, 31, 36–38, 40–45, 47
Oyama, Susan, 178

Parham, Peter, 136–137, 141, 145
Pathogens, xx–xxi, 137
Pattee, Howard H., 14–15, 162–164
Peacocke, Christopher, 309, 311, 325
Peirce, Charles Sanders, 63–65, 69, 92–95, 97–108, 110–111, 117–119, 309, 355–356
Peircean semiotics, 91–95, 97–122 (see also Peirce, Charles Sanders)
Perceptual-demonstrative identification, 311–312, 323–325, 332–334
Perry, John, 80, 315–316
Personality
  environmental influences, 375–378, 386–387
  genetic influences, 374–378
Personality factors, xxxv, 373–374, 378
Person-situation debate, 385–387, 394
Peterson, Jordan B., 373, 378–379
Phenomenological equations, 30–31
Phenomenological stoichiometry, 39
Piaget, Jean, 61, 70–72
Pietroski, Paul M., 364–365

Pointers, 336–338
Popper, Karl, 61, 66–67, 69
Posner, Michael I., 309, 313–314, 319–320, 322, 328, 331, 335
Preformationism, xxvi, 178, 192–193
Prigogine, Ilya, 13, 27, 29, 30, 32
Primary visual cortex, 291, 303
Procedural/executive theory of attention, 333–339
Proteins, 33, 162–164
Proteomics, 230–233, 239–240, 242–243
Pullum, Geoffrey K., 357–358
Pylyshyn, Zenon W., 325, 364–365

Queiroz, João, 60, 61, 62, 63, 64, 95, 105, 107, 113, 122

Rationality
  practical, 255
  theoretical, 255, 267
  valuational, 255–256, 259–264
Recursive self-maintenance, 13, 159–164
Reductionism, xiii, xv, 3, 92, 93, 153, 235, 244
Reth, Michael, 114–117, 119, 122
Retinal ganglion cells, xxx–xxxi, 291–292, 295, 301–303
RNA, 15, 34–35, 106–113, 165
  mRNA, 64, 106–112, 184–185, 230–231, 240, 242–243
  sRNA, 181
  tRNA, 107, 162
Rogers, James, 357–358
Rosen, Robert, 10, 19, 64, 66, 80
Rostral anterior cingulated cortex (rACC), 380
Ruiz-Mirazo, Kepa, 67, 160, 167, 168, 170

Schröedinger, Erwin, 29, 76–77
Segel, Lee A., 133
Self-development, xxv, 181–184, 189–197
Self-organization, 12–13, 45, 158–160, 171, 241–243

Semiotic models
  genetic sign system, xx–xxi, 107–114
  signal transduction in B cell activation, xx-xxi, 114–122
Serotonergic system, 382–383
Shanks, Niall, 134, 136
Shannon, Claude E., xxxi, xxxiii, xxxiv, 34, 45, 57–59, 63–66, 69, 72–74, 76–78, 84, 261, 289, 315–317, 354–355, 372
Shapiro, Robert, 6–7
Signs, study of. See Peircean semiotics
Singular (acts), 310–313, 331, 336–337
Strawson, P. F., 311, 325, 329, 363
Stucki, Jörg, 37–38, 41–42, 46–47
Surani, M. Azim, 184–185
Szathmáry, Eors, 13, 110–111, 160

T cells (immunity), xxii, 135, 139, 140–147, 150–151
Temporal-difference (T-D) learning algorithms, 269–271
Thermodynamics, 25–30, 32, 35, 37, 40–41, 76
  dissipative structures, 25–26, 27, 35, 47–48
  energy coupling, 35–37, 39–43
  equilibrium and nonequilibrium systems, 26–30, 32
  First and Second Laws of, xv, 25, 28–29, 35
  spontaneous vs. nonspontaneous processes, xv, 25, 28–29, 30, 32, 36
Tooby, John, 258–259. 272, 278
Torres-Padilla, Maria-Elena, 184–185
Tracking (perceptual), 311–312, 323–327, 331–332, 339
Transcription (DNA to mRNA), 43–45, 106, 109, 111–112, 181, 184, 187, 206–207, 214–220, 222, 227, 230, 236
Translation (mRNA to amino-acid sequence), 34, 106–109, 184–185, 230–231

Treisman, Anne, 317, 319, 329, 330, 333
Turing machine, 79–80, 261
Turner, Brian M., 184, 186–187
Tversky, Amos, 254, 277
Twinning. See Monozygotic twins (MZTs)

Universal grammar, 361–363

V1, xxx, 290–295, 298–303 (see also Primary visual cortex)
Van Hateren, J. Hans, 299
Valuation mechanisms, 271–275
Varela, Francisco J., 10, 158, 160
Ventral tegmental area (VTA), 268, 384
Vincent, Benjamin T., 301–303
Vitalism, xxvi, 178

Waddington, Conrad, 228, 377
Wanting versus liking, 268–269, 271–272
Ward, Patrick H., 296–297
Weaver, Warren, xxxiii, xxxiv, 34, 57–59, 62, 64–65, 69, 316, 371–372
West-Eberhard, Mary Jane, 376–377, 392, 396
Wicken, Jeffrey S., 29, 34, 68, 160
Wienands, Jurgen, 114–117, 119, 122
Wiesel, Torsten N., xxx, 289
Williams, George C., 137–138, 149, 153
Winkler-Oswatitsch, Ruthild, 8, 13, 160

Zuckerman, Marvin D., xxxv, 373, 381–382, 396–397